T0200713

# Underground Mathematics

Thomas Morel tells the story of subterranean geometry, a forgotten discipline that developed in the silver mines of the Holy Roman Empire. Mining and metallurgy were of great significance to the rulers of early modern Europe, required for the silver bullion that fuelled warfare and numerous other uses. Through seven lively case studies, he illustrates how geometry was used in metal mines by practitioners using esoteric manuscripts. He describes how an original culture of accuracy and measurement paved the way for technical and scientific innovations, and fruitfully brought together the world of artisans, scholars, and courts. Based on a variety of original manuscripts, maps and archive material, Morel recounts how knowledge was crafted and circulated among practitioners in the Holy Roman Empire and beyond. Specific chapters deal with the material culture of surveying, map-making, expertise, and the political uses of quantification. By carefully reconstructing the religious, economic, and cultural context of mining cities, *Underground Mathematics* contextualizes the rise of numbered information, practical mathematics, and quantification in the early modern period.

THOMAS MOREL is Professor of the History of Mathematics at the University of Wuppertal, Germany.

# Underground Mathematics

*Craft Culture and Knowledge Production in Early Modern Europe*

Thomas Morel

*University of Wuppertal*

CAMBRIDGE
UNIVERSITY PRESS

Shaftesbury Road, Cambridge CB2 8EA, United Kingdom

One Liberty Plaza, 20th Floor, New York, NY 10006, USA

477 Williamstown Road, Port Melbourne, VIC 3207, Australia

314–321, 3rd Floor, Plot 3, Splendor Forum, Jasola District Centre,
New Delhi – 110025, India

103 Penang Road, #05–06/07, Visioncrest Commercial, Singapore 238467

Cambridge University Press is part of Cambridge University Press & Assessment,
a department of the University of Cambridge.

We share the University's mission to contribute to society through the pursuit of
education, learning and research at the highest international levels of excellence.

www.cambridge.org
Information on this title: www.cambridge.org/9781009267304

DOI: 10.1017/9781009267274

First published 2023

*A catalogue record for this publication is available from the British Library.*

ISBN 978-1-009-26730-4 Hardback

*For my parents*

# Contents

# Figures

# Acknowledgements

This book has been a very long time in the making, building on intuitions first developed ten years ago at the end of my PhD thesis. I am indebted to many people, first of all to my *Doktorvater* Pascal Duris, whose enthusiasm and rigour shaped my historical understanding, warning me against teleology and far-fetched conclusions. This is how I learned that 'anything goes' in the history of science, as long as you find interesting archives and read them with an open mind – a piece of advice I hope I have followed here. The present project would not have even begun without Friedrich Steinle and Gerhard Rammer, who back in 2013 accepted the postdoc application of a young French graduate, dealing with a topic esoteric even for seasoned German historians of science. Thanks to them, I spent two great years at the Berlin Center for the History of Knowledge, navigating between the TU Berlin and the Max Planck Institute for the History of Science (MPIWG). Numerous discussions with Cesare Pastorino, Giuditta Parolini, and many other colleagues cemented my interest in useful knowledge and its history. During this time, I collected a good part of the archive material that eventually became the basis for the present book. Ideas and arguments slowly matured during the following years, as I tried to tie together the large, slow developments of early modern practical mathematics with a set of specific case studies about the knowledge and craft culture of the German mining states.

In the final stretch of completing the manuscript, the University of Lille granted me a much-welcomed sabbatical, while the MPIWG funded my research stay in Berlin. This is where the first draft of this book was completed, largely isolated from the outside world by a global pandemic, in the strange, at times ethereal atmosphere of a near-empty building. I would like to thank the institutions that generously provided me travel grants for archival research over the years, most notably the German Academic Exchange Service (DAAD), the Fritz Thyssen Foundation, and the International Commission on the History of Mathematics (Grattan-Guinness travel grant). My home institutions, first the Laboratoire de mathématiques de Lens and now the University of Wuppertal, provided invaluable material assistance and intellectual support. Archivists in Germany, Austria, and Slovakia were incredibly helpful in enabling me to access and understand their collections, most notably Angela Kugler-Kießling – as well as the Freiberg Institute for Industrial Archaeology, History of Science and Technology (IWTG) – Christiane Tschubel (Bergarchiv Clausthal), Patrick Kennel (University and State Library of Tirol), and Annett Wulkow Moreira da Silva (Freiberg University of Mining and Technology). Björn Ivar Berg was kind enough to send me the certificate of surveyor Gottfried Klemm from the Norwegian Mining Museum (Norsk Bergverksmuseum), while Peter Konečný's historical knowledge proved invaluable when I nearly got lost in the archive of Banská Štiavnica.

A different version of Chapter 1 appeared as 'De Re Geometrica: Writing, Drawing, and Preaching Mathematics in Early Modern Mines', *Isis* 111 (March 2020): 22–45, © by the History of Science Society. All rights reserved. I thank the University of Chicago Press for permission to reprint, as well as Floris Cohen and the editorial team for their useful comments.

Most of the following chapters have been presented in one form or another at workshops and conferences over the years, at the MPIWG and the TU Berlin of course, but also at the University of Wuppertal, during a workshop of the *Revue d'histoire des mathématiques* in Strasbourg, as well as in Barcelona, Freiberg, and Oberwolfach. Karine Chemla, Christine Proust, and all the participants of the 'Histoire des sciences, histoire du texte' seminar in Paris were very helpful, taking the time to discuss very specific points on several occasions. Antoni Malet, Jeanne Peiffer, Samuel Gessner, Tina Asmussen, Pierre Desjonquères, Jim Bennett, Sebastian Felten, and Liliane Hilaire-Perez helped me better understand the complex nature of early modern practical knowledge, during informal talks and in studying their great articles. Pamela Long and Margaret Schotte both served as unwitting writing mentors as I perused their books time and again to understand how to square the circle of writing a short and readable book without sacrificing depth or historical scholarship. A session about practical mathematics in early modern Europe, which I organized for the History of Science Society in Utrecht in 2019, was a good opportunity to present and test several hypotheses presented in this book.

Peter Konečný, Michael Korey, Ursula Klein, Wolfgang Lefèvre, and Thomas Préveraud all read various chapters of the book, and their encouraging criticisms helped me clarify my arguments on many occasions. Rachel Blaifeder and Lucy Rhymer from Cambridge University Press encouraged me with this somewhat unconventional project, helping to turn it into a more compact and clearer manuscript. Thanks to Melanie Gee, who produced the index, as well as to Narmadha Nedounsejiane and the CUP production team for their great professionalism. While numerous people helped me develop the original draft into a coherent whole, all remaining mistakes are obviously my own.

Finally, Typhaine accompanied me during the long gestation of this book, at times kindly reminding me of the existence of a real world above subterranean silver mines. Together with Margaux and Juliette, they never failed to keep me as happily occupied at home as I am at work.

# Introduction

We think we know our mine and its stones, but what this engineer is
constantly investigating here in this way is beyond our comprehension.

Franz Kafka, *A Visit to the Mine**

In 1686, mining official and tax collector Nicolaus Voigtel published
*Geometria subterranea, oder Marckscheide-Kunst*. Despite its obscure title,
the book was an instant commercial success for a very simple reason: no practitioner before him had ever printed anything about the art of underground
surveying.[1] The frontispiece was designed to catch the eye and the imagination of early modern readers (Figure I.1). Two surveyors, both wearing
improbable uniforms, surround a mirror's frame ornamented with grimacing
mascarons. Their modest hats symbolize the average miner's protecting head-
gear, indicating that they have practical experience of the matter. Similarly,
their long vests evoke the rough *Arschleder* (butt leathers) then ubiquitous
among miners.[2] Their fine stockings and fancy shoes, however, clearly reveal
a higher status and relate them to the world of learning, knowledge, maybe
even sciences. Both men are depicted in a teaching position, with the senior
surveyor using a measuring stick as a pointer towards the inside of a rocky
tunnel.

The engraver played here with the then-popular genre of the theatres of
machines, in which the world – presented as a construction site – was the central object of the frontispiece. Lured by the metaphor of the mirror, sometimes
replaced by a stage, readers were metaphorically invited behind the scenes to
observe hydraulic or hoisting machinery and envision 'a possible, mathemati-
cally-guided future'.[3] In the *Geometria subterranea*, the dark mine at the centre of the frontispiece could be taken in a literal sense, for the topic of Voigtel's

---

* Kafka, *A Hunger Artist and Other Stories*, translation by Joyce Crick (2012), p. 26.
[1] Within a generation, it went through at least three editions and nine printings. This work is studied in more detail in Chapter 4.
[2] For depictions of actual miners and surveyors, see Weigel, *Abbildung und Beschreibung derer
sämtlichen Berg-Wercks* (1721), pp. 6, 14.
[3] Keller, 'Renaissance Theaters of Machines' (1978), p. 495.

1

Figure I.1 Frontispiece of Nicolaus Voigtel, *Geometria subterranea, oder Marckscheide-Kunst* (Eisleben, 1686), engraved by Christian Romstet. Courtesy ETH-Bibliothek Zürich, Rar 271.

book was precisely to explain how to survey such obscure and sinuous tunnels. More allegorically, Voigtel presented himself as the first to divulge what had long 'been kept very secret', hidden in esoteric manuscripts and obscured by the dialect of practitioners.[4] On the ground, where an ingenuous reader might expect to find a silver ore vein or precious crystals, the younger miner hints at the instruments of subterranean geometry. One can discern a compass suspended in gimbals and a semi-circle. Both were usually suspended from a surveying chain, here represented in the miner's hand, to find one's direction underground. The implicit message was clear: the riches contained in this book were not made of silver, crystal, or even gold. They could not be extracted using a pick and a hammer, or any of the boring instruments casually represented – below the mirror, separated from the mathematical ones – at the two men's feet. The real treasure presented here was the useful knowledge needed to dig and extract precious metals. Buying this book and following the two miners' instructions, the frontispiece implied, any diligent reader could learn the esoteric art of subterranean geometry.

In the early modern period, much of the European economy relied on metals. Rulers fought for the much-needed silver coins that would fuel costly wars – *pecunia nervus belli*, as the saying went. Metal money powered the nascent capitalist economies and enabled early colonial enterprises.[5] Copper could be used for coins and engravings, or as an alloy in bronze to cast bells and cannons. Zinc, tin, and lead all had numerous uses in manufacturing and construction. Lesser-known metals such as cobalt, bismuth, or quicksilver were vital in countless crafts and processes, from medicine to glass colouring.[6] Yet these metals mostly came from a handful of mining regions in central Europe and Scandinavia, most prominently the Ore Mountains of Saxony, the Harz, and several regions of the Hapsburg dominions. Metal mines were closely monitored by local rulers whose finances crucially depended on the 'bloodstained labour and work' of miners, as Voigtel put it in his introduction.[7] A merchant from Nuremberg, a commercial metropolis known for its goldsmiths and founders, summed up this unique position in the early sixteenth century: 'that much silver cannot be found in no other land than in the Holy Empire, so that

---

[4] Voigtel, *Geometria subterranea oder Markscheide-Kunst* (1686), introduction: 'sehr geheim gehalten worden'. (NB. All translations from German, French and Latin to English are my own, unless otherwise indicated.) On this literary tradition, see the recent PhD thesis of Benjamin Ravier, *Voir et concevoir: les théâtres de machines (XVIe–XVIIIe siècle)* (2013).
[5] See Strieder, 'Die deutsche Montan- und Metall-Industrie im Zeitalter der Fugger' (1931), pp. 189–226; Graulau, *The Underground Wealth of Nations* (2019).
[6] For a very short introduction to early modern mining, see Küpker, 'Manufacturing' (2015), pp. 516–519. The seminal work on ore mining – especially in Central Europe – is the four-volume German *Geschichte des deutschen Bergbaus* (2012), here especially vol. 1, pp. 317–452.
[7] Voigtel, *Geometria Subterranea* (1686), *An den Leser*: 'mit blutsauerer Mühe und Arbeit Bergwerck bauen'.

all christian and unchristian countries have to be fed and supplied with silver from the German kingdoms'.[8] In order to meet this demand, mines became both bigger and more complex. In the late fifteenth century, the discovery of new ore deposits led to a silver rush known as the 'mining clamour' (*Berggeschrey*). Sinuous tunnels expanded, following ore veins to previously unknown depths. Capital flew from Nuremberg and Venice, later from the Low Countries, as mining cities flourished and attracted some of the finest minds of their time. Conversely, miners and assayers from various provinces of the Holy Roman Empire travelled to distant countries, helping to extract tin in Cornwall or copper in the Carpathian Mountains of Eastern Europe.[9] Skilled technicians designed the complex mining pumps, the huge furnaces, and the myriad of technical instruments depicted in early modern books such as Agricola's *De re metallica*.[10] This led Max Weber to present, in his *Economic History*, the mines as a crucible of modern rationality, a place from which 'experimentation was taken over into science'. According to Lewis Mumford, 'the mine is nothing less in fact than the concrete model of the conceptual world which was built up by the physicist of the seventeenth century'.[11] And yet, mining relied first and foremost not on mineralogy or metallurgy but on the obscure *geometria subterranea*. The usefulness of this esoteric branch of mathematics quickly became proverbial, as expressed in 1569 by the Parisian humanist Petrus Ramus: 'It is by means of geometrical hands that the Teutonic Pluto draws to itself the riches of the German soil.'[12]

The fateful interplay between geometrical practices and the nascent earth and mining sciences is an important theme of this book. From the silver rush to the foundation of modern mining academies in the late eighteenth century, humanists and later natural scientists relentlessly sought to uncover what they saw as

[8] Dietrich, *Untersuchungen zum Frühkapitalismus im mitteldeutschen Erzbergbau und Metallhandel* (1991), p. 34: 'alles silber findt man die menig in keinem anderen land denn im Heiligen Reich, sondern alle umbligene christliche und unchristliche land müssen aus teutschen landen mit silber gespeist und versehen werden' (text dates from 1523). The Holy Roman Empire produced half of the European metals by some estimates, and figures are much higher for some precious metals such as silver.
[9] Rapp, *Les origines medievales de l'Allemagne moderne: de Charles IV à Charles Quint (1346–1519)* (1989), pp. 149–152; Graulau, *The Underground Wealth of Nations* (2019), especially p. 4.
[10] Agricola, *De re metallica libri XII* (1556). While Agricola was not the only one, or even the first, to depict the mining world – see Biringuccio, *Pirotechnia* (2005) – the quality of his illustrations stands out until well into the following century. See Déprez-Masson, *Technique, mot et image* (2006).
[11] Weber, *General Economic History* (1927), p. 368; Mumford, *Technics and Civilization* (2010), p. 70: 'Did the mine acclimate us to the views of science? Did science in turn prepare us to accept the products and the environment of the mine? The matter is not susceptible to proof: but the logical relations, if not the historical facts, are plain.'
[12] Ramus, *Scholarum mathematicarum libri unus et triginta* (1569), p. 60: 'Ergo Germanicus ille Pluto geometricis manibus divitias suas Germaniae effodit atque eruit.'

the secrets of subterranean surveyors. At the same time, they often relied on the craftsmen's vernacular expertise, on their instruments and experience of practical geometry to advance their own theoretic inquiries.

What did this mysterious discipline consist of? Its German name (*Markscheidekunst*) literally reads as 'the art of setting limits'. Concretely, it encompassed all measuring operations used in the delimitation of concessions and in the daily running of metal mines. Surveyors ascertained the direction of galleries and knew how to bypass crumbled or dangerous sections. They learnt to draw mining maps and to use them to monitor local districts or to connect existing workings. Over time, subterranean geometry came to include more original tasks such as building dams to create water ponds or planning complex hydraulic systems that powered both mining and smelting machines. This artisanal mathematics gradually became an obvious tool to alleviate the uncertainty of mining, manage the technical issues, and organize the extraction of metals in a rational manner. If a few items of Voigtel's textbook belonged to the academic mathematical curriculum, the better part dealt with matters that did not follow traditional categories and classifications. The use of instruments and the numerous surveying methods were passed on from master to student, with an emphasis put on direct observation and training in the mines. In other words, the discipline fell outside of the academic geometry of its time, and vastly differed from the deductive implementation of theories that early modern scholars labelled 'applied' or 'mixed' mathematics.[13]

Subterranean geometry lays more broadly at the intersection of the history of science, culture, economy, and technology. This 'art of setting limits' hardly fits present-day classifications of knowledge, which is precisely why it has so far fallen out of view. In the last decade, however, our understanding of early modern efforts to understand and master nature has been greatly renewed, building on earlier works by Peter Dear, William Eamon, and Hélène Vérin.[14] It has become clear that mines played a decisive role in the transformation of knowledge about nature, confirming the intuitions of Max Weber and Lewis Mumford. These densely urbanized regions attracted money and skills, drawing scholars including Georgius Agricola (see Chapter 1), Jean-André Deluc (Chapter 7), Paracelsus and Gottfried Wilhelm Leibniz.[15] Pamela O. Long has

---

[13] On the distinction between pure, applied, mixed and practical mathematics, see Epple, Kjeldsen, and Siegmund-Schultze, 'From "Mixed" to "Applied" Mathematics: Tracing an Important Dimension of Mathematics and its History' (2013), pp. 657–733. And in particular Jim Bennett's contribution 'How Relevant Is the Category of "Mixed Mathematics" to the Sixteenth Century?' (pp. 677–680).

[14] See Dear, *Discipline and Experience* (1995); Eamon, *Science and the Secrets of Nature* (1994); Vérin, *La gloire des ingénieurs* (1993).

[15] Among recent works, Pamela Long has produced the most influential account on the role of early modern mines. See Long, *Openness, secrecy, authorship* (2001), esp. ch. 6, 'Openness and Authorship I: Mining, Metallurgy, and the Military Arts'; on mining experts and the circulation of knowledge, see Ash, *Power, Knowledge, and Expertise in Elizabethan England* (2004), esp. ch. 1, 'German Miners, English Mistrust, and the Importance of Being "Expert"'.

further argued that historians should devote more attention to artisans in order to understand properly the rise of new sciences. Long describes the mines as important 'trading zones' in which people and knowledge feverishly circulated and adapted to challenging environments, where scholars met and occasionally collaborated with practitioners.[16]

Building on these results, I argue that geometry decisively influenced the new technical landscape of the Holy Roman Empire, with far-reaching epistemological consequences. Despite some brilliant efforts, early modern scholars struggled to understand a culture in which geometry and arithmetic blended with mechanics, mining laws, knowledge of the earth and its minerals. Mine surveyors, for their part, had to cope with capricious economic and political landscapes. From the introduction of gunpowder blasting to the discovery of rich silver mines in New Spain, they incessantly adapted their skills, instruments, and methods. Subterranean geometry was not a distinct science but a craft culture that encompassed a vast tradition of quantification. The following studies thus rely on a broad range of sources: the few textbooks published on the topic – such as Voigtel's *Geometria subterranea* – are contrasted with a rich tradition of manuscripts circulated from masters to pupils. Sketches and survey books, mining maps, and apprentices' examinations, combined with a wide array of administrative documents, reveal the significance of this mathematical culture. Even religion came into play, with Protestant priests writing evocative sermons such as the *Spiritual Mine* or the *Mining Homilies* (see Chapter 2). Sunday sermons indeed played an important role in popularizing the discipline, praising 'the marvellous instrument, the compass, and its use in subterranean geometry'.[17]

More broadly, the present study explores the silent rise of practical mathematics in the early modern period, epitomized here by a discipline which challenges our modern understanding of practice and theory.[18] The history of subterranean geometry illustrates the growing reliance on numbers and geometric figures in civil society at large. Being a cornerstone of the all-important mining *œconomy*, the discipline elicited heated debates about the methods and values of mathematics. Sporadic exchanges with the learned spheres reflect the richness and complexity of the relationship between the science taught in universities and the crafts used underground. The inner logic of the 'art of setting limits', its historical development, and numerous ramifications blur the distinction between crafts and sciences, contributing to current debates and

---

[16] See Long, *Openness, Secrecy, Authorship* (2001); Long, *Artisan/Practitioners and the Rise of the New Sciences, 1400–1600* (2011).
[17] Eichholtz, *Geistliches Bergwerck* (1655), p. 131: 'das Wunder *Instrument* den Compaß / und dessen Gebrauch im Marscheiden'.
[18] Morel, 'Mathematics and Technological Change: The Silent Rise of Practical Mathematics' (2023).

redefinitions of what useful knowledge meant and how it was used in the past.[19] It builds and expands on a growing scholarship about technical education and the role of mathematical practitioners. This discipline, born in the mining pits of the Holy Roman Empire and performed over centuries by craftsmen, who slowly turned into engineers, exemplifies a 'vernacular conception of nature' typical of the early modern period.[20]

## Underground Mathematics

Using mathematical methods in inquiries about nature is often seen as a trait of modernity *per se*, whereas it should be historically questioned and explained. Why did geometry and arithmetic become credible tools in efforts to understand nature? How did this happen concretely in the early modern period? Who were the actors, how did they train and collaborate, and what methods were actually used? Historians dealing with the mathematization of nature have long worked to clarify the complex relationships between scholars and practitioners, sometimes questioning the relevance of such labels altogether. The once radical ideas of Edgar Zilsel (1891–1944), who argued that 'superior craftsmen' had played a key role in the shaping of the new sciences and the introduction of mathematical laws, have been revived, amended, and widely discussed, most recently by Lesley Cormack and Margaret Schotte.[21] Studies have underlined the crucial role played by mathematics and quantification in the emergence of rationality and the rise of new sciences, a position sometimes presented as a 'mathematization thesis'.[22] While there seems to be a consensus on the importance of mathematics in the early modern period, its precise influence and degree of usefulness is considerably more difficult to ascertain.

By describing the rational culture shaped by miners, its social impact and its gradual improvements, the increasing influence of mathematics can be coherently accounted for. Conversely, this book questions the scholarly depictions of early modern mathematical practices, building on Robert Halleux's seminal remark: 'Those writing books and those practising arts and crafts were not the same persons.'[23] Looking beyond academic knowledge, one can enquire

---

[19] Morel, Parolini, and Pastorino, *The Making of Useful Knowledge* (2016); Valleriani, *The Structures of Practical Knowledge* (2017).

[20] Smith, 'Science on the Move: Recent Trends in the History of Early Modern Science' (2009), p. 364.

[21] Cormack, Walton, and Schuster, *Mathematical Practitioners and the Transformation of Natural Knowledge in Early Modern Europe* (2017); Schotte, *Sailing School* (2019).

[22] On the contrary, one can find a critical account of this thesis in Cohen, *The Scientific Revolution: A Historiographical Inquiry* (1994), pp. 309–327, defending the idea that, prior to the seventeenth century, 'mathematics had impinged upon craftsmen's activities only in a few exceptional cases'.

[23] Halleux, *Le savoir de la main* (2009), p. 8.

how laws, religion, and politics converged to present mathematics as the most accurate and efficient method for understanding nature and acting upon it. While many excellent studies focus on the interactions between scholars and their patrons, the present book suggests a methodological workaround. Studying the involvement of craftsmen in human affairs and society at large, one can reconsider the multifaceted influence of quantification on early modern thought.

The historical significance of metals and money, combined with the rich history of earth crafts and sciences, has led to detailed analyses of the early developments of mineralogy, chemistry, or metallurgy. Dowsing rods and transmutative alchemy, mining trolls and popular magic have all been analysed by modern historians.[24] Mathematics, on the other hand, is surprisingly missing from historical accounts of mines, caves, and the underground world. Yet it was ubiquitous in prospecting, extracting, and assaying activities, even if nowadays exact sciences are more commonly associated with university towns or abstract theories. Past uses of arithmetic and geometry in mining are not so much ignored as taken for a static given, or purely empirical rules of thumb. Historians usually recognize that the extraction of ore implied some kind of measurements but see these as mere applications of 'general mathematical laws'.[25] Even as recent studies have recommended using actors' categories and contextualize past knowledge about nature, the growing use of geometric figures and numbers in mining is mostly mentioned only in passing and considered self-explanatory.[26]

In the sixteenth century, this book argues, an intrinsically useful culture of geometry pervaded the mining cities and conditioned their ulterior technical and scientific developments. More importantly, this culture gradually spread beyond the mining states of Saxony, Bohemia or Brunswick and was felt in all of the Holy Roman Empire and beyond. Given that both the mines and the mathematical arts are widely seen as triggering dramatic changes in early modern world-views, one can gain general insights from studying subterranean geometry, which lies precisely at the crossroad of these topics.[27] In order

---

[24] On the history of the dowsing rod and the issue of natural magic in early modern mining, see Dym, *Divining Science* (2011). On transmutative *chymistry* and magical knowledge, see Fors, *The Limits of Matter* (2015).

[25] Baumgärtel, 'Von Bergbüchlein zur Bergakademie' (1965), p. 93.

[26] Graulau, *The Underground Wealth of Nations* (2019), pp. 10–12, rightly states that 'the medieval evolution of mathematics worked to the advantage of early capitalist mining business' – the topic of her book – without engaging with the nature of this evolution or its exact influence.

[27] A series of articles on subterranean geometry was produced in the 1930s and 1940s by Walther Nehm, himself a professor of subterranean geometry. In recent times, the subject has been touched upon by Ziegenbalg, 'Aspekte des Markscheidewesens' (1984), pp. 40–49; Ziegenbald, 'An Interdisciplinary Cooperation' (1993), pp. 313–324; Ziegenbalg, 'Von der Markscheidekunst zur Kunst des Markscheiders' (1997). It is briefly mentioned in Ash, *Power, Knowledge, and Expertise in Elizabethan England* (2004), pp. 28–30.

to understand the practitioners' crafts, it is necessary to immerse oneself in the geology, laws, and religion of the time. Recent works have shown how the global contributions of mathematical practitioners, from navigation to fortification, were intrinsically linked to the specific domains they were part of.[28] To that extent, our modern ignorance of subterranean geometry simply illustrates its idiosyncratic, at times even esoteric, character. These practitioners were literally and metaphorically underground engineers, whose discipline flourished in the early modern period only to lose its global echo in the nineteenth century. Once mining academies were founded in the aftermath of the Seven Years War (1756–1763), public interest gradually shifted to the more challenging puzzles of a quickly industrializing world. The efficiency of mathematics in mine engineering, a source of wonder during the early modern period, was now simply taken for granted. A once much-discussed discipline had been all but forgotten.[29]

What makes the craft of subterranean geometry relevant is that it was routinely practised by and for 'common men' and performed publicly, as detailed in Chapter 2.[30] Long before it became a specialized profession, the *Markscheidekunst* was both an art and an administrative position that could be practised by various officials: foremen and mining masters in the sixteenth century, sworn mining officials, technicians, and engineers in later times. Its history will here be told mostly from the point of view of practitioners, many of them anonymous, including not only the surveyors, but the investors, local preachers, and the mining people at large. This choice obviously has its own challenges. Unlike university professors and clerics, most craftsmen did not publish anything – Voigtel's book being in this sense exceptional – although some of their technical accomplishments have survived to this day. The present book is therefore based on a diverse set of archival documents, handwritten textbooks, maps and sketches, travel diaries and trial reports, calculation sheets as well as the prodigiously comprehensive records of mining administrations.

Writing and publishing simply did not belong to the professional duties of mining officials, as they obeyed a logic of administration rather than a logic of

---

[28] Johnston, *Making Mathematical Practice* (1994), p. 2. Recent works in the history of mathematical practices include the domains of navigation, architecture and fortification. See Schotte, *Sailing School* (2019); Métin, *La fortification géométrique de Jean Errard et l'école française de fortification (1550–1650)* (2016); Lefèvre, 'Architectural Knowledge' (2017), pp. 247–270.

[29] At the turn of the twentieth century, the then-called *Markscheidekunde* had firmly been established as an engineering discipline with proven methods. See Wilski, 'Über die heutige Markscheidekunde' (1933), pp. 61–66.

[30] On the role of the 'common men', see Lutz, *Wer war der gemeine Mann?* (1979); Whaley, *Germany and the Holy Roman Empire* (2013), pp. 12, 187, 221. The notion of *Gemeiner Mann* refers not only to the lowest rank of society, here the illiterate miners, but to the citizen at large, whose 'participation … in the governmental process was undoubtedly an important feature of early modern German society' (Whaley, p. 12).

patronage. Johann Andreas Scheidhauer, a mining master presented in Chapter 6, carefully recorded for several decades his theoretical considerations and experiments on subterranean geometry, water wheels, and all kind of practical mathematics. These documents were known and circulated within a circle of trusted colleagues during his lifetime. However, they were not intended to be printed and thus remained after his death buried in mining archives, where their significance and originality went unnoticed. Surveyors and mining officials were mostly literate and eagerly read the vernacular booklets about geometry or mercantile arithmetic, but few of them were learned. Their lack of university education and insufficient skills in High German were repeatedly mocked by scholars. Such difficulties were compounded when they tried to describe their technical procedures using the *Bergmannsprache*, the heavy dialect spoken in mining regions. German-speaking contemporaries could not understand even a simple discourse on subterranean geometry, and would have needed to peruse the *Mathematisches Lexicon* of Christian Wolff to decipher it. On the title page of his dictionary, the famous philosopher and mathematician conveniently promised to 'describe the dialect and expressions of the subterranean surveyors, as well as of artists and artisans'.[31]

Scholars are not absent from the following case studies, as they repeatedly try to understand, systematize, and divulge what they consider to be 'books of secrets'. During the mining boom, scholars and rulers routinely complained about practitioners 'so jealous and begrudging about their art, that they do not want anybody to see it', at a time when underground surveying was mostly a hands-on know-how that could hardly be put on paper. Knowledge had to be open, scholars argued, 'so that henceforth one or two hundreds craftsmen do not have to believe a single one, without sufficient evidence' and face the potentially severe consequences.[32] These debates, as we shall see, were ultimately less about the openness of knowledge than about who had authority over it. Even the publication of Voigtel's comprehensive *Geometria subterranea*, analysed in Chapter 4, did not put an end to the criticism of practitioners. Leonhard Christoph Sturm, university professor and member of the Prussian Academy of Science, confessed that he had 'little experience of the mining language'. Having never found 'the time and patience to read the bespoken book', it was enough for him to 'leaf through it and consider the images'. Having visited a mining site once, he had witnessed surveying operations but 'could not understand anything about it'. Faced with the incommunicability of practical

---

[31] Wolff, *Vollständiges Mathematisches Lexicon* (1734), title page: 'Die Mund- und Redens-Arten derer Marckscheider auch hieher gehöriger Künstler und Handwercker, beschrieben.' For a modern analysis, see Drissen, *Das Sprachgut des Markscheiders* (1939).
[32] Reinhold, *Gründlicher vnd Warer Bericht. Vom Feldmessen, Sampt allem, was dem anhengig*, (1574), dedication: 'damit forthin nicht ein 100. oder 200. gewercken / einem allein / ohne gnungsame beweiß / mußten glauben geben.'

knowledge, the lofty professor thus reached the only logical decision: 'For that reason, I have dared to propose my method of underground surveying.' Sturm, who had never tested his proposal, confidently advised surveyors to follow it, 'although I don't know if it would coincide with the usual method'![33] By the strong reactions they provoke, mathematical practices act as a mirror that reflects the ambiguity, the ambitions, and often the limits of a learned pursuit of operative knowledge in the early modern period.

## Numeracy, Trust, and the Circulation of Knowledge

What could lead people to rely specifically on geometry and arithmetic for their inquiries and businesses?[34] Ambitious rulers, greedy investors, and the usually harsh mine foremen had no particular affinity with mathematics, which was considered to be an abstract science. It is certainly true that mathematics was experiencing formidable transformations in the early modern academic circles. The successes, however, were mostly concentrated in algebra, astronomy, and mechanical theories, areas unrelated to the daily concerns of miners, or indeed most common men. It would be a very long time before theory would trigger any rational belief in the power of numbers or geometrical figures among the masses.

The superiority granted to mathematics that developed at the time, while obvious from our modern point of view, thus warrants explanations. Miners could have settled for dowsing rods, which were firmly based on popular beliefs. Many officials, moreover, possessed a deep understanding of their local environment, knowing for instance how to ascertain the direction of ore veins from the colours and hardness of stones. Why was this tacit knowledge complemented, and in many cases slowly superseded, by a rational use of mining compasses and measuring chains? How did not only visual testimonies and visitations, but also oral and written inspection reports, come to be replaced by mining maps, on which geometrical operations were performed in an office using a pair of compasses? Visiting the Harz in the late eighteenth century, a fellow of the Royal Society still wondered: what could move 'a miner, solely upon the faith of his Geometer, and in the absolute obscurity of the entrails

---

[33] Sturm, *Vier kurtze Abhandlungen* (1710), pp. 45–47: 'Ich bin der Berg-Sprache ebenfals wenig erfahren ... habe auch sonst niemalen Zeit und Gedult gewinnen könne besagtes Buch durchzulesen / sondern habe mich bloß mit durchblättern und Betrachtung der Figuren begnüget ... nichts davon verstehen können ... Derowegen getrauete ich mir wol auf meine Art eine Marckscheider Messung zu verrichten ... weiß aber nicht / ob ich mit der gewöhlichen Methode ganz übereinstimmen würde.'

[34] Ash, *Power, Knowledge, and Expertise in Elizabethan England* (2004), p. 30, states that sixteenth-century 'German surveyors had also learned to adapt the various geometric and trigonometric techniques of their art'.

of the earth', to undertake 'a labour that is to cost him years in daily boring through a rock'?[35]

This book argues that the answer has little to do with the epistemological certainty of exact sciences and their growing prestige in the learned world. The dynamic of practical mathematics at the time was mostly unrelated to the recent breakthroughs of algebra or, later, to the burgeoning calculus. The authority of subterranean surveyors built on an earlier, broader, and more systematic use of quantitative knowledge, a tradition notoriously underrated in historical analyses. The formidable efficiency of the four operations, basic geometrical methods, and a skilled use of rudimentary tools was further supported by a combination of economic, cultural, and religious factors.[36]

In other words, successful and authoritative uses of mathematics in the mines were not conditioned by their stage of theoretical development, but rather by their cultural impact and the wide recognition they enjoyed. If anything, the influence was the opposite: the regular experience of the senses and the concrete accomplishments based on arithmetic and geometry might well, in the long run, have heightened the public trust in abstract theories. At a time when many cities and merchants were still using Latin numerals, counting boards, and jetons for their operations, a culture of mathematics quickly spread in the mining towns of the Empire.[37] The archetypal example of the arithmetician (*Rechenmeister*) Adam Ries who left the city of Erfurt and its prestigious university to work as a mining clerk and open a reckoning school in the booming mining city of Annaberg, will be presented in Chapter 3. A written culture of bookkeeping, concession setting, and public accounting developed, soon amplified by the nascent Protestant Reformation.[38] In order to grasp the deep influence of mathematics on early modern thought, the skills and methods of mine surveyors should be understood along with the general culture of their time.

In response to 'the wildfire spread of literacy' in early modern society, surveying operations were publicly displayed and open to legal challenges and debates, as detailed in Chapter 2.[39] Subterranean surveyors performed their measurements in ceremonies so as to ensure their social acceptability. Protestant preachers purposely incorporated the mathematics of mining in their sermons about the credo of reason. Witold Kula has brilliantly shown, in

---

[35] Deluc, 'Barometrical Observations on the Depth of the Mines in the Hartz' (1777), p. 424.
[36] Recently, the concept of 'everyday mathematics' has seemed to open new perspectives on this often-neglected part of sciences, see Wardhaugh, *Poor Robin's Prophecies* (2012).
[37] Rüdiger, 'Zur Rolle der Lateinschulen' (2008), p. 329. A more general reference about the spread of Hindu-Arabic numerals is Danna, 'Figuring Out' (2021), pp. 5–48.
[38] On the influence of Protestantism on mining culture, see Knape, *Martin Luther und der Bergbau im Mansfelder Land* (2000).
[39] Porter, *Science, Culture and Popular Belief in Renaissance Europe* (1991), p. 7.

his studies on metrology and measuring practices, how widespread the 'distrust of counting and measuring' could sometimes be in medieval Europe.[40] When this body of knowledge made its way into most metal mines of Europe and the American colonies, it was still carried out by or under the supervision of German-speaking specialists. One encounters these surveyors almost everywhere, from the great copper mine of Falun in Sweden to the quicksilver mines of Almaden in Spain.[41] In another cultural context, however, the fragile consensus built around the use of geometry could quickly falter. French officials routinely complained about the 'foreigners that we bring in our Mines at great cost'. In some instance, impostors even produced stolen maps and pretended to be 'oracles of subterranean geometry'. Seducing and deceiving officials with their 'strange accents that strike the ear', they caused costly accidents.[42]

In the mining cities of the Holy Roman Empire, on the contrary, surveyors never appear as mystery-mongers but mostly as known and trusted artisans. Common miners learned to respect measuring procedures because they witnessed the laws of geometry on a daily basis, in the mines, in churches, and even during trials. This broad diffusion introduced new values in mining societies and enhanced the status of mathematics, ensuring in turn the acceptability of more risky and elaborate methods, described in the last three chapters of the present book. Along with well-known centres such as Nuremberg, Paris, and London, a greater variety of places thus deserves our attention if we are to understand the silent rise of practical mathematics.[43] As mining operations grew more complex and technical during the seventeenth century, surveyors improved their instruments and methods, assimilating theoretical developments – such as map-making and trigonometry – on their own terms. A curriculum of surveying techniques was written down in a handwritten corpus widely circulated among practitioners, as described in Chapter 4.[44] To cope with the peculiarities of mining problems, a formal system of companionship appeared. Young people trained for several years, learning on site how

---

[40] Kula, *Measures and Men* (1986), p. 14.
[41] On exchanges with Spain and the Habsburg Empire, see Morel, 'Circulating Mining Knowledge from Freiberg to Almaden' (2016). On German miners, and most specifically subterranean surveyors, employed in early modern Scandinavia, see Hillegeist, 'Auswanderungen Oberharzer Bergleute nach Kongsberg/Norwegen im 17. und 18. Jahrhundert' (2001).
[42] The complainant was Jean-Pierre-François Guillot-Duhamel, a teacher at the Parisian *Ecole des Mines* founded in 1783. See Duhamel, *Géométrie souterraine* (1787), introduction: 'des étrangers que nous faisons venir à grands frais dans nos Mines [.] ils s'annoncent comme les oracles de la Géométrie souterraine ... on est séduit par l'accent inconnu qui frappe l'oreille.'
[43] On crafts and practical mathematics in Nuremberg, see Smith, *The Body of the Artisan* (2004), ch. 2, especially pp. 66–72, 89, as well as Iwańczak, *Die Kartenmacher* (2009).
[44] On the corpus of *geometria subterranea*, see Morel, 'Five Lives of a *Geometria subterranea* (1708–1785)' (2018), pp. 207–258.

to perform measurements and collect the relevant data. Back in their mining offices, they were taught the calculations necessary to draw maps and work out solutions on paper. Long before any international standardization of engineering curricula was implemented, the specificity of mining geometry implied many exchanges among technicians. Long-distance circulations and a diplomacy of skills linked various states of the Holy Roman Empire with Scandinavia, Russia, and other European powers. This gradually led to the creation of structured institutions delivering reputed certificates. In Chapter 6, we shall see how mining schools slowly matured in the eighteenth century, trying to standardize and improve existing practices without getting lost in abstract or unpractical solutions.

## Tools of Visualization and Control

Mining sites were among the biggest engineering projects of the early modern period. As mines went deeper, digging and extraction became capital-intensive and involved massive machines. Such projects required meticulous management operations in a world full of unknowns and hugely unpredictable. In a continent plagued by wars, mining sites were obvious targets for ambitious monarchs – the ravages of the Thirty Years War described in Chapters 4 and 5 being the most striking examples. Moreover, the economy was subject to severe and long-lasting downturns: when the gold and silver of the New World led to widespread inflation in the second half of the sixteenth century, many continental mines suddenly became unprofitable.[45] At a local level, any well-crafted plan could be made obsolete if investors went bankrupt, if the local ruler died or simply decided to buy extravagant items rather than to fund much-needed machines or tunnels.

Rulers and mining officials needed tools to cope with uncertainty. In that context, geometry and arithmetic soon became ubiquitous instruments, rare islets of certitude in an ocean of instability. Mines gradually turned into sites of numerate expertise and practices. One should not, however, fall into a positivist illusion. Its efficiency notwithstanding, subterranean geometry was no exact science – just as modern surveying isn't – and its goal was not to reach the perfection of a scientific demonstration. That being said, a surveyor using wooden stakes to delineate the path of a planned tunnel on the surface made an abstract and daunting task suddenly palpable. Information about the invisible was also a powerful tool to ensure the equity (*Gerechtigkeit*) of mining. Given

---

[45] On the economic conjecture and its influence on mining, see Dietrich, *Untersuchungen zum Frühkapitalismus im mitteldeutschen Erzbergbau und Metallhandel* (1991); Tenfelde, Berger, Seidel, Bartels, and Slotta, *Geschichte des deutschen Bergbaus* (2012), vol. 1, pp. 317–354, 453–475.

that 'no one can see through stones', preacher Mathesius warned, 'a *quarter* of favour often weighs more than a *centner* of equity'.[46] Using geometry, it became possible to produce mining maps representing the subterranean world. In that context, measurements and computations provided valuable data for making technical decisions.

In the mining states of central Europe, practical mathematics served both as a technical and as a political tool, as highlighted in Chapters 3 and 5. The tight control exerted by rulers over surveying and mapping activities mirrored in many cases the control that regalian monopolies exerted on mint or justice.[47] Subterranean geometry was practised under extreme conditions, where information was hard to grasp and major errors were clearly not an option. Working with instruments that often were self-designed and self-repaired, surveyors were nevertheless able to achieve incredible results. The rough conditions of metal mines directly shaped their methods, which were designed with robustness in mind. Measures were repeated to avoid mistakes, while frequent visitations during digging operations ensured that miners' deviations could be corrected in real time.

The broader conception of practical geometry underwent important transformations during the early modern period, and this topic has recently become a lively field for historical analysis. What it meant to make good use of numbers, diagrams, and plans changed tremendously.[48] The idea of a fully controlled world, or of a perfect mathematical model, were not simply out of reach in the early sixteenth century but, just like mining maps, such ambitions did not even exist. It took several generations for the mining people to learn to adapt charts and diagrams to their daily concerns. To understand how new forms of visualization of caves, ore veins, and the underground emerged, our analysis must be broadened and include visitation reports, jurisprudences, and a wide range of geometrical manuscripts. By doing so, we will understand both how mining maps developed from a technical point of view, and how they were subsequently produced, stored and used to design ever larger and more complex projects.[49]

Recent studies have tried to move beyond the image presenting mathematical practices as a static given to understand their materiality, uneven progress,

---

[46] Mathesius, *Sarepta oder Bergpostill* (1562), p. 204v: 'weil niemandt durch den stein sehen kan / und offt ein quintet gunst mehr gilt / denn ein centner gerechtigkeit'.

[47] This argument is examined in detail in Chapters 3 and 5. On the idea of an early modern 'geometry of power', see Korey, *The Geometry of Power* (2007); Dolz, *Genau messen, Herrschaft verorten* (2010).

[48] On early modern visualization and diagrams, see a summary and bibliography in Bigg, 'Diagrams' (2016), pp. 557–571.

[49] Stephen Johnston has offered a similar analysis in his study of practical geometry, plats and design of Dover Harbour. See Johnston, *Making Mathematical Practice* (1994), ch. 5.

and ubiquitous development.[50] At the turn of the sixteenth century, artisanal geometry used to be mainly a set of efficient, versatile, and yet rudimentary tools to obtain quantitative information about nature. Early successes and incremental improvements widened the range of technical problems that a skilled *Markscheider* could hope to solve and allowed for bolder ambitions, which in turn transformed the very essence of the work. Around 1800, mine surveyors were full-fledged engineers who could design and monitor major engineering projects from an initial idea to the final breakthrough.

## A Sinuous Path

This book explores various sides of subterranean geometry to understand how a culture of accuracy and artisanal practices of mathematics developed in the mining pits of the Holy Roman Empire. A first chapter introduces the discipline from the point of view of Renaissance scholars. Early modern humanists were fascinated by the underground world of metal mines. The richness of the geometry contained in Georgius Agricola's *De re metallica* (1556) and Erasmus Reinhold's *On Surveying* (1574) is presented. I further show that these books, despite their lifelike descriptions and illustrations, did not limit themselves to straightforward, faithful depictions of actual practices. Early modern readers were presented with rational reconstructions and pseudo-technical procedures. In spite of a thorough knowledge and a genuine interest for the underground world, scholars mainly used their writings on mines in a patronage context, or to display their interpretation of Euclidean geometry.

The second chapter, by contrast, delves directly into early modern metal mines. The birth of a vernacular culture of geometry is described, detailing the daily work of craftsmen and insisting on the materiality of measuring practices. Surveys, carried out in public during solemn ceremonies, were a keystone of mining laws. A central hypothesis of this book is that mathematical accuracy acquired a dual meaning at the time: measurements had to be precise enough to solve intricate technical problems, while respecting procedures codified in mining customs and laws. Far from being a mere tool, geometry was meant to ensure trust; it was ubiquitous and pervaded many aspects of a miner's life. In the early years of the Protestant Reformation, Lutheran pastors actively fostered the rise of practical mathematics. Mathematical and religious rationality were equated, making subterranean geometry accurate in a third way, this time as an expression of divine will. The omnipresence of measurements, combined with their legal and religious recognition, ultimately conferred a higher status to the discipline.

---

[50] Streefkerk, Werner, and Wieringa, *Perfect gemeten* (1994); Morera, 'Maîtriser l'eau et gagner des terres au XVIe siècle' (2016), pp. 149–160; Jardine, 'Instruments of Statecraft' (2018), pp. 304–329; Schotte, *Sailing School* (2019).

Chapter 3 is set both in the Ore Mountains and at the court of Dresden under August of Saxony (1553–1586). It offers a broader picture of the vibrant intellectual life in mining cities, illustrated here by the example of Annaberg. Local officials and technicians developed remarkable skills in arithmetic and geometry, on which rulers came to rely to map their realms and tame their capricious landscapes. I focus on the careers of two dynasties of practitioners, respectively subterranean surveyors (the Öders) and reckoning masters (the Rieses). After both patriarchs contributed to the economic rise of the city, their descendants became versatile engineers and courtiers of the Saxon Electors. They collaborated with university professors and instrument-makers, using their skills all over the Electorate and beyond, temporarily turning the court of Dresden into a centre of practical mathematics.

Mining greatly evolved during the troubled time of the Thirty Years War (1618–1648) and so did subterranean geometry. Balthasar Rösler (1605–1673) introduced numerous innovations, and his teaching was disseminated by his students among mining regions, in a series of beautifully illustrated and hitherto unstudied manuscripts. The fourth chapter analyses the birth of this technical genre, its evolution and uses within the training system of mining regions. In 1686, Nicolaus Voigtel published the first practical textbook on the topic, to which belonged the frontispiece presented above. Surprisingly, the craftsmen's manuscripts weathered the rise of the printed press. I argue that authoring and publishing books failed to supersede the authority of practitioners precisely because their know-how was embedded in a peculiar technical and cultural setting. Subterranean geometry would remain an underground knowledge for another century, as most innovations arose within this handwritten tradition.

Chapter 5 focuses on the visualization of the underground, arguing that its development was strongly linked to broad changes in the political structure of mining regions. Drawing mining maps and working on them became widespread in the second half of the seventeenth century, gradually replacing alternative tools such as written reports of visitations, wood models, or annotated sketches. In Saxony, Captain-general Abraham von Schönberg (1640–1711) put his weight and reputation behind the new cartographic technology, hoping that its acceptance would in turn help him advance his reform agenda. At-scale representations were instrumental in justifying new investments, while offering technical road maps to implement them. Johann Berger (1649–1695) spent years producing a monumental cartographic enterprise, the *Freiberga subterranea* (1693) to support his patron's ambitions. As surveyors finally realized the old dream of 'seeing through stones', the administrations rapidly seized their skills to reform and police their subterraneous cities.

Chapter 6 offers a reappraisal of the foundation of mining academies. Subterranean geometry merges here with broader questions about technical education in the eighteenth century. Early attempts to replace the guild-like training and to establish brick-and-mortar institutions prompt a familiar debate between professors and practitioners. Who could best formalize and improve a century-old corpus? Moreover, what was the right way to teach it? Major mining centres, I argue, offered varied solutions to improve theoretical teaching, of which mining academies were but the ultimate step. I focus here on the biography of Johann Andreas Scheidhauer (1718–1784), mining master and autodidact mathematician. His vast geometrical production – unpublished and long forgotten – looms large in the early projects of mining academies, not least through the influence of his student Johann Friedrich Lempe (1757–1801), emblematic professor of the *Bergakademie* Freiberg.

The last chapter relates the digging of the Deep-George draining tunnel (1771–1799), named after George III, King of Great Britain and Hanover. During the planning phase, surveyors designed this engineering project with a previously unknown level of detail. Jean-André Deluc, Fellow of the Royal Society and reader to Queen Charlotte, visited the Harz mines three times while the operations were under way. Deluc's geological and meteorological inquiries led him to perform barometric experiments in these mines. He relied on practitioners' data to test and calibrate his instruments, marvelling about their precision in the *Philosophical Transactions* and the *Journal des Sçavans*. Scholars and amateurs – from Goethe to Watt – but also merchants and their wives rushed to visit the project. Year after year, journals reported to the public how the various sections of the tunnel connected seamlessly. In the late eighteenth century, the two worlds of natural scientists and mine engineers met one last time; this time around common issues of precision, data gathering, and instrumentation.

Subterranean geometry is an original lens through which we can consider the rise of numerical information, the evolution of engineering, and finally the cultural significance of early modern mathematics. Setting aside – for a moment – the bold issue of the rise of the new sciences and entering the metal mines of the Holy Roman Empire, one suddenly encounters craftsmen whose geometry was firmly rooted in popular beliefs and daily practices.[51] This seemingly invisible discipline illustrates the gradual evolution and local challenges of practical mathematics, its irregular improvements, wide circulation, and evolving relationships with related arts and crafts. By studying these surveyors on their own terms – 'as it essentially was'[52] – one understands how geometry quietly found its way into public ceremonies, customs, and legal disputes.

---

[51] Pumfrey, Rossi, and Slawinski, *Science, Culture and Popular Belief in Renaissance Europe* (1991), p. 6; Cooter and Pumfrey, 'Separate Spheres and Public Places' (1994), pp. 237–267. More recently, see Smith, 'Science on the Move' (2009), pp. 345–375.
[52] Ranke, *Geschichten der romanischen und germanischen Völker* (1824), vol. 1, p. vi: 'wie es eigentlich gewesen'.

One witnesses how it ultimately became indispensable to the administration of nascent modern states.

From the sixteenth century onwards, the ubiquitous activities of mathematical practitioners exerted an immense influence on civil life and technological change, in the mines and elsewhere. Everyday mathematics was increasingly employed in the public sphere, mostly without being mediated by academic knowledge. The relevance of scholars, and later natural scientists, is not refuted; yet I argue that it should be qualified and thoroughly compared with other forms of knowledge, instead of being universalized. Subterranean surveyors wrestled with the chaotic world of the underground. Facing issues of instrumentation and visualization, they used measurements to objectively assess the present and anticipate future developments. In doing so, they would contribute to a new quantifying culture in the mining regions of the Holy Roman Empire and advance a concrete mathematization of nature.

# 1    Of Scholars and Miners

Subterranean geometry developed as a set of measuring practices and rules of thumb, performed by mining officials in several regions of the Holy Roman Empire. Highly sought after, German-speaking technicians were recruited by virtually all European powers in the early modern period, for their skills were unmatched. They knew how to prospect, extract, and smelt the ore better than anyone else, and their machinists were unrivalled. They were also familiar with a complex system of mining laws and customs that relied on very precise surveying operations. Knowing one's position at the surface and underground was both crucial and priceless: using cords and compasses, these experts were able to ascertain concession limits and find the direction of mining veins with an efficiency that would both astonish foreign observers and comfort worrying investors.

Given the lack of written sources from craftsmen, however, the study of early modern mines today usually involves dealing with scholarly works such as Agricola's *De re metallica*. Georgius Agricola (1494–1555), a Saxon physician and the author of numerous books about the mining world in his time, is often depicted in the historiography as the 'father of metallurgy'.[1] A polymath with a strong humanist inclination, he is mainly remembered as the author of the *De re metallica* (1556), containing twelve books. These twelve books 'on the nature of metals', presenting at length and in great technical detail the renewal of ore mining, are rightly seen as 'the most famous mining treatise of the sixteenth century'.[2] Other famous works about early modern mineralogy and metallurgy include Vannoccio Biringuccio's *De la Pirotechnia* (1540), Lazarus Ercker's *Treatise on Ores and Assaying* (1574), and the collection of *Bergbüchlein*, small booklets published earlier in the century, although none quite matched the influence of Agricola's work. The mining literature deeply

---

[1] Kessler-Slotta, 'Die Illustrationen in Agricolas *De re metallica*' (1994), p. 55; Hannaway, 'Georgius Agricola as Humanist' (1992), p. 553. German historians tend to present him rather as the 'father of mining sciences [*Montanwissenschaften*]', see Ernsting, *Georgius Agricola, Bergwelten 1494–1994* (1994).

[2] Long, *Openness, Secrecy, Authorship* (2001), pp. 183–184.

influenced the way useful knowledge was formalized in the early modern period; these books 'set out many technical processes in written form, rationalizing the disciplines of mining and metallurgy, using and creating precise technical vocabulary'.[3]

In order to understand the esoteric discipline of underground surveying, this mining literature seems to offer a good starting point. Eric Ash, for instance, praises the *De re metallica* in his lively analysis of expertise in modern England as 'the most important single work on mining operations published during the early modern period'. He uses it as a true description of what actually happened in the mines, notably for subterranean geometry, following most historians, both local specialists of mining history and cultural historians of science, in that respect.[4] Despite Agricola's stature and the importance of the underground surveying, there is no thorough analysis of the mathematics presented in his *De re metallica*. Other early modern scholars addressing these themes, such as Erasmus Reinhold the Younger (1538–1592), court physician and son of the famous astronomer, are all but forgotten. Reinhold published in 1574 a surveying treatise entitled *Vom Marscheiden kurtzer und gründlicher Unterricht* (A Short and Rigorous Instruction to Underground Surveying). Both works deal extensively with the use of geometry and instruments in mining; they also contain comparisons of numerous measuring units, both ancient and modern, together with the use of cords or various measurement methods.[5]

Scholarly books offer a wealth of information to understand the learned world's growing curiosity regarding the underground world, the generation of metals, and the progress of mechanical arts. One should, however, be aware of a major caveat, summed up concisely by Robert Halleux: '[T]hose writing books and those practising arts and crafts were not the same persons'.[6] Amazed as we are by the beautiful engravings of pumps and water wheels presented in the *De re metallica*, we know little about the machinists who patiently designed, built, and improved the original models – and it is very difficult to tell how accurate these beautiful figures indeed are. The same

---

[3] Long, *Artisan/Practitioners and the Rise of the New Sciences, 1400–1600* (2011), p. 112. On the *Bergbüchlein* tradition, see Darmstaedter, *Berg-, Probir- und Kunstbüchlein* (1926); Koch, *Geschichte und Entwicklung des bergmännischen Schrifttums* (1963).

[4] Ash, *Power, Knowledge, and Expertise in Elizabethan England* (2004), pp. 22–30; Déprez-Masson, 'Richesse minières et traités techniques' (1989), p. 184; Déprez-Masson, *Technique, mot et image* (2006), p. 29; Long, 'The Openness of Knowledge' (1991), pp. 318–355; Graulau, *The Underground Wealth of Nations* (2019), p. 13.

[5] Some mining engineers, writing about modern surveying techniques, mention Agricola in passing: see Wilski, *Lehrbuch der Markscheidekunde* (1929), pp. 117–119. The only work to seriously engage with books 4 and 5 is Déprez-Masson, *Technique, mot et image* (2006), pp. 194–208.

[6] Halleux, *Le savoir de la main* (2009), p. 8.

holds for the mathematics of deep-level mining. In this chapter, the geometry of mines included in humanist books of the sixteenth century will be analysed in detail, to understand precisely both the endeavour of Agricola and Reinhold, and how they presented the underground world to their respective audiences. I then focus on the ambiguous relationships between written works and actual practices, arguing that literary descriptions of surveying practices are interesting and relevant, but not quite in the way one might expect. A significant part of their content had in fact little to do with actual measuring methods – it was borrowed from a learned corpus of the medieval period, the *practica geometriæ*, and adapted both to the context of the German mining boom and to their authors' humanist aspirations. Their limitations notwithstanding, these important works contributed to contemporary debates among scholars on the legitimacy of mining, the role of mechanical arts, and the usefulness of mathematics.

## Mining and Geometry in the Middle Ages

In the Middle Ages and up to the late sixteenth century, the major regions for the extraction of precious metals were located in the Holy Roman Empire and more generally in Central Europe. This fact would have come as a surprise to most Ancients, who thought precisely the opposite to be true. The most extensive antique description of the region is found in Tacitus' *Germania* (originally entitled *On the Origin and Situation of the Germans*). Written around 98 CE, this work was rediscovered in the fifteenth century and was then much discussed by German humanists.[7] Tacitus described a region deprived of precious metals: 'Silver and gold the Gods have denied them, whether in mercy or in wrath, I am unable to determine', wisely adding: 'Yet I would not venture to aver that in Germany no vein of gold or silver is produced; for who has ever searched?'[8] Modern archaeology has confirmed that metallurgy in Central Europe was largely confined to the self-production of iron and 'was by far not as important as, for example, in the Spanish peninsula'.[9] Little is known about the exact surveying methods employed in antiquity, especially in the extensive Roman operations. Roberto Rodriguez, an archaeologist, has shown how instruments such as the water level, the *groma*, and *dioptra* must have been used, and one might thus speculate that

---

[7] Cochlaeus, Langosch, and Reinhardt, *Kurze Beschreibung Germaniens = Brevis Germanie descriptio (1512)* (2010).

[8] Tacitus, *On Germany* (1910), part 1, adding: 'For the use and possession [of gold and silver], it is certain they care not'. See Tenfelde, Bartels, and Slotta, *Geschichte des deutschen Bergbaus* (2012), vol. 1, pp. 95–110.

[9] Tenfelde, Bartels, and Slotta, *Geschichte des deutschen Bergbaus* (2012), vol. 1, p. 121. See also Suhling, *Aufschließen, Gewinnen und Fördern* (1983), pp. 38–67.

ancient mine geometry derived from the vast agricultural corpus written by the official surveyors, known as *agrimensores*.[10]

In the late Middle Ages, mining operations still concentrated on relatively small sites around the Mediterranean Sea but also in the mines of the Rammelsberg in Western Germany. The gradual 'German eastward colonization' of the inhospitable forests and mountains of Central Europe soon revealed promising veins, and pockets of German-speaking miners settled in mining regions from the Alps to the Carpathians. Mining towns were founded in the late twelfth century, for instance Freiberg in Saxony and Kutná Hora in Lower Hungary (now part of Slovakia).[11] Although the extent to which ancient surveying techniques were transmitted is unclear, current interpretations suggest a break during the Middle Ages.[12] At this point, two brief comments should be made. First, these methods were no mere bookish knowledge, and might indeed have survived the decline of the *agrimensores* tradition. In any case, they were fundamental enough to be either passed down through the generations or independently rediscovered. A good example is the water level – a long piece of wood hollowed and filled with water in order to establish precisely the horizontal plane – that was independently developed in many cultures. Second, more advanced mining techniques could not be simply transferred from one society to another, or from Mediterranean workings to the pits of Northern Europe. Local geographies and the appropriation of knowledge necessarily implied a degree of accommodation, as Chandra Mukerkji has shown in regard to the adaptation of Roman hydraulic techniques in the French Pyrenees.[13]

Political evolution in the late Middle Ages altered the development of mining and, crucially, the role played by surveyors.[14] Authority over mining districts was highly contested, for it conditioned the appropriation of underground riches. The Golden Bull of 1356 – an elaborate constitutional frame for the Holy Roman Empire – gave the ownership rights of metallic ore to local rulers, who in turn tried to lure investors by promulgating *Bergfreiheiten*

---

[10] Rodríguez, 'Ingeniería Minera Romana' (2004), p. 180, who indicates that these works 'required the application of topographic techniques of a certain precision, similar to those used in other Roman public works, with instruments such as the water level, dioptra, groma or chorobates' (my emphasis). A synthesis about which ancient texts on surveying have been preserved can be found in Reimers, 'Wurzeln des Markscheidewesens im Spiegel gelehrter Schriften' (2021), pp. 93–127.

[11] See Pounds, *An Economic History of Medieval Europe* (1994), pp. 327–333 and Suhling, *Aufschließen, Gewinnen und Fördern* (1983), pp. 72–78; see also Graulau, *The Underground Wealth of Nations* (2019), pp. 30–72.

[12] Alberti, 'Entwickelung des bergmännischen Rißwesens' (1927), p. 9.

[13] Mukerji, *Impossible Engineering* (2009), pp. 134–153.

[14] A well-documented example of mining laws and operations in the Late Middle Ages is Massa Marittima, analysed in Pfläging, *Bergbuch Massa Marittima: 1225–1335* (1977). The role of surveyors (*arbitrator*) is detailed on pp. 60–64.

(mining liberties).[15] Prosperous districts were formed by many prospectors opening small concessions aligned along ore veins. This environment made a precise
setting of boundaries necessary and an 'art of setting limits' (*Markscheidekunst*,
the term that later came to designate underground geometry) developed in order
to secure property rights. Even in France, where the mighty kings readily used
dubious claims to seize natural resources, surveying was becoming central.
When Charles VII prosecuted his financier Jacques Cœur in 1453, taking control of his silver mines in Pampailly, one of the first decisions was to hire specialists – among them the *niveleur* (leveller) Claus Smermant – to achieve a
gallery draining groundwater. All of them belonged to small communities of
German-speaking experts circulating between mining regions.[16]

In the Middle Ages, mathematics was thus not 'applied' to mining in the
way that a modern science would be coherently transferred to a technical issue.
Various tools and measurement techniques were used in a piecemeal fashion to
solve seemingly unrelated issues. In other words, experts in mine surveying –
*Markscheider, niveleur,* or *Schiner* – did not conscientiously acquire geometrical
skills they would later use in their daily work.[17] In tackling issues such as concession setting and following ore veins, they improved their know-how based
on experience, trials, and numerous errors. At the time, most of the extraction
happened in small-scale workings, located either at the surface or with limited
depth, which made rough approximations and estimates sufficient. Doing so,
miners gradually came to rely on counting and measuring, among many other
skills related to stones, minerals, and lodes. Mathematics was amalgamated
with the rest of mining knowledge to the point where it was often invisible to
the untrained eye. A casual early modern observer likely saw in the expression
'teaching of cracks and veins' a vague allusion to rocks and cavities, while mining officials immediately knew that it referred to a geometrical procedure to find
the direction of a mining vein.[18] Altogether, this eminently pragmatical use of
measurements, blended with crafts and know-how without any overarching theory, led to long-lasting prejudices towards subterranean geometry. As late as the
mid-eighteenth century, a university professor could write that the discipline was
'badly regarded by *mathematicis*, and held for a bastard in friendship', adding
that no proper scholar could have 'respect' for it.[19]

[15] This was detailed in the ninth chapter of the Bull (*Of Gold, Silver, and Other Kinds of Ore*). See
also Suhling, *Aufschließen, Gewinnen und Fördern* (1983), pp. 78–83; Tenfelde, Bartels, and
Slotta, *Geschichte des deutschen Bergbaus* (2012), pp. 191–192; Weber, *General Economic
History* (1927), pp. 178–189.
[16] Benoit, *La mine de Pampailly: XVᵉ–XVIIIᵉ siècles* (1997), pp. 78–80.
[17] Gautier Dalché and Querrien, 'Mesure du sol et géométrie au Moyen Âge' (2015), pp. 119–121.
[18] See, among many others, Agricola, *Vom Bergkwerck: xij. Bücher* (1557), preface: 'ein grundtli-
che erfarung der gengen / klüfften / schichten / und absetzungen des gesteins'.
[19] Jugel, *Geometria subterranea* (1773), p. 7: 'sie bishero von den *Mathematicis* schlecht angesehen
und für einen Bastard in der Freundschaft gehalten, und aus deren Achtung ganz ausgeschlossen
worden', although Jugel himself tried to explain why this position was, in this view, not justified.

This does not mean that no such thing as a practical geometry existed in the Middle Ages. In fact, the very term of *practica geometriæ* was introduced in the twelfth century by Hugh of Saint-Victor (d. 1141), head of the Saint-Victor abbey in Paris.[20] He coined it to distinguish between the theoretical or 'speculative' part of geometry and its practical or 'active' part. The latter typically dealt with instruments or procedures, while the former focused on abstraction and demonstration. His *Practica geometriæ* successively presented the science of length, surfaces and the sphere and gave birth to a literary genre that blossomed in the Renaissance, as will be shown later in this chapter.[21] Suffice it to say that this so-called practical geometry was influential in scholastic circles but hardly known to practitioners.[22] The instruments mentioned in this textual tradition, most prominently the astrolabe and the quadrant, have little in common with the tools that mathematical surveyors, gaugers, or *Markscheider* used on a daily basis. Broadly speaking, these geometrical texts had a theological goal, aiming at 'enlarging the scope of meditation on spiritual subjects'.[23] Their relationships with actual practices mirror those of the recipe books that developed in monasteries in the same period. As William Eamon has aptly noted, 'the medieval craft recipe books have to be considered, strange as it may seem, as literary creations within a complex literary tradition'.[24] Similarly, the *practica geometriæ* were literary objects assembled by scholars without contact with the actual surveyors, who at the time rarely wrote down their know-how and relied instead on oral teaching. To sum up, while there undoubtedly were active mine surveyors during the Middle Ages, as well as scholars interested in practical geometry, there is nothing to indicate that a relationship existed between the two. Fruitful exchanges would in any case hardly have been possible, given the current development of *scientia* and mining techniques. Things began to change in the late fifteenth century, when the sudden mining boom attracted some of the brightest scholars of the time in the Ore Mountains of Saxony.

---

[20] Baron, 'Sur l'introduction en Occident des termes "geometria theorica et practica"' (1955), p. 288: 'Omnis geometrica disciplina aut theorica est, id est speculativa, aut practica, id est active'. A further discussion of the *geometria practica* can be found in Shelby, 'The Geometrical Knowledge of Mediaeval Master Masons' (1972), pp. 309–405, who studied the medieval masons, and came to the similar conclusion that 'this geometry scarcely resembles either the classical geometry of Euclid and Archimedes, or the medieval treatises on *practica geometriae*'.

[21] On the development of this tradition in the High Middle Ages, see Victor, *Practical Geometry in the High Middle Ages* (1979).

[22] Hugh of Saint-Victor, *Practical Geometry* (1991), p. 17.

[23] Zaitsev, 'The Meaning of Early Medieval Geometry' (1999), p. 552. A recent review has concluded that the exact methods described in the *Practica Geometriæ* corpus had little to do with actual practices, and more generally that 'technologies are never applied sciences'; see Gautier Dalché and Querrien, 'Mesure du sol et géométrie au Moyen Âge' (2015), pp. 97–139.

[24] Eamon, *Science and the Secrets of Nature* (1994), p. 30.

## Georgius Agricola and the Humanist
## Tradition in the Ore Mountains

An ample body of literature has been devoted to Agricola and the humanist tradition to which he belonged. Born in Glauchau, then part of the Saxon Electorate, in 1494, he studied at the University of Leipzig from 1514 to 1518 and was subsequently appointed rector of the Latin school of Zwickau.[25] A few years later he left Saxony for Italy, living in the cities of Bologna and Venice, where he was involved in the publication of new editions of the great Greek physicians Galen and Hippocrates. Back in the mining states, he became a physician and apothecary in the then booming town of Saint Joachimsthal, in the southern part of the Ore Mountains of Saxony (now Jáchymov in the Czech Republic). There he could directly observe mining operations and subsequently published his first major work in 1530, a dialogue on mining entitled *Bermannus* (The Miner). Finally settling down as a town physician in Chemnitz, a medieval city in the northern Ore Mountains in which many mining investors and officials lived, Agricola continued to gather information and publish on mining until his death. His masterpiece, the *De re metallica*, was posthumously published the following year.

As explained above, Agricola's work is often seen today as a true depiction of what he saw, a direct testimony of sixteenth-century mining, 'free from any speculation'. While it is true that he had ample contact with mining officials and possessed first-hand knowledge of mining, extracting, and smelting methods, the context in which he wrote should not be forgotten. As a learned scholar who benefited from the patronage of the Saxon Elector, Agricola composed his works with specific purposes. Pamela Long has shown how, together with a handful of fellow contemporaries, they 'created out of mining and metallurgy a learned discipline worthy of a well-born readership'.[26] One way to achieve this was to write in Latin, and another was to emulate classical works such as Vitruvius's *Architectura*. Agricola thus gathered the crafts of miners and smelters and, by giving his book a coherent structure, endeavoured to bring these closer to the 'liberal arts', to which geometry obviously belonged.[27]

---

[25] Among the many biographies of Agricola, see Agricola, Halleux, and Yans, *Bermannus le mineur* (1990). For a brief yet informative English biography, see Long, *Openness, Secrecy, Authorship* (2001), pp. 183–187. See also Flachenecker, 'Zwischen Universalität und Spezialisierung: Agricola als Humanist' (1994), pp. 101–103; Hannaway, 'Georgius Agricola as Humanist' (1992), pp. 553–560; Beretta, 'Humanism and Chemistry' (1997), pp. 19–26.

[26] Déprez-Masson, *Technique, mot et image* (2006), p. 184; Long, *Openness, Secrecy, Authorship* (2001), p. 191. See also Long, 'The Openness of Knowledge' (1991), pp. 334–341.

[27] On Agricola's endeavour to 'give the metallic art the pedagogical structure of a liberal art', see Halleux, *Le savoir de la main* (2009), pp. 75–77.

Agricola was a skilled writer who could provide vivid and naturalistic descriptions of the mining regions and the underground world. Erasmus of Rotterdam himself, in a preface to the *Bermannus*, acknowledged this skill: 'I thought that I wasn't reading about valleys, hills, mining pits, and machines, but that I was seeing them'.[28] Agricola thus wrote as a humanist in the frame of the European Renaissance. His writings possess a sense of renewal typical of his time. His *De re metallica libri XII* were to mining what Vesalius's *De humani corporis fabrica* (1543) was to anatomy, or Münster's *Cosmographia* (1544) to geography. All three books exerted a long-lasting influence on European scholars, contained impressive woodcuts, and were published in a centre of humanistic learning, the city of Basel, Switzerland. These works gave contemporaries a concrete sense that the Ancients did not know everything about anatomy, geography, or mining.

Agricola spared no effort to emphasize the novelty of mining enterprises, especially given that the discipline had been despised in the past and had even been considered suspect for its lack of certainty and the risks it implied. In his *Bermannus*, he took the time to quote Tacitus' *Germania*, in which the author wondered why gods had punished Germany by depriving it of metals.[29] To early modern readers, who were familiar with the contemporary silver rush, the absurdity was patent, putting Agricola in the position of presenting his innovations in all matters. In a quote that will be important for my analysis of subterranean geometry, he further emphasized the novelty and challenges associated with deep-level mining in contrast to the previous overexploitation of superficial veins (*Raubbau*). Comparing the state of machinery in antiquity and in his time, Agricola wrote:

BERMANNUS: In our times, however, plenty of similar and different machines have been invented, that would let the Ancients' ones far behind and maybe entirely differing from them.

NAEVIUS: I cannot argue with you about these things that I do not know about.

BERMANNUS: I am convinced that Vitruvius faithfully transcribed all machines that he found in the works of the Greeks. However, these absolutely do not stand comparison with ours. The depth of our shafts forced us to some extent to think of machines that would match their extension.[30]

---

[28] Agricola, *Bermannus* (1530), p. 3: 'Visus sum mihi valles illas et colles, et fodinas et machinas non legere, sed spectare'.

[29] Ibid., pp. 22–23. About German humanism and the specificity of northern Europe in that respect, see Smith, *The Body of the Artisan* (2004), pp. 63–67.

[30] Agricola, *Bermannus* (1530), pp. 36–37: 'BER: [Forte inueniretur], multæ tamen machinæ, si quid ego sentio, a nostris hominibus inuentæ sunt, quæ ueteres illas non parum artificio uincunt, aut multum ab eis differunt. NAE: De re quam ignoro non contenderim tecum. BER: Equidem quod de machinis apud Græcos authores fuit, a Vitruuio transcriptum puto: quod si iam nostra cum illis conferes, tum quid intersit iudicare poteris. Puteorum certe profunditas necessitate quadam nostros adegit ut tot et tanta instrumenta tractoria excogitarent'.

Agricola was not bragging, as the extension of mining operations in the sixteenth century surprised even the most bullish observers. Mining had existed for centuries in central Europe, but only recently had the operations expanded to such a degree. In the older way of mining, an ore vein would generally be dug to a dozen *Lachters* (ca. 25 metres) and then either abandoned or connected to a mining gallery to drain water down the mountain. These practices had recently been supplemented by new, capital-intensive extraction processes, where vertical shafts were dug in rich ore veins to a depth of up to 200 metres. These shafts would in turn require new machines to extract the ore and drain water, leading to the beautiful pictures represented in the sixth book of the *De re metallica*.

The *De re metallica* thus contained words and pictures unlike anything previously published, even by Agricola himself, making this work literally incomparable. Its success among humanist circles, attested by subsequent Latin editions, contributed to spur enquiries on the generation of metals, their chemical nature, and even the study of fossils.[31] This should not prevent historians from underlining the ambiguities of his method. Although historical research has shown discrepancies between what scholars wrote and what practitioners did, and despite all the precautions taken when studying other authors writing about arts and crafts, many historians have come to take Agricola's words quite literally. The most obvious reason is the numerous illustrations contained in *De re metallica*. They are not only technically superior to contemporary works such as Biringuccio's *Pirotechnia*, but are also presented as actual observations of the mines, an expensive and time-consuming task financed by Agricola's publisher.[32] Many of them were realized by Hans Rudolf Manuel Deutsch (1525–1571), who also carved gorgeous woodcuts for Münster's *Cosmographia*. As the art historian Elisabeth Kessler-Slotta wrote, Agricola developed his own method to 'combine knowledge with empiricism and didactic', according to which pictures 'should explain mining tools, instruments and technical procedures as exactly and as ostensibly as possible'.[33]

Was Agricola then aiming at a perfect yet passive description of the mining world? In a carefully crafted and deliciously ambiguous passage, the humanist scholar described his method in the Latin preface of the first edition:

---

[31] The Latin version went through four editions in 1556, 1561, 1621, and 1657. About Agricola's influence on the development of earth sciences and chemistry, see Beretta, 'Humanism and Chemistry' (1997), pp. 36–43.

[32] For an English translation, see Biringuccio, *Pirotechnia* (2005).

[33] Kessler-Slotta, 'Die Illustrationen in Agricolas *De re metallica*' (1994), p. 64. See also Hannaway, 'Reading the Pictures' (1997), pp. 49–66.

I have omitted all those things which I have not myself seen, or have not read or heard of from persons upon whom I can rely. That which I have neither seen, nor carefully considered after reading or hearing of, I have not written about. The same rule must be understood with regard to all my instruction, whether I enjoin things which ought to be done, or describe things which are usual, or condemn things which are done.[34]

This statement sounds like a bold empiricist claim, but a closer reading underlines the ambiguities of his method. Excluding at first everything he has not directly seen, Agricola then puts direct observation on the same level as testimonies from reliable persons. At first, this seems to infer that he gathered experience from trustworthy miners. However, these 'reliable' or 'probable' sources could in the Renaissance include all sorts of relevant works written by fellow humanists or some of the Ancients. Another passage, introduced in the German preface of 1557, did little to remove the ambiguity by saying that one could 'get to a proper understanding of mining ... by reading these books'.[35]

Should the content of the *De re metallica* thus be taken at face value? There can be no single answer, for the author discusses many arts and disciplines. His descriptions and pictures of machines, especially in the sixth book, are probably the most studied part of this work. According to a leading historian of the topic, 'one can be fairly sure that the drawings and woodcuts are pictorial representations of devices that were actually employed on the contemporary mining and smelting sites of continental Europe'.[36] On the other hand, countless examples from the early modern period show that both scholars and craftsmen used seemingly benign descriptions, from *pharmacopoeia* to *alchymia*, to introduce either fantastic objects or their own ideas, without aiming at practicability.[37]

## The Scope of Geometry in the *De re metallica*

The issue is more difficult to address when dealing with the geometrical parts of Agricola's work for several reasons. First of all, the author was a humanist physician with a special focus on minerals, mining, and metallurgy. We know little about his mathematical background except that he was an expert on ancient metrology, another typical endeavour of the humanist movement; he even devoted a complete volume to the topic entitled *De mensuris et*

---

[34] Agricola, Hoover, and Hoover, *De re metallica* (1912), pp. xxx–xxxi.
[35] Agricola, *Vom Bergkwerck: xij. Bücher* (1557), Introduction: 'auß sölchen büchern / so sie die selbigen mitt fleiß durchläsendt / zu rechtem verstandt des bergwercks kommen mögen'.
[36] Lefèvre, 'Picturing the World of Mining in the Renaissance: The Schwazer Bergbuch (1556)' (2010), p. 9.
[37] Hall, 'Der meister sol auch kennen schreiben und lessen' (1979), pp. 57–58.

*ponderibus Romanorum atque Græcorum* (On Weights and Measures of the Romans and Greeks).[38] Moreover, describing a surveying method in which technical operations and abstract geometrical considerations are intertwined is inherently different from describing the work of a machine or the underground world. Finally, the hypothetico-deductive model of classical geometry, as introduced in Euclid's *Elements*, seems to have little to do with empiricism and observation.

Agricola's *De re metallica* contains twelve books following the architectonic process of a mining enterprise, beginning with the discovery of an ore vein and ending with smelted metal coins. When modern writers mention this work, they usually discuss the preface, the sixth book on mining machines, or the ninth book on ore smelting, which happen to contain the impressive and well-known machine engravings. Let us re-place the surveying and geometrical parts in the context of the first half of this work, describing the activity of mining from the very first approach to the extraction of the ore. The first three books include an introduction in which the author defends mining against those who think it is 'a kind of business requiring not so much art as labour'.[39] Enumerating all the disciplines a miner, or more likely a mining official, should know, he names most of the mathematical sciences of his time: astronomy, geometry, arithmetic, construction, and draftsmanship. Disliking the division of knowledge into 'parcels', he was also disappointed with those who 'hide the art of underground surveying' – an art mastered by few people that should be shared.[40]

Books 2 and 3 describe the theory of mountains and ore veins, often mentioning in passing the art of measuring and the limits of concessions. Agricola insists on the respect of these boundaries and presents mining laws as the pillar of a sustainable mining state. The mining master, for example, has 'to give the right to mine to those asking and to confirm it, to survey the mines and to set limit stones'.[41] This crucial point has received too little attention: mathematics and geometry were quite literally keystones of the mining states. Contemporaries routinely expressed the relationship between legal, mathematical, and biblical laws, such as this sermon emphasizing the role of geometry to precisely set concession limits:

---

[38] Agricola, *Libri quinque De mensuris et ponderibus* (1533).
[39] Agricola, *Vom Bergkwerck* (1557), p. 1: 'ein solchs geschefft / das mehr arbeit dan kunst bedörffe'. Hoover uses 'skills' to translate 'Kunst', but what Agricola means here clearly includes art informed by 'intellectual understanding', to use the characterization of *Kunst* in Smith, *The Body of the Artisan* (2004), p. 72.
[40] Agricola, *Vom Bergkwerck* (1557), p. 2: 'ein anderer verbirgt die kunst des marscheidens'.
[41] Ibid., pp. 68–73, here p. 70: 'Uber das ist auch sein ampt / denen so begeren die gerechtigkeit zur grubenn / zu geben / und die selbige zu bestätigen / die gruben zu marscheiden / und jnen marstein zusetzen.'

One should thank God for this / and praise the industrious people / who achieve this with the triangle. … This is why God gave [us] the liberal arts / so that one can recognize him / and thank him / and the fairness / and that the people are served by these / profitably and with good conscience.[42]

Contrary to what has sometimes been written, Agricola's opinion on dowsing or divining rods was pretty severe, concluding that in order to locate an ore vein, a pious miner 'shouldn't make use of an enchanted twig' and should instead rely on his observation, be 'skilled in the natural signs', and know his surveying methods.[43] The next two books present the geometry of the mines and form the single most important scholarly presentation of mining mathematics at the time. Book 4 describes the concession allocation and, conversely, various offices of the mining administration. Concessions were ordinarily made of squares with a side of seven fathoms (ca. 14 metres) that could be extended in rectangles following an ore vein, using precise rules.[44] Setting property limits was a crucial and complex duty, since boundaries set on the surface had to be respected underground. A remarkable feature of this fourth book is that Agricola did not include, in his otherwise extensive description of mining offices, the subterranean surveyor or *Markscheider*. Given the importance of the matter and the abundant references to surveyors in the previous and following books, this discrepancy needs to be accounted for. Because of the nascent stage of mining administration and the lack of highly skilled officials, this duty had often been attributed to mining masters. Agricola's description thus matched the reality at the turn of the sixteenth century, neglecting the importance geometry had gained within a few decades. While it might seem fussy to argue about a couple of generations, I think that this observation is crucial. When describing machines and legal cases, the author was quick to highlight as archaic a practice 'from the time of our ancestors' and to describe a 'way of making pumps that has been invented ten years ago' or a dividend system 'that has changed fifteen years ago'.[45] Given the prodigious

---

[42] Spangenberg, *Die XIX. Predigt Von Doctore Martino Luthero* (1574), pp. 64–65: 'dafür man billich Gott dancken / unnd fleissige Leute / so solches durch den Triangel zu wege gebracht … Denn darzu hat Gott die freyen Künste gegeben / daß man ihn daran erkennen / und jm dancken soll / und der gerechtigkeit / unnd den leuten damit nutzlich / und mit gutem gewissen gedienet werde.'

[43] Agricola, Hoover and Hoover, *De re metallica* (1912), p. 41. For another point of view, see Dym, *Divining Science* (2011).

[44] On the system of *Kuxen* (mining shares) presented in details by Agricola, see Laube, *Studien über den erzgebirgischen Silberbergbau von 1470 bis 1546* (1974); Dietrich, *Untersuchungen zum Frühkapitalismus im mitteldeutschen Erzbergbau und Metallhandel* (1991); Asmussen, 'The Kux as a Site of Mediation' (2016), pp. 159–182.

[45] Agricola, *Vom Bergkwerck* (1557), pp. 65, 149, 67: 'zu der zeit unserer vorfaren … von der zeit an vor zehen jaren erfunden … ist nuhn von fünffzehen jaren här so verendert worden.'

development of deep-level mining and smelting machines in the previous few decades, he had good reason to pay attention to chronology. The disparity of the material gathered by Agricola on surveying and geometry will be analysed in detail later.

The fifth book describes both the digging operations of shafts and galleries and how subterranean surveyors should proceed to direct them. Several figures explain what vertical shafts (*Schachten*) and horizontal galleries or tunnels (*Stollen*) are and describe various kinds (see Figure 1.1). The natural theory of signs, introduced at the end of the third book, is developed for the underground. The use of shafts and galleries for ventilation, water drainage, and ore extraction is detailed to give the reader a sense of the technical constraints on both planning and surveying. Agricola then turns to subterranean geometry:

> I have completed one part of this book, and now come to the other, in which I will deal with the art of surveying. Miners measure the solid mass of the mountains in order that the owners may lay out their plans, and that their workmen may not encroach on other people's possessions. The surveyor either measures the interval not yet wholly dug through, which lies between the mouth of a tunnel and a shaft to be sunk to that depth, or between the mouth of a shaft and the tunnel to be driven to that point which lies under the shaft. ... Or in some cases, within the tunnels and drifts, are to be fixed the boundaries of the meers [concessions], just as the *Bergmeister* [mining master] has determined the boundaries of the same meers above ground. Each method of surveying depends on the measuring of triangles. A small triangle should be laid out, and from it calculations must be made regarding a larger one.[46]

### Similar Pictures of Similar Triangles

Agricola's introduction immediately presents the reader with interpretative issues. First of all, the vocabulary of subterranean geometry traditionally belonged to a specific German sociolect, the so-called language of miners (*Bergmannsprache*). Agricola mastered it but decided to write his book in Latin; the subsequent German translation, although realized only one year after the original, is only moderately more helpful because it was made by a fellow humanist from Basel, Philippus Bechius (1521–1560). Agricola did not use a specific Latin term for the art of setting concession boundaries (*Markscheidekunst*), using the Latin *ars mensorum* or *mensor* for the surveyor. Was this term used for convenience, in order to be understood by scholars, or for lack of a better word? As a learned humanist, Agricola knew of the corpus of the Latin surveyors, the *agrimensores*. In the previous (fourth) book he had mentioned in passing a collection of Roman surveying texts printed in Paris in

---

[46] Agricola, Hoover, and Hoover, *De re metallica* (1912), pp. 128–129.

*iugum* A. *Iugi pertica* B. *Puteus* C. *Primus funiculus* D.
*Primi funi. uli pondus* E. *Secundus funiculus* F. *Idem in terram*
*infixus* G. *Caput primi funiculi* H. *Os cuniculi* I. *Tertius fu-*
*niculus* K. *Tertij funiculi pondus* L. *Menfura prima* M.
*Menfura fecunda* N. *Menfura tertia* O. *Triangulus* P.

Figure 1.1    Measuring the depth of a shaft and length of a gallery using similar triangles, according to Agricola's *De re metallica* (1556). The surveyor is presented alone, with a plumb line as the only instrument. He points at the small triangle, while the sides of the bigger triangle are the shaft, the gallery and the inclined cord. Courtesy Max Planck Institute for the History of Science, Berlin.

1554, the *De limitibus agrorum* ('On Fields' Limits'), even borrowing some of the diagrams for his own book.[47]

When discussing square-shaped concessions or using triangles, his text referred to various triangles by their Greek names. If his description and examples matched what can be found in Roman surveying works, one could assume that he simply borrowed his theory from the Latin corpus. However, a philological analysis has shown that Agricola's vocabulary, while proving his knowledge of classical surveying texts, was distinct from this tradition.[48] When analysing the figures and methods used by Agricola, one should thus keep in mind that he was knowledgeable about both practitioners' methods and ancient surveying literary tradition, but decided to produce his own synthesis.

Agricola presented his first and main instruction on underground surveying as based on the use of similar triangles (see Figure 1.1). It seems fairly easy to understand, and the picture is described at length, insisting on the method of the surveying process. The goal of this operation is to find the distance to dig before a shaft and its gallery finally meet. The gallery and the shaft are seen as two sides, *cathetus* and *basis*, of a large right triangle whose *hypotenusa* is represented by the inclined cord HG. Using the wooden infrastructure and the base point H, the surveyor constructs a smaller, yet similar, triangle, shown by the person in the figure. With the help of proportionality, one could then deduce from the length of the line segments O and N the depth of the shaft and length of the gallery, as Agricola explains: 'Just as many times as the length of the first cord [M] is contained in the whole length of the oblique descending cord, shall you take the second cord to find the distance between the mouth of the gallery and the [point to which] the shaft will be dug.'[49]

The method is mathematically simple and sound, but this does not mean that surveyors necessarily used it. Indeed, there is much evidence to indicate that this figure does not represent actual mine surveying. First of all, there is no mining infrastructure (*Berggebäude*), in stark contrast to most other pictures of his work. Moreover, most pictures in the *De re metallica* involve several protagonists: aiming at realism, the artist depicted many individuals performing operations as in real life, with an illustrative character in mind. These images should depict 'mining tools, machines and technical procedures as exactly and

---

[47]  In Agricola, *De re metallica* (1556), p. 60: 'De limitibus agrorum' is inspired from Flaco, *De agrorum conditionibus & constitutionibus limitum* (1554). Agricola also borrowed several diagrams (esp. pp. 227–230).

[48]  Déprez-Masson, *Technique, mot et image* (2006), pp. 197–201.

[49]  Agricola, *Vom Bergkwerck* (1557), p. 93: 'Derhalben / wie viel erste meß in der gantzen Schnur die flach hinunder gehtt / gfunden werden / so viel zeigendt die anderen meß an was die lucken so da ist zwüschen dem mundtloch des schachts / und zwüsche dem mittell boden des stollens.'

Figure 1.2    Mine surveying as presented in the *Schwazer Bergbuch*, 1554.
The surveyor is surrounded by aides, using a compass to ascertain the direc-
tion of the gallery and recording it on wax rings. Courtesy Bergbau-Museum
Bochum, Entwurfexemplar 872, fol. 31v – detail.

as concretely as possible'.[50] Oddly enough, the *Markscheider* depicted here is
not using any instrument besides the cords symbolizing mathematical lines. He
is depicted alone, not even working but merely showing the reader the fundamental point of Agricola's reasoning; this is all the more surprising for a task
that necessarily involved several people. Fortunately, a good point of comparison can be found in the *Schwazer Bergbuch*, a mining manuscript written
in the same decade, which also represents surveyors at work. In these images,
one clearly sees several people working together, the *Markscheider* and his
apprentices or aides (see Figure 1.2).[51]

While the accuracy of this second depiction might also be questioned, for
the *Schwazer Bergbuch* was written in a courtly context, let us now turn to a
third document.[52] This drawing is one of the oldest working documents by a
Saxon *Markscheider*, Hans Dolhopp, to have been preserved (see Figure 1.3).
It was made around 1603 and was obviously not meant for publication, but
rather to track the progress in digging several shafts and a gallery, since

[50] Kessler-Slotta, 'Die Illustrationen in Agricolas *De re metallica*' (1994), p. 64.
[51] Bartels, Frey and Bingener, *Das Schwazer Bergbuch* (2006), vol. 2, pp. 61–62.
[52] SächsStA–F, 40 001, Oberbergamt Freiberg, Nr. 608, f. 26.

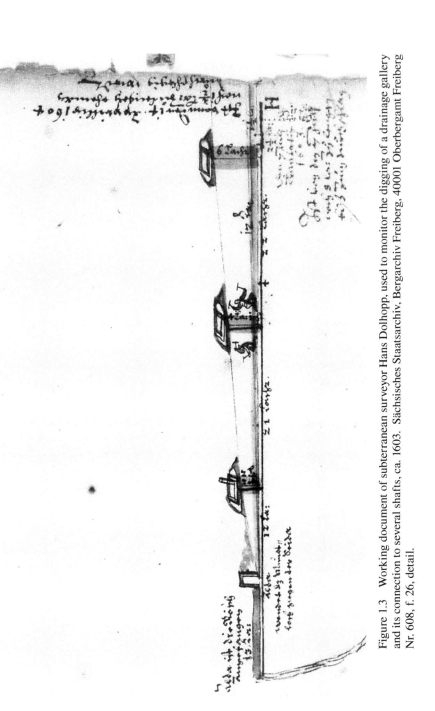

Figure 1.3 Working document of subterranean surveyor Hans Dolhopp, used to monitor the digging of a drainage gallery and its connection to several shafts, ca. 1603. Sächsisches Staatsarchiv, Bergarchiv Freiberg, 40001 Oberbergamt Freiberg Nr. 608, f. 26, detail.

these enterprises took years to complete. Here we can see an actual occurrence of the problem Agricola was supposedly addressing in his picture. This sketch is representative of the geology of the Ore Mountains, where mining operations mostly take place in a hilly countryside. As can be seen on the drawing, the fundamental step used by Agricola – measuring the *hypotenusa* formed by the mountainside – was hardly practicable barring exceptional circumstances. Most of the time, one simply could not draw any oblique cord that plays this crucial role because of the geological configuration of these rotund hills.

However, my main objection is that this method does not solve any meaningful problem that would arise in an actual sixteenth-century mine. To work in low-depth shafts of about 20 metres, a formalized use of similar triangles was probably not crucial. Accordingly, the drawing of Hans Dolhopp was in all likelihood not the basis of any elaborate reasoning, but rather a fieldbook used to record surveying information. As we can see from the annotations, the surveyor regularly visited the mine working, measured the progress of the enterprise, and then took notes on the plan: 'As of July 7 there is 8 fathoms to dig until breakthrough' (Figure 1.3, bottom right corner). It is certainly possible that methods involving proportionality would be used to perform back-of-the-envelope calculations, but not to 'demonstrate the depth of the shaft', as Agricola wrote.[53] In such operations, the need for geometrical formalization is not obvious. Frequent annotations to this map suggest that Dolhopp checked the progress every couple of months, eventually refining his estimates based on observations, not computations. After all, this shaft would only be 12 metres deep!

In the sixteenth century, the use of geometry could only be decisive for the new and deeper shafts called *venæ profundæ* in the *De re metallica* and described as 'very deep, sometimes as much as sixty, eighty, or one hundred fathoms' (ca. 200 metres). Agricola went on to explain that it was for those 'vertical shafts' that the new water wheels were put into use, in order to turn a profit 'when a vein is rich in metal'.[54] The impressive machines and new methods of the *De re metallica* thus deal precisely with this deep-level mining. But in that case, it would be geologically impossible to use a cord as *hypotenusa* to measure the mountainside, since operations would happen well beneath the surface. In other words, the first and main geometrical example given by Agricola could not address the most acute and interesting problems of his time.

Finally, the real conceptual difficulty, when digging a shaft that would eventually meet the gallery following a silver vein after years of work, was related

[53] Agricola, Hoover, and Hoover, *De re metallica* (1912), p. 144.
[54] Ibid., p. 122.

to the horizontal direction. The risk was to dig deep enough, but then not meet the gallery because its direction would have changed in the meantime: the real geometrical problem arises when one takes into account the third dimension. This is precisely the reason why the surveyors represented in the *Schwazer Bergbuch* (Figure 1.2) can be seen observing a compass, to assess the direction of the gallery and use wax rings to record it. On the famous illustration presented in the *De re metallica*, the real question should not be *when* the shaft will meet the gallery, but *whether* they will meet at all. Although numerous pictures in this book use perspective, the drawing technique used in that case is an elevation. Despite the skills of the engravers, the way this picture is drawn precisely masks the real issue, that is, that the two excavations are not being dug in the same vertical plane.[55]

### *Geometria Theoretica*, *Practica*, and *Subterranea*

This famous triangular method, which is the main illustration about subterranean geometry to be found in the *De re metallica* and as such the most often reproduced in the historiography, is thus very problematic. It depicts a procedure highly unrepresentative of the actual issues faced by sixteenth-century miners, and yet Agricola clearly said that he 'omitted all those things which I have not myself seen'. In that context, his ambiguous addition takes on a new meaning: 'or have not read or heard of from persons upon whom I can rely'. Unlike the art of extracting and smelting, about which little had been written at the time, the tradition of classical geometry was already more than two millennia old. By its very nature, this geometry relied on reading or hearing demonstrations, and a sound proof was enough for a scholar to 'rely on'. In fact, a theory of similar figures was already laid down in detail in the sixth book of Euclid's *Elements*, where the second, third, and fourth propositions deal with similar triangles.[56] The *Elements* were a touchstone of the humanist curriculum, both for their mathematical content and as example of a method that should be applied to all sciences.[57]

In the first part of this chapter, I have depicted the implication of Agricola in humanist networks. Turning to one of his contemporaries, the French humanist Oronce Finé (1494–1555), who published *Protomathesis* in 1532, one can indeed identify a problem not unlike Agricola's[58] (see Figure 1.4). The

---

[55] Hannaway, 'Reading the Pictures' (1997), p. 52, rightly states that the woodcuts of Agricola's *De re metallica* 'do not provide a [mere] depiction of scenes or a panorama of the workings of a mine; they are composite pictures that have to be read'.
[56] Euclid, *The Thirteen Books of Euclid's Elements* (1908), p. 3.
[57] Hoppe, 'Die Vernetzung der mathematisch ausgerichteten Anwendungsgebiete mit den Fächern des Quadriviums in der frühen Neuzeit' (1996), especially pp. 5–7.
[58] On Oronce Finé, see Axworthy, *Le mathématicien renaissant et son savoir* (2016).

Figure 1.4    Measuring the depth of a well according to Oronce Finé, *Protomathesis* (Paris, 1532), f. 72. Knowing the three sides of the small triangle produced by the *carré géométrique* (quadrant), one deduces the depth of the well. Note the position of the observer pointing at the smaller triangle. Courtesy ETH-Bibliothek Zürich, Rar 9724.

mathematics underlying these two methods is rigorously identical, as is the fundamental idea of constructing a smaller triangle in order to compute an unreachable length. The similarity of the lone observer, his position and hand showing the smaller triangle is striking, as can be seen comparing the settings of Figures 1.1 and 1.4. To be clear, my hypothesis is not simply that Agricola borrowed his figure from Finé's *Protomathesis*. The Euclidean geometry had by then disseminated in a vast scholastic literature, the textual tradition of *practica geometriæ* initiated by Hugh of Saint-Victor. Agricola's geometrical approach was rooted in these humanist traditions, so that he could similarly have taken it from closer humanists such as Johannes Stöffler (1452–1531), Petrus Apianus (1495–1552), Walther Hermenius Rivius (ca. 1500–1548), or many others, all of whom display similar pictures.

The sixteenth century was indeed a period of great transformation in the scholarly use of mathematics to understand the world. Martin Waldseemüller (1470–1520), who also worked near the important mining centre of Saint-Dié-des-Vosges, improved an ancient instrument of Latin *agrimensores*, the *dioptra*. Apianus, Thomas Digges (1546–1595), and Christoph Puehler (ca. 1500–?) all proposed various quadrants as well as other instruments to calculate lengths using ratios and lengths of triangles. By then, the theory of similar figures was frequently mentioned in Latin geometry treatises, with instruments – generally a quadrant – or without. The two most popular examples were measuring the height of a tower or the depth of a well, *puteus* in Latin. It is hardly surprising to realize that *puteus* means not only a well but also a (mining) shaft. The classical wording of the method, by Hugh of Saint-Victor, is the following:

Here there are two similar triangles, one larger, the other smaller, one in the instrument, the other in the physical figure. The larger perpendicular drops from the *mensor*'s eye along his height and the well depth to its bottom; its base is the width, or well diameter. The hypotenuse is the sight line dropping obliquely from the mensor's eye to the far side of the bottom, and taking the entire bottom as base. The other triangle, set up orthogonally under the same base, is similar to the first.[59]

In other words, this situation depicted by Agricola did not arise from a practical necessity, but reproduced a typical problem of the existing literature. How much of this scholarly tradition had been transmitted to actual mine surveying is difficult to say. As can be seen in Figure 1.3, the peculiarities of these underground and vertical settings make a direct transfer of Euclidean geometry unlikely.[60] What is clear is that Agricola, by reproducing and adapting this method to introduce his *geometria subterranea*, reveals himself as an attentive and astute reader of both the Roman surveying corpus and the *practica geometriæ*. He does not simply copy these works but adapts them to the mining context. These methods, however, still belonged to a scholarly frame of thought. Historians studying the *practica geometriæ* have come to the

---

[59]  Hugh of Saint-Victor, *Practical Geometry* (1991), p. 54.
[60]  On this topic, see Morel, 'Bringing Euclid into the Mines' (2017), pp. 154–181. The specificity of studying the underground has recently been underscored by historians in what they called 'vertical scientific practices', but mostly refers to modern sciences. See Reidy, 'The Most Recent Orogeny: Verticality and Why Mountains Matter' (2017).

conclusion 'that the works on practical geometry were not aimed at operators, surveyors or "site managers"', a conclusion corroborated by the present case.[61]

In short, Agricola's work provides a sharp view on the subterranean world but raises serious questions about its actual use. While surveyors were busy solving actual mining problems with piecemeal solutions, Agricola must have immediately seen the analogy between mining operations and a standard problem of his learned tradition. Although he knew how the surveyors were working on the surface and underground, his explanation likely relies on his reading of fellow humanists. As Floris Cohen noted in another context, early modern scholars could use such practical experiences as an illustration of their principles; in such cases, 'craft experience is important yet secondary; it helped concentrate wonderfully a mind already well-prepared through its own efforts'.[62] Instead of pure observations, Agricola could in good conscience have promoted a scholarly method that he had, in his own words, 'carefully considered'.

This does not mean that Agricola was not interested in or that he ignored what practitioners were doing. One can indeed find a small cluster of plausible surveying methods, together with several surveying instruments, interestingly packed at the end of Book 5. Most of these methods also rely on cords but use these instruments in a completely different conceptual frame. Instead of constructing similar triangles, that is, entering the Euclidean theoretical frame, cords were prosaically used to record an underground length. The surveyor would then 'lay them out in the same way on the surveyor's field in the open air', producing a rough sketch of the underground situation for all to see. This process was typical for surveying practices, not only in subterranean geometry but in many technical settings such as polders in the Low Countries.[63] However, Agricola only sketches these processes and does not illustrate them with figures, which made it difficult to put them into practice. This part might have been written separately from the rest of the geometrical part, as the writing style and rhythm differ slightly; as such, it may have been a later addition to the first draft of 1550.

I thus think that Agricola's main interest when discussing geometry rests in the principle of similar triangles and its underlying theory. Providing guidelines for actual surveying operations was neither his intent nor what his readership expected, so that actual procedures were only summarily covered, without

---

[61] L'Huillier, 'Practical Geometry in the Middle Ages and the Renaissance' (2003), p. 189.
[62] Cohen, *The Scientific Revolution* (1994), p. 348.
[63] Agricola, Hoover, and Hoover, *De re metallica* (1912), pp. 144–148, here p. 147. This method is attested in Tirol in the 1520s, see Krumm, 'Visualisierung als Problem' (2012), pp. 294–313. On surveying in the Low Countries, see Streefkerk, Werner and Wieringa, *Perfect gemeten* (1994).

sufficient analysis. While efficient, these methods were based on estimates and not susceptible to demonstration. My hypothesis about the aim of the geometrical part is supported by metrological arguments. In this domain, Agricola was an expert, having published an entire volume on the history of measuring units – a work so valuable that it has been reissued and is still used by modern historians.[64] When explaining how to perform the computation on similar triangles, he gave numerical examples such as 'a depth of 12 fathoms, 2 feet and a half, 1 hand and the fifth of a half-finger'. This unit system using fingers, hands of four fingers, and feet of four hands was the common system of the time, as can be seen in contemporary surveying books.[65] However, subterranean surveyors of the time used another unit system, where a *Lachter* (fathom) was divided into three and a half *Ellen* or *Rute* (cubits), then divided in *Zölle* (inches), as can be seen, for example, in Figure 1.3.[66] When discussing examples, Agricola did not use these units; this undoubtedly conscious choice shows that his intended audience was not mining technicians and his aim was not any direct reproducibility.

In fact, a rare humanist actively trying to bridge the gap between the university culture and practical concerns seems to have been Petrus Ramus (ca. 1515–1572). Teaching in Paris, Ramus frequently visited the local merchants and goldsmith. When he toured the Holy Roman Empire in 1568, he spent several days in artisan workshops in Nuremberg, engaging with instrument-makers.[67] Ramus attributed the great number of chairs for mathematics in German-speaking territories to its usefulness, both for wars and in the mines. It is certainly true that most of the mines and many of the wars were at that time concentrated within the Holy Roman Empire. Furthermore, I will detail in Chapter 3 how some rulers tried to foster cooperation between scholars and practitioners. For most university professors, however, the practical uses of mathematics continued to be purely rhetorical. Their practical geometry had little to do with actual issues in navigation or wine gauging, and one should thus not mistake these printed descriptions with what actually happened. The discrepancy might be harder to notice when discussing the underground geometry, but it is just as relevant.

---

[64] See Agricola, *Schriften über Maße und Gewichte* (1959).
[65] Compare the measuring units used in Agricola, *Vom Bergkwerck* (1557), pp. 95–96, and Stöffler, *Von künstlicher Abmessung aller grösse* (1536), p. Aiij. Historians now think that Agricola referred more specifically to the lengths units used in Chemnitz, see Agricola, *Schriften über Maße und Gewichte* (1959), pp. 432–435; Witthöft, 'Die Metrologie bei Georgius Agricola' (1996), pp. 19–27.
[66] The use of measuring units was not fully standardized in mining regions in the sixteenth century. See Nehm, 'Georg Öder und seine markscheiderische Tätigkeit auf dem Rammelsberg' (1934), pp. 64–72; Kula, *Measures and Men* (1986), chapter 10.
[67] On Petrus Ramus, the reference work is Hooykaas, *Humanisme, science et réforme* (1958), here pp. 81–85 and 91–96.

## Courtly Instruments and Scholarly Methods

A second important work dealing with subterranean geometry was published in the late sixteenth century. It was written by Erasmus Reinhold the Younger (1538–1592) and entitled *Vom Marscheiden kurtzer und gründlicher unterricht* (A Short and Rigorous Instruction to Underground Surveying). His father, E. Reinhold the Older, had been a leading astronomer and author of the *Prutenic Tables* – which greatly contributed to the success of the Copernican heliocentric theory.[68] The younger Reinhold had studied mathematics under his father at the University of Wittenberg before turning to medicine. He was subsequently appointed court and town physician in Saalfeld, by then a Saxon protectorate. In 1575, Tycho Brahe visited him to peruse his father's copy of Copernic's *De revolutionibus* and consult its numerous annotations.[69]

Reinhold's *Instruction* on subterranean geometry was written as the second part of a surveying book that – its introduction claimed – had been drafted by his famous father and completed by himself. The first part dealt with general surveying and bemoaned the 'gross mistakes' of ordinary surveyors. In contrast, the author acknowledged that subterranean surveyors 'hit their mark' and that 'the measures they produce can sometimes be fine'.[70] Strikingly, the first section of this part opens with a problem that will be familiar to the present reader: 'how to find properly how deep a shaft can be brought at a given place, and how far it needs to be dug'. The illustration confirms the similarity to Agricola's picture, in spite of its sketchy character and the inferior quality of its engraving (see Figure 1.5).

The descriptions of the method by the two humanists have the same structure, both being variations on the 'depth of the well' exercise found in the corpus of *practica geometriæ*. Reinhold focuses, as Agricola did before him, on the principle of using cords to construct a smaller triangle similar to the 'physical figure' formed by the mining shaft and the mouth of the gallery, to use the words of Hugh of Saint-Victor. There is no reason to think that this was plagiarism, since both the wording and numbered examples differ. Reinhold's illustration and wording are more abstract and strongly hint at the geometrical concept of similarity, probably because of his mathematical education, whereas Agricola depicted a more evocative mining landscape. Both authors

---

[68] See Gingerich, 'The Role of Erasmus Reinhold and the Prutenic Tables in the Dissemination of the Copernican Theory' (1973), pp. 43–62, 123–126.

[69] On his biography, see Wilkening, 'Erasmus Reinhold, der Verfasser der ersten deutschen Markscheidekunde' (1960), pp. 13–15, 58–74; Wilkening, 'Aus dem Leben des Erasmus Reinhold' (1963), pp. 11–12.

[70] Reinhold, *Gründlicher vnd Warer Bericht* (1574), dedication: 'so da gar genahe zum zweck schiessen / und ihnen ihr abmessen bißweilen fein zutrifft'.

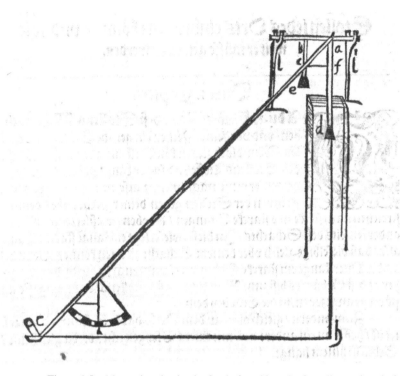

Figure 1.5    Measuring the depth of a shaft and length of a gallery using similar triangles, according to Erasmus Reinhold the Younger, *Vom Marscheiden*, 1574, f. c verso. Courtesy ETH-Bibliothek Zürich, Rar 4257: 3.

thus chose to begin their mine geometry with the same situation, which – as we have shown – does not correspond with actual surveying practices. One can conclude that their choice reflects a similar humanist goal 'to extract certain underlying principles from their practical concerns'.[71]

Reinhold's prose reveals a slight embarrassment with the practicability of his method. While defending this manipulation of cords – 'in itself correct and true' – he acknowledges that 'something can easily go wrong' and abruptly turns to another technique: using a quadrant, as represented on the bottom-left part of the picture. This quadrant was no standard object but a specific instrument designed by Reinhold himself, 'which to [his] knowledge, is not yet used by miners'. It was meant to be used with specific trigonometric tables that the

---

[71]  Cohen, *The Scientific Revolution* (1994), p. 131.

author included in the book.[72] This tool should have combined the functions of a compass and of a quadrant, allowing to record both vertical angles (as shown on the picture) and horizontal angles. It was in all likelihood not broadly used in a real context, for reasons similar to those raised in analysing Agricola's case.[73]

I think that Erasmus Reinhold was consciously displaying his versatile skills in a context of patronage.[74] What mattered for him was to prove his ability to design instruments based on mathematical theories. The book's dedication to the powerful ruler of Saxony emphasized the novelty of the instrument, before conveniently flattering Elector August for his 'special affection and inclination for the arts [Künste]'. Criticizing the subterranean surveyors for 'keeping secretly hidden' their methods, his description of the quadrant aimed at underlining his command of the 'irrefutable principles of the Geometriae'.[75] Reinhold's publication was quickly rewarded when the Elector appointed him in the following year mining master (Bergvogt) in the district of Saalfeld.[76]

Reinhold's and Agricola's writings on subterranean geometry are thus scholarly depictions on a practical matter. Both toggle between description and prescription, giving reasons to doubt that actual underground surveying practices resembled their accounts. When Reinhold explains how to build his compass, he seems at first very down to earth: 'you should begin by taking a very dry board', he advises his reader, going as far as to recommend walnut as 'the best of all' trees. Having drawn the circle, the reader should 'let a carpenter … cut it out'.[77] His description is perfectly realistic until one notices that his instrument is to be graduated in degrees, whereas mining compasses and all legal texts always recorded directions according to different unit systems (mostly dividing a circle not into 360 degrees, but twice twelve hours). The pseudo-concrete setting was rather a way for Reinhold to exhibit his command

---

[72] Reinhold, Gründlicher vnd Warer Bericht (1574), p. cii recto: 'Aber doch dieweil gar leichtlich etwas kan verrückt werden.' Dedication: 'habe auch hierzu einer sondern art eines Bergk Compasses meldung gethan / welche meines wissens hiebevron / bey Bergkleuten nicht gebreuchlichen gewest'.

[73] This instrument required at least some understanding of trigonometry, which could hardly be expected from an average sixteenth-century mine surveyor. In fact, the tables given by Erasmus Reinhold, possibly inspired by his father, were not standard and could not easily be reconstructed if lost. Finally, such an instrument could not legally have been used as such during an official ceremony, as described in Chapter 2.

[74] On the 'quest for patronage and advancement' in mining, see Long, Openness, Secrecy, Authorship (2001), pp. 188–192.

[75] Reinhold, Gründlicher vnd Warer Bericht (1574), dedication: 'sondere lieb / und neigung zu den künsten … heimlich verborgen halten … unwidersprechlichen fundamenten der Geometriæ'.

[76] Wilkening, 'Aus dem Leben des Erasmus Reinhold' (1963), p. 12.

[77] Reinhold, Gründlicher vnd Warer Bericht (1574), p. b verso und bii recto: 'Du must anfenglich ein gar dürre bret … unter welchen allen Nußbeumen / das beste ist … laß dir es einen Tischer sein geheb demselben Circkelriß nach / abschneiden.'

of trigonometry and to demonstrate the scope of his mathematical skills. When describing his quadrant, it is the Elector of Saxony that he truly had in mind, not the mine surveyors.[78] Once again, this argument should not be overstated by confusing the work of a scholar with his understanding of the subject. I do not mean that Erasmus Reinhold failed to master the art of setting limits. Other sections of his *Instruction* clearly demonstrate his abilities and knowledge of actual practices, for instance the penultimate chapter dealing with the ubiquitous issue of 'bringing a point to the day', that is, locating the spot at the surface precisely above a given point in the mine.[79] As in Agricola's *De re metallica*, however, such parts were confined to the last portion of the book and hardly understandable for their brevity and lack of illustrations. More importantly, a casual reader lacked means to understand that both works were composed of heterogeneous material, and could not distinguish between technical description and scholarly prescription.

Surveying methods presented by scholars can thus generally not be seen as a mere account of what they had undoubtedly witnessed in the ore mines of the Holy Roman Empire.[80] When dealing with mathematics, humanists had to strike a balance between observation and rationalization, between the empirical methods actually used and the demonstrative principles they unknowingly relied on. There was an essential tension between the worlds of scholars and practitioners, and yet Agricola thought he could both 'describe things which are usual' and 'enjoin things which ought to be done'. Reconciling these two goals proved especially difficult when dealing with mathematics, where the powerful Greek tradition had been reinvigorated in the realm of the *practica geometriæ*. Their accounts on subterranean geometry and the prominent place given to impracticable methods illustrate how, according to Jim Bennett, 'surveying is another area where rhetoric might stretch beyond the realistic limits of practice' during the early modern period.[81]

The issue was by no means unique to geometry and could be generalized, *mutatis mutandis*, to many scholarly works dealing with mining sciences. R. Halleux has shown that, in the long run, Agricola's Latin vocabulary was admired by scholars but generally did not replace the technical language of miners. The same holds for subterranean geometry, where the vernacular and heavily dialectical expressions continued to be used and can be found in mining or mathematics dictionaries in the following centuries. Using Latin did

---

[78] Reimers, 'Wurzeln des Markscheidewesens im Spiegel gelehrter Schriften' (2021), pp. 93–127.
[79] Reinhold, *Gründlicher vnd Warer Bericht* (1574): 'Das Sechzehende Capittel. Die örtung aus der Gruben heraus an Tag zu bringen.'
[80] For another example, this time related to Georgius Agricola, see Morel, 'Subterranean Geometry and its Instruments' (2017), pp. 80–83.
[81] Bennett, 'Practical Geometry and Operative Knowledge' (1998), p. 207.

not create an international community around mining practices, and people continued to visit the German-speaking mining regions to learn this vernacular knowledge and expertise on site, and to recruit German surveyors all over Europe.[82] Agricola's influence on actual practices should not be overstated either. Recent analyses of chemical processes, most notably in mercury production, show that his scholarship might have been further apart from actual practices than previously thought.[83] Even concerning machines, the masterpiece of the *De re metallica*, 'Agricola's work contributed less than is usually thought' to technical evolution.[84]

Far from being mere descriptors, Agricola and Reinhold sought to present a learned approach to practical problems. Their mathematization of nature was not exactly false: their texts are indeed theoretically sound and coherent, until one realizes that such methods and instruments would hardly have fulfilled their purpose in a concrete setting. In other words, they were influenced by their mathematical culture, and their writings were mediated by and oriented toward their own humanist interests. Such books were not written to be deceptive, because their fundamental aim was not practicability. Reinhold was writing in a patronage context and was accordingly rewarded by a position in the mining administration. Describing in detail the know-how of seasoned practitioners would not only have been a daunting task, it was above all not something scholars could even think of, for most of the early modern scholars 'assumed without questions that their own verbal ingenuity far outstripped the material ingenuity of the artisan', as Pamela Smith has argued.[85] Agricola and Reinhold sought to present a higher truth in applying what they saw as paramount, that is the Euclidean frame, in a practical context, even if this meant setting aside some recent developments of surveying for deep-level mining. Both authors presented subterranean geometry as a variation of the Latin tradition of practical geometry. This only reminds us of a basic truth about scholarly writings of the Renaissance dealing with craftsmen and artisans: these texts should not be taken at face value.

*

In 1550, the geographer Sebastian Münster (1488–1552) published the third edition of his *Cosmographia Universalis*, an encyclopaedic work describing the whole world known in his time. German states were being transformed by the silver rush and Münster, revising the corresponding part, now 'found

---

[82] Lüschen, *Die Namen der Steine* (1968), p. 67; Agricola, Halleux, and Yans, *Bermannus le mineur* (1990), p. xxix; Garçon, 'Réduire la mine en science' (2008), pp. 317–336.
[83] Raphael, 'Producing Knowledge about Mercury Mining' (2020), pp. 95–118.
[84] Ernsting, *Georgius Agricola, Bergwelten 1494–1994* (1994), p. 168.
[85] Smith, *The Body of the Artisan* (2004), p. 82.

worthy of a chronicle to describe the great and fantastical mines that have emerged in our times'.[86] He thus decided to 'ask many scholars [*gelerter leüt*]' to send him their contributions on mining and to explain why the subject had not already been treated, but could not find a proper answer:

So, I have to answer myself my question on the reason why there is so few things written on mines. I think that an important reason is that no one can write about something they have not experienced or seen. Mining uses numerous instruments and peculiar names that are not easily known to writers and scholars, and which are so fantastic that no one could fully master them, however far one may wander.[87]

Münster's savvy intuition was seemingly disproved when Agricola published his *De re metallica* just six years later. For the first time, a scholar who had spent decades among miners wrote down and published his observations. On a deeper level, however, Münster was right: works written by humanists were not aimed at an audience of practitioners but at scholars. At the time, there was indeed no way to adequately write down the know-how and specific skills of miners and craftsmen. Agricola chose to write his masterpiece in Latin, a language alien to most early modern practitioners. A German translation was admittedly published one year after, in 1557, but was a commercial failure and proved to be 'unsaleable'.[88] While the Latin *De re metallica* was regularly re-edited, the unsold stock of the first German edition was substantial enough to be bought by another publisher one generation later. Brought on the market with minor alterations and a new title page, the repackaged version hardly sold better. In a strikingly similar fate, and despite its superior treatment of underground surveying, Reinhold's book was soon forgotten and only rediscovered in the mid-eighteenth century.[89]

Given the importance of Agricola's work, historians tend to inquire into early modern mining through this lens, even as we know that bookish knowledge was not yet in a position to influence surveying practices. The abundant Latin literature testifies of an intense curiosity from scholars about the

[86] Münster, *Cosmographei* (1550), p. dxxviii: 'Chronick würdig sein zu beschreiben die grossen wunderbarlichen Bergwercken / so zu unsern zeiten auffkommen seind.'
[87] Ibid., p. dxxviii: 'vil gelerter leüt gefragt ... Dise mein frag / was die ursach / das man so wenig von Bergwercken geschrieben findt / muß ich mir selbs auff lösen / und halt das für ein grosse ursach / die weil nie mans wol von eim ding schreiben kan / der es nit selbs erfaren oder gesehen / so hat dis Bergwerck vilerley gebreüchlich instrumenten und eygnenamen / die nit bald ein schreiber oder gelerter weißt / auch so wunderbarlich / das keiner / wie weit er wandert gar auß lernen kan.' Münster ultimately visited the mines of Leberthal himself (in 1545) and reported fondly about it. See McLean, *The Cosmographia of Sebastian Münster* (2013), pp. 154, 309.
[88] Koch, *Geschichte und Entwicklung des bergmännischen Schrifttums* (1963); Bartels, *Vom frühneuzeitlichen Montangewerbe zur Bergbauindustrie* (1992), pp. 279–281.
[89] Reinhold's *Instruction* was admittedly re-edited in 1615, but neither edition is referenced in manuscripts or by later authors until Christian Wolff listed it in his *Mathematical Lexicon* in 1734.

underground world, as the examples of Erasmus, Agricola and Münster illustrates. These works certainly played a role in heightening the status of mining disciplines, and might have encouraged the nascent vernacular literature of the *Bergbüchlein*, or the formalization of hitherto customary mining laws. Still, there was an important gap between the crafts and the books that were written about it, even when these texts claimed to present actual practices. Humanists presented this geometry as a variation on an existing corpus, the practical geometry of the Renaissance scholars. Their legacy, however, was largely restricted to academic circles, while the practical geometry of the mines followed its own dynamic course.

As one understands that scholarly books were no descriptions of surveying processes – and that they were in all likelihood neither read nor used in mining offices – one should set them aside for a moment and turn to practitioners themselves. From a conceptual point of view, it equates to changing our perception of the history of science. Instead of referring to the work of individual – admittedly brilliant – scholars to study the inner logic of practical knowledge, we should learn about the craftsmen's collective culture of mathematics, its nature, and the values associated with it. What were their problems, methods, and instruments? Instead of asking to what extent the content of the *De re metallica* depicts real operations of underground surveying, let us turn the question around and try to find out what the major operations of subterranean geometry were. Only in this way can we properly follow the path taken by the actors themselves towards a rational use of quantification. By entering the early modern silver mines and observing the subterranean surveyors at work, one gets a very different perspective on the art of setting limits.

## 2 A Mathematical Culture
### The Art of Setting Limits

– Music –
Measure sets a goal and limits
Measure shows, how far?
Measure prevents disputes
Measure ends quarrel
Measure instructs a miner
Where he can extract ore

Mining song from the Harz Mountains*

In the Late Middle Ages, long before it became a mathematical art, the demarcation of mining concessions originally followed a set of rules and material procedures. As soon as ore outcrops were discovered, prospectors competed to secure the most promising spots. Local mining masters, required by law to delimit concessions for all applicants, devised methods to properly set the marking stones that symbolized the concession boundaries. The term used by medieval miners to denote this activity, *Markscheidekunst*, belonged to the so-called *Bergmannsprache* (language of miners) and literally translates as the 'art of setting limits'.[1] Far from a geometry produced in the quiet of an office, the necessary measurements were made in inhospitable locations. As mines went deeper, limits were increasingly set underground, in 'narrow and arduous shafts, in which one often has to operate at the risk of one's body and life', in the words of a surveyor.[2]

From the end of the fifteenth century onwards, a favourable legal setting combined with the discovery of new mining sites led to a silver rush known as the *Berggeschrey* (mining clamour) in several regions of the Holy Roman Empire and Central Europe. The scale of extractive operations was transformed, raising new

---

* Henning, *Programma de historia recentiori Hercyniæ* (1726), p. c2: 'Messen setzet Ziel und Schrancken / Messen zeigt, wie weit? / Messen beugt vor dem Zancken / Messen hebt den Streit / Messen weisst den Bergmann an / Wo er Ertze fördern kan.'
[1] Several spellings of the term exist, such as *Marscheiden, Mark-Scheiden, Marckscheidekunst*, etc. The subterranean surveyor – the *Markscheider* – was in Tirol named *Schiner*, and his instrument the *Schinzeug*. See Bartels, Frey, and Bingener, *Das Schwazer Bergbuch* (2006), vol. 2, p. 58r.
[2] Rösler, *Speculum metallurgiae politissimum* (1700), p. 87: 'viel enge Strecken und beschwerliche Schächte / darinnen man offt mit Leibes- und Lebens-Gefahr darbey handlen muß'.

technical problems that led in turn to innovative surveying methods. Constructive geometrical procedures had been integral to mining activity from its medieval origins onwards.[3] In order to set the marking stones, sticks and cords had regularly been used. Mathematics, however, was not originally seen as the only legitimate, objective way to ensure fairness among competing miners. The length of a parcel given to prospectors could sometimes be set 'as far as they can throw their pick', hatchet, or mining hammer from the central pit. Anthropocentric procedures also abounded, involving steps and jumps to fix or extend concessions.[4] Such methods had several obvious drawbacks, the first of which was the lack of a unified framework. Moreover, there was no obvious way to check later on if boundaries had been moved over time, or even properly set in the first place. Well into the eighteenth century, when subterranean geometry was well established, mistrust was sometimes evident and local custom still advised prospectors to place coal, glass, and brick splinters under marking stones, to deter cheating and dishonesty.[5]

The aim of this chapter is to retrace how a culture of accuracy and quantification developed in early modern mines, and to highlight both its richness and significance. One should not expect from practitioners a coherent doctrine based on Euclid's *Elements*, for this classical geometry was not available – and would hardly have been useful – to mine surveyors. The art of setting limits used in metallic mines developed within a specific technical, administrative, and economic context, examined here in detail. We dissect the actual procedures of underground geometry in order to understand the encroachment of mathematics into human affairs. Since measurements played an important role in technical operations, this culture grew to become an integral part of early modern thought in mining regions. The surveyors' skills with instruments were blended with knowledge of the generation of metals and accumulated experience of the behaviours of ore veins. Moreover, subterranean geometry was constrained by the legal frame in which mining operated. Unlike the abstract sciences, it had to follow strictly codified rituals that are reconstructed here using a purposely diverse set of sources.[6] Mining laws, customs, and rulings, as well as sermons and town chronicles, reveal striking details about the materiality of measuring operations and ceremonies.[7]

---

[3] On the concept of 'constructive geometry', see Shelby, 'The Geometrical Knowledge of Mediaeval Master Masons' (1972), pp. 395–421.

[4] Zeidler, *Erbvermessen der Silber-Gruben Catharina* (1714), p. 18. See also Clauss and Kube, *Freier Berg und vermessenes Erbe* (1957), pp. 123, 129; Ziegenbalg, 'Aspekte des Markscheidewesens mit besonderer Berücksichtigung der Zeit von 1200 bis 1500' (1984), p. 44.

[5] Clauss and Kube, *Freier Berg und vermessenes Erbe* (1957), p. 110.

[6] This 'methodological opportunism' has been theorized by Pascal Duris in 'Faire feu de tout bois' (2013), p. 243: '*Faire feu de tout bois*, that is resort without prejudice to various kind of material and conceptual sources.'

[7] Eric H. Ash rightly mentions both mathematics and the legal frame as decisive advantages of German miners over their English counterparts, but stops short of assessing how intertwined both factors were. See Ash, *Power, Knowledge, and Expertise in Elizabethan England* (2004), p. 30.

If we are to understand the inner logic of subterranean geometry, we have to set aside for a moment 'illustrious names, marvellous machines, seducing courts, and scholarly discourses'. Early modern humanists, as the previous chapter has shown, relied on general principles and hardly addressed the interaction between geometry and technology. It is noteworthy that the administrative, religious, and technical sources used below were written for and by practitioners, to address their current, practical concerns. As Bertrand Gille phrased it in his studies on Renaissance engineers, 'our means of investigations are limited, our knowledge fragmentary'. The conservation of these surveying procedures up to today was at best an unexpected by-product, but certainly not the authors' original intent. A lot of information was preserved only through working sketches or administrative documents focused on the extraction of ore. They contain very few detailed illustrations, and their geometrical vocabulary can be scattered or wrapped up in religious discourse or visitation reports. Understanding them is 'a matter of difficult steps and long journeys' but through systematic cross-referencing the modern reader can get much closer to actual surveying practices.[8]

Unearthing the art of setting limits reveals how common miners and superior craftsmen considered measurements in the sixteenth century. This appreciation gradually turned into a full-fledged culture of geometry, which gained pre-eminence in the mining states of central Europe. Although surveying methods did not impress contemporaries as immediately as miraculous ore findings or mechanical machines, multiple factors nevertheless converged to enhance the status of geometry and quantification in the long run. The legal frame of extracting activities came to establish mathematics as a guarantee of objectivity, in order to foster prospecting and investments. Quantified information, official receipts, and marking stones served as evidence in trials about property and boundaries. Furthermore, these transformations unfolded as the Reformation took off in many states of Central Europe. While historians have pointed out that Protestantism valued the academic teaching of mathematics, a close reading of Lutheran pastors such as Johannes Mathesius (1504–1565) allows us to show how religion also contributed to its wider popularization. Mining sermons, teeming with concrete details, praised underground geometry as an expression of divine perfection. Pastors consciously promoted mathematics to foster rationality, trust in surveying procedures, and ultimately obedience to a higher law. Geometrical rules were gradually conflated with the rules of law and religion, fostering a popular culture of accuracy.

---

[8] All quotes are from Bertrand Gille, who discusses the Renaissance engineers in Gille, *Les Ingénieurs de la Renaissance* (1964), p. 41.

## Geometry and the Rules of Law

Everyone will get his due share
No one breaks any laws
Everything is well ordered and reserved
Here one uses justice, not violence
Equity is maintained for everyone

Mining poem from Saint Joachimsthal, Bohemia, 1523[9]

Early modern mining was a highly regulated activity. Local rulers sought to maximize profits from the exploitation of underground riches, while at the same time attracting much needed technicians and investors.[10] Ore extraction was a perilous enterprise that required a lot of capital for an uncertain result. Precious metals were located in narrow and sinuous ore veins, surrounded by hard waste stone, and silver concentration varied greatly within a single vein. Regulations could take different forms: mining laws, customs, or jurisprudences. Mining laws (*Bergordnungen*) were enacted by rulers, such as the Elector of Saxony, the Duke of Brunswick, and the King of Bohemia, in order to lure capital and workers; several were printed and widely circulated in Central Europe. These laws granted a remarkable degree of freedom, in the form of the *Bergfreiheit* (mining freedom): every miner could prospect wherever he wished, regardless of who owned the land.[11] Once a concession (*Muthung*) was granted, it was crucial to set its boundaries quickly, even if in a preliminary way, because new miners would immediately prospect around the most promising spots. Mining freedom was offset by duties, which also aimed at stimulating extraction. Any locator had to actively operate his site and pay his workers, or he would lose his digging rights within days or weeks.

In this context, it was only natural that surveying would also be regulated, and mining laws dealt with measurements at length. Geometrical methods played two crucial and complementary roles. First, they were used to secure concessions' boundaries, so that, in the words of a contemporary, 'the investors

---

[9] Rudhart, *Antzeigung des Nauenn Breyhberuffen Bergwerks Sanct Joachimsthal* (1523), reproduced in Lempe's *Magazin der Bergbaukunde* in 1794, vol. 8, p. 194: 'Einez iezlichen wirdt das seine genuglich vorricht/ Daselbist auch nymands keines rechten gericht / Ist alles wol vorordent und bestalt / Man gebraucht da rechtes und nyt gewalt / Der billigkeit tut man menniglich pfflegen.'

[10] Graulau, *The Underground Wealth of Nations* (2019), pp. 1–29, summarizes the problem: 'mining required capital' but 'lords had no capital for mine improvement' (p. 9 and p. 17).

[11] On mining laws, see Löscher, *Das erzgebirgische Bergrecht des 15. und 16. Jahrhunderts* (2003); Braunstein, 'Les statuts miniers de l'Europe médiévale' (1992), pp. 35–56. If a miner was not satisfied with local rules, he could leave the district without asking permission, a rare privilege at the time.

can make their calculation' and anticipate future profits.[12] Another mining official enjoined rulers to avoid arbitrary decisions, 'to dispute neither the mining freedom nor the mining law, but to let the yes be yes and the no be no'.[13] Surveyors were thus charged with the crucial task of providing a factual basis on which legal decisions could be reached. Second, geometry was expected to provide information about the direction of ore veins, the depth of underground galleries, and the probable duration of digging operations. However, there was initially an understandable unease with the esoteric aspect of mathematical procedures, since mining operations relied on something few people understood. Its reliability had to be secured, and ideally publicly displayed, or local rulers would be perceived as exploiting it for their own interests.[14] Surveyors were justifiably viewed with defiance because their mistakes could 'dig into the investors' purse'. Using geometry was meant to avoid the situation where 'one or two hundred craftsmen have to believe a single one, without sufficient evidence', if only because 'we are all humans', and humans sometimes make mistakes.[15]

To fulfil his role, a subterranean surveyor thus had to measure accurately. The notion of accuracy involved in these operations, however, differs significantly from the definition that any contemporary scholar could have given. Subterranean geometry could not be demonstrated using axioms and logical reasoning, as surveyors' computations could generally neither be corroborated nor refuted. The most diligent determination of a vein direction would have to be corrected if the ore strata later turned out to be fractured or deviating from the original direction. In many cases, the complexity of real-world conditions posed a major challenge: Intrinsic geological and technical aspects of most problems made the idea of a perfect method illusory. In fact, subterranean geometry did not need to be perfect, as long as it was socially accepted: measurements had to be – as the etymology *accuratus* indicates – performed with care. This meant, on the one hand, surveying as precisely as possible while, on the other hand, conforming to legal and customary practices. Only if these two conditions were met could a survey be considered valid. Procedures were not checked against an abstract standard of validity, in the way a geometrical proof was evaluated in universities, according to the Euclidean frame. Surveys had

[12] Reinhold, *Gründlicher vnd Warer Bericht* (1574), introduction: 'damit die gewercken ihrer rechnung machen konnten'.

[13] Löhneysen, *Bericht vom Bergwerck* (1650), p. 45: 'weder die Bergk Freyheit oder die Bergk Ordnung *disputiret*, sondern Ja/Ja und Nein/Nein seyn lassen'.

[14] While local rulers, most notably the Elector of Saxony and the Habsburg monarchs, benefited immensely from local underground riches, this was mostly through their own mining shares (*Kuxen*) and tithes.

[15] Reinhold, *Gründlicher vnd Warer Bericht* (1574), dedication and introduction: 'ihr fehlen den gewercken in die Beutel greiffet ... damit forthin nicht ein 100. oder 200. gewercken / einem allein / ohne gnungsame beweiß / mußten glauben geben ... wie wir dann alle Menschen seind.'

to be checked against common sense and, ultimately, on the concrete outcome of the mining operation.

The *Schwazer Bergbuch*, a sixteenth-century treatise on Tyrolean laws and customs, observes that a surveyor should have 'good and fair' instruments and avoid 'unjust tools'. The word used by the author, *ungerecht*, denotes both something *inaccurate* and something *unfair*, implying that such instruments would produce measurements both imprecise and incorrect. Failing to observe this would 'bring harm and prejudice'.[16] A careful surveyor should, for example, bear in mind that 'water – for many places are very wet – causes the cord to be wrong', as surveying cords were usually made of hemp at the time. He should not forget that 'some places are ferriferous, moving the magnet needles from their places and making the survey wrong'. In spite of countless local specificities, a surveyor should strive to 'give to each pit its due measure and equity'.[17] The impossibility of a perfect survey also had its upsides. Some rules, for instance, could easily be abstruse and even arbitrary, and yet be accepted because they were customary. In some regions, the length of a concession was traditionally 28 *Lachters* (ca. 56 m) while in others it was 40 *Lachters* (ca. 80 m), for the simple reason that it had always been so. What mattered most was social agreement on the dimensions of concessions and respect for surveying procedures. The accuracy of measuring operations was therefore charged with a moral character.

The pace of technological change and the flourishing of mining towns made some mining laws quickly obsolete. Mining masters and local courts thus had to cope with numerous unexpected situations, giving birth to a vast jurisprudence: the mining customs or *Berggebräuche*. In Saxony, Simon Bogner (ca. 1500–1568) was a mine entrepreneur who also owned a smelting hut. From 1541 onwards, he was mining master in Freiberg and contributed to the establishment of a general mining administration, collecting rulings and jurisprudence in his own manuscript. These legal sources, together with town chronicles that provide lively examples of disputes and trials, open a window onto the work of subterranean surveyors that complements, often in precise detail, the better-known mining literature of the *Bergbüchlein* and other works written in a courtly context.[18]

---

[16] Bartels, Frey, and Bingener, *Das Schwazer Bergbuch* (2006), vol. 2, p. 339: 'mit guetten, gerechten … niemand durch solichen ungerechten Werchzeug zu Schaden oder Nachtail kome, sonnder yeder Grueben jer gepurennd Maß und Gerechtigkait erfolge'.

[17] Rösler, *Speculum metallurgiae politissimum* (1700), p. 87: 'das Wasser (dann es manches Orts sehr naß ist/) beschwerlich / verursachet dass die Schnure falsch ist … an manchem Orte Eisenschüssig … die die Magnet-Nadeln von ihren Ort verrücken / und den Zug falsch machen'.

[18] Moeller, *Theatrum Freibergense chronicum* (1653), pp. 408 and 458. The custom written by Bogner is presented in Löscher, *Das erzgebirgische Bergrecht des 15. und 16. Jahrhunderts* (2003). On the *Bergbüchlein* tradition, see Long, *Openness, Secrecy, Authorship* (2001), pp. 177–178.

## Claiming a Concession

Let us now follow the acquisition of a concession, in order to see precisely how mining laws and customs relied on underground surveying. While each region had its particularities, the use of geometry followed a roughly similar process in the main mining centres.[19] Attracted by rumours or geological signs, miners would begin prospecting new hills. They would be looking for natural signs indicating ore, such as metal-tasting sources – for 'these are not far from the veins' – or discoloured trees, 'since the exhalation of veins consume them', according to mine surveyor Balthasar Rösler.[20] Miners then dug prospecting shafts and, if signs were encouraging, tried to secure the area in a first round of preliminary surveying. The mining master was obliged to perform this measurement: 'If a pit is furnished with workers, mining buckets, and rope, and the investors want the mining master to set their measure provisionally, he shall not refuse', just as surveys should not be influenced by 'favor, friendship, gifts or presents, hate, envy, or enmity'.[21] This first area was called the *Fundgrube* (founder's shaft), a square whose side was roughly oriented along the direction of the ore vein. 'The mining master shall take the cord and place it on the middle of the winch', mining laws indicate, and from there build a square, as represented in Figure 2.1.[22] This figure could then be extended by adding rectangles or squares knows as *Maaße* (measures) following the lode. Lured by the rumours of newfound riches, competitors would soon prospect along the ore vein and acquire neighbouring concessions.

The first survey was only provisional and indicative, as no meaningful quantity of silver had usually been extracted at that point. It was performed with a simple cord of the wanted length, without taking slope into account, so that people spoke of a 'lost measurement' or a 'lost cord' (*verlornen tzuges* and *verlohrne Schnur*). It certainly appears rather rudimentary, but judging practical

---

[19] One can argue that mining laws induced a common pattern of surveying method, just as the printed *Künstbüchlein* contributed to the standardization of crafts procedures, as in Eamon, *Science and the Secrets of Nature* (1994), p. 113. Moreover, many workers in the new mining regions originally came from Saxony, contributing to the diffusion of Saxon customs. Furthermore, the *Bergordnungen* were printed from the late fifteenth century onwards, ensuring a swift circulation of knowledge.

[20] Rösler, *Speculum metallurgiae politissimum* (1700) p. 11: 'diese sind von den Gängen nicht weit … weil die Exhalation von Gängen denselben verzehrt'.

[21] Ermisch, *Das sächsische Bergrecht des Mittelalters* (1887), p. 171, transcribes the Annaberg mining law of 1509, article 24: 'So eine tzeche iren schacht belegt, kewbel und seyl einwirfft und die gewercken am bergkmeister begern yre maß zu uberschlahen, das sall er nicht weygern.' *Bergk-Ordenung* (1589), p. 14v: 'Gunst / Freundschafft / Gifft oder Gabe / Haß / Neid / Feindschafft.'

[22] *Ursprung vnd Ordnungen der Bergwerge inn Königreich Böheim* (1616), p. 7: 'So soll der Bergmeister die Schnur nehmen / und sol sie legen mitten uff den Rundbaum.' Figure 2.1 comes from Beyer, *Gründlicher Unterricht von Berg-Bau* (1749), table IX, fig. 20; see the explanations given by the author in the text, pp. 189–191.

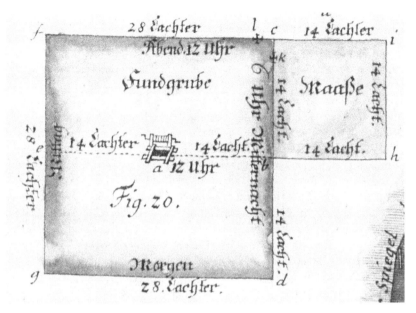

Figure 2.1 Preliminary survey of a mine concession. The large square represents the founder's shaft (*Fundgrube*) and the smaller square on the right a first extension (*Maaße*). Both are oriented along the dashed line representing the direction of the ore vein, as determined by the subterranean surveyor. Courtesy of the TU Bergakademie Freiberg/Universitätsbibliothek II 410 4.

procedures by our modern criteria, or even by the standards of theoretical mathematics at the time, would be missing the point. A lost measurement was meant to be an approximation, purposely vague so as not to be legally binding. A popular saying stated that 'the lost cord of the mining masters proves nothing', while a legal compendium defined 'measuring with a lost cord' as when 'one lets the subterranean surveyor pull [cords] for one's own advice'.[23] For this preliminary survey the mining master 'should not take more than one *groschen*' as salary, just enough to buy a *Groschenbrot*, a typical loaf of bread of ca. 5 kg.[24] In any case, a geometrical procedure must be assessed in light of

[23] TU BAF – UB XVII 18, f. 58v: 'Verlohrne Schnur der Bergmeisters beweiset nichts'; Schönberg, *Ausführliche Berg-Information* (1693), p. 102: 'Mit verlohrene Schnur vermessen lassen / heist / wenn einer zu seiner Nachricht durch den Marckscheider sein Feld ... abziehen.'

[24] Annaberg mining law of 1509, article 5, transcribed in Ermisch, *Das sächsische Bergrecht des Mittelalters* (1887), p. 165: 'von einer muttung nicht meher den einen groschen nemen'. The price of bread loaves was set to avoid using change, but their weight was subject to fluctuations regulated by tables computed by reckoning masters (see Chapter 3).

the context in which it was used. During the context of the silver rush, the aim was primarily to dampen conflicts and ensure that prospecting would not lead to endless brawls.

Once a concession had been surveyed, a mining clerk wrote the location down in his mining book, together with the day, sometimes even the hour, 'so that another after him' – meaning after the first concession holder – 'who claims the neighbouring field or *Maaß*, can begin to work and mine'.[25] Boundaries were simply marked by cheap wooden stakes in the ground. This survey was sometimes labelled *minus solemnis* (less formal) or even *superficiaria*. This latter term underlined both that measurements were only made on the surface of the mountain and that they had little legal value. Nevertheless, it was strictly forbidden to perform even this inconsequential operation without the knowledge of the administration! As a matter of principle 'no one should perform mine surveying unless it has been approved by our captain or mining master'.[26] The administration exerted a monopoly on surveying, however rudimentary the procedure was. In the Holy Roman Empire, it was considered a sovereign function belonging to rulers, just as justice or coinage, and abusing it was severely punished.

Real work began after the first ore had been extracted and assayed, if meaningful quantities of silver were found. When the results were encouraging enough to warrant further investments, a second survey was made to properly establish the concession. Depending on geological configurations, it was distinctly possible to find small pockets of highly concentrated metallic ore, so that a difference of a few feet could lead to fortune or to ruin, which incidentally was the reason why mining was sometimes deemed immoral.[27] During this operation, named either *dimensionem solemnem* (formal survey) or *Erbbereiten*, an orchestrated ceremony publicly displayed mathematics as the guarantee of fairness. Legal compendia describe this operation at length. An announcement had to be publicly shouted in town 'three Saturdays in a row', after which the ceremony would be performed in the presence of the local mayor and mining officials. It also included the investors and workers, who were given a free meal while a celebration was organized with music and dance.[28] Emphasizing the importance of the ritual conferred legitimacy to the survey and acted as a kind

[25] Rösler, *Speculum metallurgiae politissimum* (1700) p. 32: 'damit ein anderer nach ihm / der das folgende Feld oder Maassen gemuthet / anzusitzen und zu bauen weiß'.

[26] Ermisch, *Das sächsische Bergrecht des Mittelalters* (1700), p. 194, article 90: 'Es sal sich auch nu hinforder uff vilgemelten unßerm bergkwergk nimandt marckscheidens understehen, er sei dan von unßerm heuptman und bergkmeister tzugelaßen.' On the term '*superficiaria*', see Beyer, *Otia metallica* (1751), vol. 2, p. 231.

[27] Agricola himself defended mining against such accusations, as analyzed in Agricola, Halleux and Yans, *Bermannus le mineur* (1990), pp. xviii–xix, 28–30.

[28] Schönberg, *Ausführliche Berg-Information* (1693), pp. 28–30, here p. 28: 'drey Sonnabende nacheinander'. See also Clauss and Kube, *Freier Berg und vermessenes Erbe*, p. 94.

of public record within the local community. In the same spirit, 'young miners were slapped and their hair pulled to imprint the event' in their memories, so they could act as witnesses even decades later, if needed.[29]

The subterranean surveyor had performed his duty 'in advance' and already described the concession in the corresponding mining book.[30] In the case of a formal survey, this meant ascertaining the direction of the vein with his mining compass, before carefully aligning the concession along this imaginary line. All measurements were nevertheless re-enacted once again in public to ensure general agreement. Rulings warned against surveying without 'public announcements' and were very clear that 'such a thing cannot rightfully happen', consistently nullifying the results.[31] Any miner was allowed to contest each of the steps and complaints were handled on the spot, but within a strictly civil and codified frame. Touching the surveyor's chain, for instance, was fined with 20 *marks* of silver, an absurdly high sum tantamount to being banned from the district.[32] The practical geometry a surveyor had to use in an official measurement was by no mean trivial. Mountains are not usually flat and numerous obstacles could hinder his survey. The *Markscheider* had to work out a solution and 'mark out a figure that would enclose *geometrically* just as many square-*Lachter*, as the founder shaft contained *arithmetically*', according to a mining clerk. The *Markscheider* would then check if the winch was indeed placed in the middle of the square by materializing the diagonals with cords; this visual proof that the winch was located in the 'middle' of the concession was psychologically important to people.[33]

Surveyors had to adapt to particular cases depending on the situation of the vein or local customs. If the field was new, for instance, the prospector could extend his concession from the length of a 'backward jumping', an age-old custom still alive in some districts at the end of the seventeenth century.[34] If previous concessions existed, the surveyor was forced to respect the existing boundaries and negotiate to enclose a satisfactory area. Mining official Adolph Beyer (1709–1768) indicated in his *Otia Metallica* (Metallic Leisures) that real-life surveys were always a matter of compromise between the principle of

[29] Clauss and Kube, *Freier Berg und vermessenes Erbe* (1957), p. 110: 'beteiligten Bergjunden versetzte man eine Ohrfeige und raufte ihnen das Haar, um diesen Zeuges das Ereignis nachhaltig einzuprägen'.

[30] For a typical example of such book (*Vermeßbuch*) of the early seventeenth century, see SächsStA–F, 40015 Bergamt Schneeberg, Nr. 1309, Vermessbuch 1597–1632.

[31] Span, *Sechshundert Bergk-Urthel* (1636), § 272, case from 24 September 1615: 'weder offentlich verruffen … so kan dasselb. zu recht nicht Passiren'.

[32] Schönberg, *Ausführliche Berg-Information* (1693), pp. 28–30.

[33] Beyer, *Otia metallica* (1751), vol. 2, p. 261.

[34] Schönberg, *Ausführliche Berg-Information* (1693) p. 28: 'worauff die Geschwornen / nach des Lehenträgers rückwerts gethanen Sprung / das Feld verlochsteinen'. Clauss and Kube, *Freier Berg und vermessenes Erbe* (1957), p. 129.

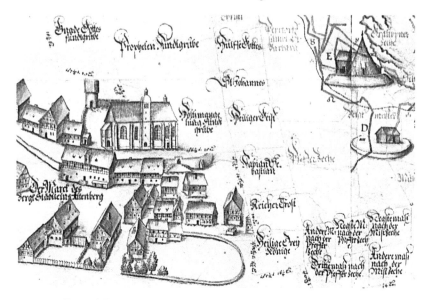

Figure 2.2  Mining concessions on the outskirts of Altenberg (Ore Mountains of Saxony). In practice, concessions were irregular quadrilaterals following the direction of ore veins, whose sides' lengths were written on the plans. Dots represent the marking stones and each concession was given a name, mostly religious (here *Holy Spirit, Saint Johann, God's Help* or *Shaft of the Prophet*). Detail from a mining plan by Balthasar Rösler, 1668, Sächsisches Staatsarchiv, Bergarchiv Freiberg, 40040, Nr. B624.

rectangular forms and real-life conditions, adding: 'the shapes on mining maps indicate how poorly this quadrangular form is or can often be respected'.[35] Looking closely at such plans, one can indeed grasp the challenge of setting limits in uneven landscapes that were already crowded not only with other concessions, but also houses and mine buildings (see Figure 2.2.) In all cases, a formal survey had a *forma probante*, meaning that it had legal value and could hardly be contested later. Boundaries were officially marked by heavy stones, known as *Lochteine* or *Markscheide*, that would in turn be depicted on the maps. The price of this second kind of surveying was understandably much higher, including the legal fees, a tax to be paid to the mining administration, as well as a banquet for dozens of people. The use of geometry was orchestrated into a costly public performance, primarily to ensure its legitimacy.[36]

---

[35] Beyer, *Otia metallica* (1751), p. 261: 'Wie schlecht aber diese viereckige Figur offte beybehalten wird oder kan, bezeugen die *Figuren* derer Marckscheider-Risse.'
[36] The expression '*in forma probante*' can be found in numerous mining customs; see, for instance, Schönberg, *Ausführliche Berg-Information* (1693) p. 28.

Geometry was thus used as a reliable tool, although the price of accuracy was indeed high. A formal survey represented a large investment – routinely higher than the annual salary of a miner – but it brought real legal security. Two Saxon investors choosing an *Erbbereiten* over a preliminary survey justified their decision saying that 'they would like to be in rightness [*in Richtigkeit*] in regard to the neighboring extensions'.[37] At any rate, accuracy was less costly than its absence. A lack of measurement, or a rushed observation, could induce anything from major delays to mine flooding or crumbling. Preliminary surveys, on the other hand, were designed to be easy and cheap, purposely lacking legal value to avoid lengthy procedures and endless controversies.

What if a miner or a company did not accept the outcome of a formal survey? As with all procedures designed to settle property issues, it is inevitable that some people contest an unfavourable outcome. Mining laws anticipated this scenario and took pains to design mechanisms that would ensure the results achieved consensus. In the eventuality that consensus was not reached, a disgruntled prospector could ask that a third kind of measurement be performed, the *Wehrzug* or *dimensio decisiva*. The plaintiff was then authorized to summon a subterranean surveyor from a neighbouring district, or even another region, albeit 'at his own cost'.[38] The foreign surveyor could use whatever method he wished in order to reach a decision: his impartiality was to be guaranteed by his status as a foreigner and by his geometrical skills. Without overstating the modernity of such control processes, it prompts two general remarks about early modern practical mathematics. Calling a foreign geometer for a *decisive survey* allowed for comparisons of procedures, ensuring the circulation of this esoteric know-how between mining regions.[39] Secondly, the procedure of *dimensio decisiva* illustrates how practical matters were not susceptible to demonstrations akin to Euclidean proofs. One could not make one's case simply with a paper and a pen by reasoning within a hypothetical-deductive framework. Surveying with different methods until agreement was reached illustrates how mining administrations strove for social consensus instead of abstract proofs.

During the early modern period, the convincing power of geometry was in the making, and its legal display was an important factor in that process, a process that Witold Kula has similarly observed in the public accessibility of standards and measures outside town halls.[40] Surveying was a ubiquitous

[37] SächsStA–F, 40015, Bergamt Schneeberg, Nr. 139, 'weile sie wegen derer benachbarten Maaßner gerne in Richtigkeit seyn mögen'.

[38] *Bergkordnung des freyen königlichen Bergkwercks Sanct Joachimsthal* (1548), article 12: 'auff sein kost alhieher zubringen'.

[39] A concrete example of *dimensio decisiva* dated from 1612–1613 can be found in Herttwig, *Neues und vollkommenes Berg-Buch* (1734), pp. 72–74.

[40] Kula, *Measures and Men* (1986), pp. 79–80.

phenomenon, which came to be socially accepted because it was openly demonstrated and provided evidence that could then be used during trials. We are dealing here with a prime example of what Max Weber has labelled 'rational technology'. The attribution of concession was rational insofar as laws 'reduced [it] to calculation to the largest possible degree'.[41] Mathematics contributed to the reduction of uncertainty in mining and made the quantification of space visible.

### The Ubiquity of Measurements

Delimiting mining concessions by means of preliminary, formal, and decisive surveys was only one of the many duties of a surveyor. As deep-level mining developed during the sixteenth century, most of the technical issues now arose underground. A pressing issue was the locating of concession limits within mine workings. As shafts went deeper, operations were pursued on different levels simultaneously. The marking stones delimiting a concession were first set on the surface during the solemn measurement described above. In a second step, they had to be fixed underground by means of symbols engraved in the galleries, as can be seen in Figure 2.3. Mining laws specified that 'all limit marks have to be brought from the surface into the mines by the subterranean surveyor, the mining master being informed, and noted and described in a specific book'.[42] The task amounted to locating underground the spot exactly perpendicular to a given marking stone set on the hillside, or more rarely with a given angle. In practice, surveyors used mining compasses to assess the direction of the ore vein. In the sixteenth century, miners used fairly regular compasses, except for the units of measurement.[43] While courtly instruments were made of brass or ivory and nicely ornamented, practitioners had to work with simple wooden compasses.[44]

Once the direction was ascertained, the surveyor measured the distance between the winch and the limits of the concession. He then went down the

---

[41]  Weber, *General Economic History* (1927), p. 277. Mining laws were as detailed as possible, ensuring a 'calculable law', in Weber's words (ibid., p. 277), while free labour ensured the circulation of men and techniques.

[42]  *Bergkordnung des freyen königlichen Bergkwercks Sanct Joachimsthal* (1548), p. I, § 12: 'Es sollen auch alle Lochstein / so vom Tag hinein / inn die Gruben gebracht werden / so inn der Gruben / durch die Marscheyder verbracht werden / beim Bergkmeyster ordentlicher weiss / inn ein sonderlich Buch eingeschrieben / und verzeichnet werden.'

[43]  In mining compasses, the full circle was divided into twice twelve hours, complemented by the four cardinal points, each hour then being divided into quarters for greater precision. Directions were indicated in the following way: 'ME.12.¼.' means: 12 hours and a quarter, to be read in the southern (*Meridiens*) quarter of the compass. Several exemplars are described and studied in Michel, 'Boussoles de Mines des XVIᵉ et XVIIᵉ siècles' (1956), p. 617.

[44]  In his *Instruction*, Erasmus Reinhold recommended in 1574 that surveyors worked with carpenters to produce their own instruments, suggesting that casual instruments were not made by professional instrument-makers. See Reinhold, *Gründlicher vnd Warer Berich.* (1574), p. bii.

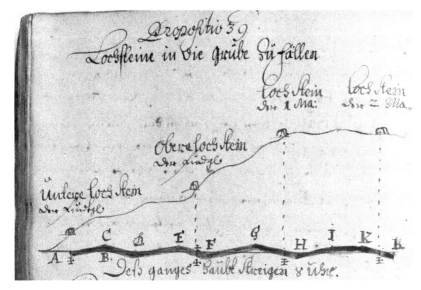

Figure 2.3 Setting of concession limits in the gallery, as presented on a surveyor's manuscript. The marking stones (*Lochsteine*) separating concessions at the surface have to be vertically 'brought' or 'felled' into the galleries and pits. TU Bergakademie Freiberg/Universitätsbibliothek XVII 333, f. 58v.

mine and assessed where, if at all, this distance had effectively been reached. The obvious difficulty was that mine galleries followed the sinuous paths of ore veins, changing their course and alternatively ascending and descending. Surveyors sometimes relied on an instrument composed of concentric wax rings to record the varying directions and adjust their measurements accordingly (see Figure 1.2). They noted every change of direction in the mine, together with the distance covered, and immediately reconstructed on the surface the gallery path they had followed underground. If the end point of the gallery appeared to lay outside of the concession, underground operations would come to a halt. The surveyor would go back underground and carve in stone a sign indicating the boundary of the concession. If the end point of the gallery still laid within the concession, extraction could proceed unimpeded. A rare description of this process can be found in the German *Methodus geometrica* of Paul Pfinzing (1554–1599), a cartographer from Nuremberg, also active in the city council. Despite having no known relationship to subterranean geometry, he must have gathered testimonies about the procedures of mine surveyors.[45]

---

[45] On Pfinzing, see Schmidt, 'Pfinzing and Friends' (2020), pp. 364–386.

Pfinzing, whose book mostly describes standard cartographic and surveying practices, modestly remarked:

The compass can also be very usefully applied in the mines, to set boundaries, [and] ascertain the direction of galleries and veins underground. For just as one can measure and record altogether each mountain and valley at the surface, one can also perform this underground with a compass, going through the gallery and the vein, reproducing the direction, and thereby find out where everyone is in its field, of which practice and experience will give a better instruction than I could write upon.[46]

Setting concession limits underground was a daunting task for surveyors with little formal knowledge of mathematics. Moreover, this kind of measurement could hardly be demonstrated in public, since it did not take place on the surface. Mining master Enderlein emphasized the issue of survey accuracy in his handwritten work on customs: 'in measuring, and especially while setting the boundaries on an ore vein, subterranean surveyors should use and keep one and the same method [*Ordnung*]'. If asked, they would 'have to disclose and show this method' to anybody.[47] In most cases, a surveyor was accompanied by two (sometimes four) mining jurors to compensate for the absence of the public. They followed him and gave their approbation before he finally carved the sign into the gallery's wall. This was meant, I think, to ensure the accuracy of the procedure. Even if nobody, including the mining master, was skilled enough to replicate the survey, people were entitled to witness it and to testify that things had been done with care, that is, according to local customs.

Once the boundaries had been fixed, challenging the decision was very difficult, even when it could be proved that the marks had been incorrectly placed. In 1720, George Zecher, supervisor of the *Bestowed Luck* mining pit, complained to his local mining master about a mistake made during an earlier formal survey of the mine twenty-two years before: the marking stones 'had not all been brought or carved on the main branch vein', so that 'one stands too far in the floor, the other too far in the roof' of the gallery.[48] The administration

---

[46] Pfinzing, *Methodus Geometrica* (1598), p. 44r: 'Nachmals ist auch der Compass sehr nützlich zu gebrauchen / zu den Bergwercken / Marscheidung / Stollen und Gäng zu finden / wie sie undter der Erden streichen / Dann gleich wie man ob der Erden / alle Berg und Thal / kan zusammen messen und Eintragen / Also kan man auch mit dem Compasten / undter der Erden / mit abgehen der Stollen unnd streichung der gäng / solches verrichten / und dadurch wissen / Wo ein jedes im Landt außkombt / Wie dann die ubung und Erfahrung mehr anleitung geben wirdt / dann ich darvon schreiben kann.' Note the beautiful illustration on f. 44v.

[47] Mining customs of Matthes Enderlein, in Löscher, *Das erzgebirgische Bergrecht des 15. und 16. Jahrhunderts.* (2003), p. 114: 'sollen in ihrem ziehen und sonderlich im hineinbringen der erbstufen alweg auf einem gang einerley und gleiche ordnung halten und brauchen'; 'ime zu eroffnen und anzuzeigen schuldig sein'.

[48] SächsStA–F, 40010 Bergamt Freiberg, Nr. 326, f. 1r: '[die] gesezten Lochtseine, welche beym Erbbereiten nicht alle ufs Thrum gekommen oder gesezet worden … einer zu weit in liegende, der andere zu weit in hangenden stehen'.

turned to August Beyer, the surveyor who had performed the original measurement, for answers. Beyer stood by his previous survey, interestingly using two complementary arguments. First, he claimed to have measured the right distance in what then seemed to be the correct direction of the metallic vein. In 1698, however, the lode was split in different branches and it was not clear which one would turn out to be the *Hauptgang* (main vein). Back then, and this was his second argument, neither the local supervisor not the mining authorities had protested, because nobody could foretell 'that the marking stones would not be set on the correct main vein'. What retrospectively seemed incorrect could thus not 'make the whole official surveying flawed'.[49] In other words, what mattered most was that the original procedure had been correctly performed according to local customs: the survey was thus accurate. Beyer repeatedly underlined that there had been no 'carelessness', pointing to the fact that a consensus had been reached at the time. His geometry was sound, the surveyor implied, even if the geological configuration had turned out to be different than expected.[50] We witness here how mining laws took pains to distinguish between correct mathematical procedures and the uncertainty inherent to mineralogy and geology.[51]

Setting concession limits might sound arduous enough, but surveyors routinely coped with much more complex problems. Most of the ore extraction took place deep underground, with many miners competing in relatively small areas. During the silver rush, the mining book of the Bohemian town of Saint Joachimsthal indicated that 462 concessions were granted for the year 1518, with 102 for the month of November alone – more than three preliminary measurements per day![52] Mining cities were thus surrounded by crowded fields – as can be seen in Figure 2.2 – leading to legal issues that amounted to complex three-dimensional geometry problems. What if two outcropping veins, located in two different concessions on the surface, 'nevertheless merge[d] together in the depth', that is, turned out to meet at some point underground?[53] In that case, rulings recommended that the oldest concession gets a right of *Vierung* (intersection). This meant that it owned property on everything in a square of 7 *Lachter* (ca. 14 m) centred on the gallery. Sixteenth-century surveyors were deprived of crucial tools to theoretically tackle this kind of problem, lacking both maps and

---

[49] SächsStA–F, 40010 Bergamt Freiberg, Nr. 326, f. 3r: 'daß die Lochsteine nicht auf den rechten Haupt-streichen gesetzt worden wären'; f. 4r: 'das ganze Erb-Vermeßen keines weges vitiös machen'.

[50] SächsStA–F, 40010 Bergamt Freiberg, Nr. 326, f. 3v: 'Nachlässigkeit'.

[51] SächsStA–F, 40010 Bergamt Freiberg, Nr. 326. In that particular case, there were numerous written discussions between the administration and the plaintiff, but the marking signs do not appear to have been moved.

[52] Matthes, *Das erste Bergbuch von St. Joachimsthal 1518–1520* (1965), pp. 7–9.

[53] Annaberg mining law of 1509, § 92, as transcribed in Ermisch, *Das sächsische Bergrecht des Mittelalters* (1887), p. 195: 'am tag genug ferne von einander weren und doch in der teuffe tzusamne fielen'.

trigonometric tables. In other words, the fairly common case of two distinct concessions colliding underground led to an intricate problem with no obvious solution. In such cases, the Annaberg mining law of 1509 recommended to stop any mining activity as soon as a point of contact between competing galleries had been reached. Litigated ore was stored for as long as its ownership was not settled. The mining master summoned 'the jurors and other trusted knowledgeable miners' and tried to reach a compromise. Jurisprudences representing decades of legal precedents were only slightly more specific: a surveyor should be allowed to visit the two concessions and conduct a 'visual observation' (augenscheinlich) about the specifics. Accompanied by jurors, he would then 'in their presence open the intersection and mark it as far as it goes'.[54]

Compared to the precise indications of previous methods, one perceives here that sixteenth-century subterranean geometry still relied on experience and compromises in the most complex situations. The use of geometry was not meant to impose a definitive and static truth, but to provide a common framework, whose possibilities and limits were constantly debated. Long lists of rulings indeed reveals that these decisions were often challenged. When mathematics was unable to produce a satisfying answer, a costlier method was needed and mining officials went back to medieval tried and tested solutions. As a mid-fourteenth century mining law states: 'should one have to survey underground an ore vein, one should do it with open breakthrough'.[55] In that case, the two galleries had to be fully connected in an offener Durchschlag (open breakthrough), which could mean weeks of digging work, until the physical intersection could be fully excavated.[56] The surveyor then used cords to compare the marking stones on the surface and underground. This solution was fine for wage workers, as it provided weeks of work and common miners were entitled to a free 'breakthrough beer' when the operation was completed. For investors and the mining administration, however, such methods were both expensive and time-consuming. Without geometry, one could only rely on the direct experience of senses.[57]

### The Birth of a Profession

In the early years of the mining rush, around 1500, most surveyors were not specific professionals. They held other official positions in the mining

---

[54] Ibid., pp. 100–101: 'in ihrer kegenwart die vierung wie fern sich die erstreckt, eroffnen und vorstufen'.

[55] Ibid., p. 54 (Freiberg mining law B from ca. 1346, § 20): 'Ist daz man marscheyden zal yn der tueffe uff den gengen, daz mus man thun myt offen durchslegyn.'

[56] Ibid., pp. 100–101: 'offene durchschlege'. For an actual example taking place in 1564, see Hake and Denker, Die Bergchronik des Hardanus Hake, Pastors zu Wildemann (1981), pp. 74–75.

[57] Jacobsson, Technologisches Wörterbuch (1781), article 'Durchschlagbier', literally 'breakthrough beer' – 'a beer, offered to miners for having overcome the danger of a breakthrough'.

administration – such as mining master, juror, or clerk – and simply received a fee when performing measurements. If nobody was available or able to complete a measurement, someone from a neighbouring district would usually be called upon. Surveying tasks, however, came to involve more complex issues over time. Full-time availability was needed to maintain drainage galleries or connect the now extensive networks of mine workings. Underground surveying finally became a distinct profession as surveyors standardized their own training. In the late sixteenth century, the subterranean surveyor working in Freiberg, Hans Dolhopp, already received a fixed salary known as *Wartegeld* (waiting allowance). This annual sum, fifty thalers, was roughly equal to the salary of a miner and was supplemented by surveying fees for every operation.[58] Such a modest fixed salary represented a first step towards a professionalization of the *Markscheider* activities. A second step was the oath swearing, in which practitioners pledged to fulfil their duties. The accuracy of their measurement was guaranteed by law and a mistake could have severe consequences. In the Saxon mining law, the official oath ended with an unveiled threat: 'If the indications of the bespoken subterranean surveyor are inaccurate, he shall be properly punished.'[59]

The professionalization of subterranean geometry illustrates a wider dynamic of specialization at work in early modern practical mathematics. The geometric skills of the *Markscheider* were highly specific and indispensable within their domain of expertise, as were those of other superior craftsmen, for instance the Dutch draining engineers. Without them, mining enterprises could be costlier or simply impossible. In the mid-sixteenth century, the ancient imperial city of Goslar (Harz region) tried to revive the legendary – and long-drowned – Rammelsberg mines by achieving a long-abandoned drainage gallery. In the absence of a dedicated surveyor, political authorities could not even assess the extent of the problem and had to ask the Saxon Elector to send mining experts (*Bergverständiger*).[60] Mining master Wolf Schleusing and surveyor Georg Öder the Second thus travelled across the Empire to, in their own words, 'visit, inspect, and survey the mine'. They described pumps and workings that had 'fallen into decay, crumbled, so that one cannot stop the flood of water'.[61] The city council asked Öder to 'bring the gallery to the day', that is, to locate on the surface its proposed path. Using his compass and surveying cords, the

---

[58] SächsStA–F, 40001 Oberbergamt Freiberg, Nr. 3477, f. 25r.
[59] *Bergk-Ordenung / des Durchlauchtigsten / Hochgebornen Fürsten und Herrn / Christianen / Hertzogen zu Sachssen* (1589), article 17: 'Wann alsdann desselben Marckscheiders angeben nicht zutreffen würde / sol derselbige gebürlich gestrafft werden.'
[60] See Nehm, 'Die ersten Ansätze des Markscheidewesens auf dem Rammelsberg' (1933), pp. 79–88; Nehm, 'Georg Öder und seine markscheiderische Tätigkeit auf dem Rammelsberg' (1934), pp. 64–72.
[61] Ibid., p. 66: 'des orths sehr vorbrochen, vorgangen, das mahn dy Waßer ... in fluten nicht halten kan'.

surveyor was able to make the underground pathway visible, representing it with wooden stakes, and indicated the procedure to follow in order to reclaim the gallery.[62] This is one of countless examples of travelling subterranean surveyors, and Georg Öder the Second himself visited other mining regions to solve complex issues, often related to the path of drainage galleries.[63]

In the early modern period, crafts could only circulate with the practitioners themselves, for it was not enough to send ready-made formulas, tables, or instructions to be given to local officials. These useful skills and know-how were so deeply specific that seasoned experts had to travel in order to survey and work out locally adapted solutions. The ensuing contacts sometimes triggered exchanges of knowledge or comparisons between local and foreign procedures. This fostered in turn the diffusion and standardization of methods and instruments. Surveying was becoming a differentiated profession, as the increasing complexity of measurement led some individuals such as Hans Dolhopp or Georg Öder – whose career and family will be discussed in the next chapter – to a high degree of specialization.

When accepted by all, geometry had proven to be a cheaper alternative to the medieval procedures. Instead of making the underground visible with brute-force solutions that required years of hard work and thousands of thalers, a skilful surveyor could produce precise measurements within days or weeks. As issues grew more complex, a major new challenge emerged: determining the best course of action.

In practical mathematics, issues often escape formal comparisons, let alone proof. Likewise, small differences or incremental improvements for instruments or procedures were hard to pinpoint. In other words, subterranean surveyors had to ensure the unity of the discipline while avoiding two pitfalls: stagnation or practical solipsism. If everyone relied on one's own set of procedures, geometry could not play its role of social consensus any more – even if it proved efficient. This is why mining laws and customs emphasized the accuracy and the openness of surveying procedures. Customs directed surveyors to 'use and keep one and the same method' to ensure that every miner or company was treated in the same way. Conversely, the complexity of mining prevented the mere generalization of existing methods, and new procedures had to be devised to cope with technical challenges.[64] The *geometria subterranea* did not operate within a Euclidean

---

[62] On the Tiefer-Fortunatus-Stollen, see Nehm, 'Georg Öder und seine markscheiderische Tätigkeit auf dem Rammelsberg' (1934), pp. 68–72; Pitz, *Landeskulturtechnik* (1967), pp. 64–67: 'bergkwergeß … zw befarn und zw besichtigen und zw margkscheiden … den Stolln an Tag zu brengen idem'.

[63] Reichert, 'Die Kurfürstlich-sächsischen Markscheider Georg Öder die Jüngeren sen. und jun' (2014), p. 155.

[64] Enderleins' *Joachimsthaler Berggebräuche*, §103, in Löscher, *Das erzgebirgische Bergrecht des 15. und 16. Jahrhunderts* (2003), p. 114: 'einerley und gleiche ordnung halten und brauchen'.

frame, as its academic counterpart, but within a very specific legal context. Its methods had to be not only effective, but also standardized enough to be recognized and respected. Measurements ultimately became ubiquitous and fairly standardized not because mathematics was seen as infallible or demonstrative, but mostly because it could be presented as objective. Rulers and mining masters pushed to replace anything that could be seen as arbitrary by the ever-developing art of setting limits.

In the early modern mining world, even the workers who could write, count, and read had no particular reason to trust mathematics. Surveys, compasses, and cords were thus presented as neutral and efficient tools, and their ubiquitous use gradually lent these values to the discipline of subterranean geometry itself. Practical mathematics, in other words, came to be seen as objective and conclusive because it obeyed clear, precise, and codified rules. Simultaneously, surveying became specialized to the point where it became a differentiated profession for which an oath had to be taken. Uses of subterranean geometry in the sixteenth century illustrate how the ubiquity of measurements slowly gave specific meaning and values to mathematics. In mining cities, it came to be associated with accuracy, fairness and efficiency. Geometry was not primarily respected for its demonstrative power but because it was trusted as right – meaning both correct and just. To further understand how subterranean geometry became a fundamental pillar of the mining ethic, we should also consider the religious context of early modern central Europe.

## 'A Trustworthy Geometer on the Mountain of Our Lord'

It is difficult to think of a place more concerned with religious matters than early modern Central Europe; miners were prone to adhere to new theories, from the Ultraquist movement to the Protestant Reformation.[65] The importance of religion in the daily life of miners was considerable, and mining towns were at the forefront of the movement when the Reformation gained steam in the 1520s. Their wealth and autonomy led most rulers, even Catholic ones, to begrudgingly tolerate the new faith.[66] Mining cities used their newfound wealth to erect huge churches, and the most important pits even included small underground chapels.[67] Given the huge importance and rapid development of these regions, they attracted many gifted young Lutheran ministers, such as

---

[65] On mining and the Reformation, the reference work is Knape, *Martin Luther und der Bergbau im Mansfelder Land* (Leipzig, 2000).

[66] Brown, *Singing the Gospel* (2009), pp. 40–41; Wolf, 'Die "Himmlische Fundgrube" und die Anfänge der deutschen Bergmannspredigt' (1958), pp. 347–354.

[67] On St Anne's Church in Annaberg, where the famous mining *Bergaltar* is preserved, see Bürger, 'Die Annaberger St. Annenkirche' (2013), pp. 353–376.

Johannes Mathesius (1504–1565) and Cyriacus Spangenberg (1528–1604), who actively promoted the mining culture.[68] Mathesius was a central figure of the Reformation who compiled and published notes from Martin Luther's *Tischreden* (table conversations). After some personal experience of mining in his youth, Mathesius had the opportunity to study theology in Wittenberg. He went on to become director of the Latin school of Saint Joachimsthal precisely when Georgius Agricola – whom he called 'my dear friend' – worked there. He was later ordained pastor of the same town and published his Sunday sermons under the title *Sarepta or Mining Homilies*.[69]

The *Mining Homilies* aimed at presenting biblical episodes to common miners in a familiar way. Presenting himself as 'a spiritual miner', Mathesius praised Agricola's Latin writings but chose to express himself in vernacular 'with clear, distinct, and German mining words'. He explicitly endeavoured to 'match' the religious parables of the Old and New Testament with 'this mining town and surrounding' pits. True to the Protestant credo, Mathesius thought that doing so would help miners understand the Bible: 'a miner will accept guidance if someone can show him, with mild and good reason, or with open breakthrough, something better, more correct, and certain'.[70]

The idea of using everyday examples to convince miners of God's greatness was fairly straightforward. Interestingly, Mathesius not only recommended to use one's 'mild and good reason' to convince miners, but also relied on 'open breakthrough' (*offenen durchschlegen*). Unintelligible outside of the mining regions, this dialectal expression was instantly clear for miners. It referred to the operation described above, in which concession holders disagreeing on the outcome of a survey dug until their sections were fully connected. Mathesius framed his down-to-earth approach of religion, scattered with concrete examples taken from the mining world, as similar to the open breakthrough used by surveyors. In his eyes, both displayed tangible, manifest truths and were, as such, obviously convincing. From the very introduction of his sermons collection, one sees how Mathesius used subterranean geometry as an example of certitude.[71]

Spangenberg was even more explicit. Son of a protestant preacher, he had studied in Wittenberg too and became teacher, eventually switching to

---

[68] A combined biography of both men can be found in Boettcher, 'Martin Luthers Leben in Predigten: Cyriakus Spangenberg und Johannes Mathesius' (2000), pp. 163–188.

[69] Sarepta was a mining city mentioned in the Old Testament, and the Hebrew word became a synonym for smelting operations. Mathesius, *Sarepta oder Bergpostill* (1562), introduction: 'mein lieber freundt / Doctor Georg Agricola'. Mining sermons were held during the carnival season, as explained in the introduction and again on p. 34.

[70] Ibid., introduction: 'Ein geystlicher Bergkman … im alten und newen Testament … deutlichen / vernemlichen / deutschen Bergkworten ausspreche.'

[71] Ibid., introduction: 'ein Bergkman gerne sich wil weysen lassen / da jemand mit glimpff und gutem grunde oder offenen durchschlegen / was bessers / richtigeres oder gewisseres im wirdt anzeygen können'.

preacher and prolific writer in the Mansfeld mining county, where his mentor Luther had been raised. His most famous work is his twenty-one sermons on the life of Luther, written to promote the Reformation.[72] In 1574, he published the nineteenth of the series, entitled *On Dr. Martin Luther. How He Was a Trustworthy Subterranean Surveyor on the Mountain of Our Lord God* (see Figure 2.4). Mathesius and Spangenberg are two examples of a larger tradition of sermons aiming at audiences of miners, which also includes Johann Schreiter's *Decimae Metallicae* or Peter Eichholtz's *Spiritual Mine*.

These printed books, despite their modest format, were certainly too expensive for the average worker. The intention of the authors was rather to circulate biblical material adapted for miners that local preachers could buy and use in their parishes. As Roger Chartier has taught us, 'relationship with the written word did not necessarily imply individual reading' in the sixteenth century. Books were often used in a collective context, be it in the public setting of a church sermon or in smaller private gatherings.[73] Their content, always written in a vernacular German full of mining idioms, therefore reflects more accurately than scholarly books the mining knowledge and the mathematical culture of a common *Bergmann*. Remarkably, all mining sermons were written by Lutherans. Catholic authors were certainly aware of the cultural importance of mining: Johannes von Paltz, preacher of indulgences and a papal legate, had in fact preached and published a sermon entitled *The Celestial Concession* as soon as 1490. However, references to mining were actually very scarce and superficial, with no technical or legal description, let alone instructions about geometry.[74]

There was indeed a specific inclination towards mathematics in most of the Protestant universities in which these preachers had studied. Philip Melanchthon (1497–1560), who after Luther was the intellectual leader of the movement, used his considerable influence to promote sciences.[75] He addressed the mathematical sciences in his sermons and wrote prefaces to academic textbooks, among others to a Latin version of Euclid's *Elements*. Melanchthon recommended that all Lutheran universities hire not one, but two professors for mathematics. This involvement came from a deep belief that

---

[72] On C. Spangenberg and his *Chronicles* of Mansfeld county, which provide plenty of information about early modern mining, see Tenfelde, Bartels, and Slotta, *Geschichte des deutschen Bergbau* (2012), vol. 1, pp. 341–342. For his mining sermons, see Berndorff, '"Und da habe ich müssen mach ihrer sprach reden". Einsichten in die lutherischen Bergmannspredigten des Cyriacus Spengenberg' (2000), pp. 189–203.

[73] Chartier, *Cultural Uses of Print*, here quoted in Eamon, *Science and the Secrets of Nature* (1994), p. 123.

[74] Von Paltz, *Die himmlische Fundgrube* (1490).

[75] On Melanchthon and mathematics, see Berhardt, *Philipp Melanchthon als Mathematiker und Physiker* (1865); Reich, 'Philipp Melanchthon (1497–1560) und sein Einsatz für die Mathematik' (2017), pp. 1–14.

Die XIX. Predigt:

Von Doctore
Martino Luthero / wie
er so ein getreuwer Marscheider
auff vnsers HERRN Gottes
Berge gewesen/
Durch

M. Cyriacus Spangenberg.

Franckfurt am Mayn/ 1574.

Figure 2.4 Title page of Cyriacus Spangenberg, *On Dr. Martin Luther: How He Was a Trustworthy Subterranean Surveyor on the Mountain of Our Lord God*, with a portrait of Martin Luther, Frankfurt-am-Main, 1574. The original sermons were held in the early 1570s in the mining county of Mansfeld. Courtesy of the Staatsbibliothek zu Berlin – Preußischer Kulturbesitz (4 an: Cn 4980-15/21).

it could lead to a better understanding of God's perfection. After all, had he not 'ordered all things in measure and number and weight', as stated in the Old Testament?[76] The interplay between Lutheran culture and mathematics has been thoroughly studied by scholars. However, little is known about the influence of the reformed religion on the perception of geometry by common men, miners, or craftsmen.

A first goal of these sermons was to give miners a basic understanding of the instruments used by the mining master, the assayer, or the subterranean surveyor. The compass, in particular, had become a ubiquitous tool, and yet, as Mathesius acknowledged, 'it is not mentioned in the Holy Scriptures'. Miners, who likely had never seen a sea, were nevertheless informed that sailors used it to find their way on the oceans.[77] And since God was said to rule the earth just as he ruled the seas, miners could use the instrument to know 'in which hour [of the compass] are the directions of ore veins / and where you should head'.[78] Several pages were dedicated to the precise description of the instrument. The preachers' goal was, in my view, less to actively train miners to use a compass than to have them understand its principle, and respect decisions taken using it. The *Bergkompass* was presented by Peter Eichholtz as a pillar, not only of mining, but of society:

Without equity among one another ... we are dispersed, and go crazy, just as wild animals. This is why all miners respect and hold in high regard this wonderful instrument, the compass, and its use in subterranean geometry, since its proper use saves not only many expenses, but also conserves much equity [*viel Gerechtigkeit*].[79]

Geometric explanations routinely exceeded the immediate practical use in mining to approach other aspects of civil life. When describing the quadrant used to record angles, Mathesius digressed from subterranean geometry. To relate to the daily experience of miners, he explained that it could be used to 'scale gauging rod' and thus helped to 'measure how much barrels contain'.

---

[76] Book of Wisdom 11:20 (King James Bible). This famous (apocryphal) excerpt from the Bible is mentioned and analyzed both by Spangenberg and by Mathesius; see Spangenberg, *Die XIX. Predigt Von Doctore Martino Luthero* (1574), pp. 39, 45, 48; Mathesius, *Sarepta oder Bergpostill* (1562), p. 204v.

[77] Mathesius, *Sarepta oder Bergpostill* (1562), p. 137: 'I say this to explain / both the words and the nice sayings of Zachariah / Now you know how foremen use their level or their plumb line / how you miners use you compass, cords, quadrant, and plumb to survey the underground / and how seamen use their plumb line to measure the depth of the sea.'

[78] Ibid., p. 201v: 'In der heyligen Schrifft ... / wirdt des Magneten nicht erwehnet.' 203v: 'und auff welche stunde die gänge streichen / und wo ihr zufahren sollet'. A long description of mining instruments is also given in Spangenberg, *Die XIX. Predigt Von Doctore Martino Luthero* (1574), pp. 56–58.

[79] Eichholtz, *Geistliches Bergwerck* (1655), p. 130: 'Ohne Gerechtigkeit untereinander ... so werden wir zerstrewet / und gehen gleich in der irre / wie die wilden Thiere. Darumb wie alle Bergleute hoch achten und halten / das Wunder Instrument den Compaß / und dessen Gebrauch im Marscheiden / denn durch desselbigen rechten Gebrauch nicht allein bey den Bergwercken viel unkosten mag ersparet / sondern auch viel Gerechtigkeit damit erhalten werden.'

Going further, he claimed that 'one could use the quadrant to learn how far is a city from another one'. Mathesius even hinted at the general method: 'when one can have two sides of a triangle, the third can be sought by the art of measuring or reckoning'.[80] It is striking to see a Lutheran sermon, delivered in vernacular German in early modern mining cities, describe the basic principle of triangulation, less than a generation after its first publication in the Low Countries by Gemma Frisius, of course in Latin.[81] While the descriptions of instruments served a didactic purpose, the religious aspect was never far away. These ingenious tools all relied on 'the liberal and natural arts' offered by God, and the audience was repeatedly reminded that he had created 'the compass, the [wax] rings, the quadrants, and the hard-working people' using them.[82]

Mining sermons were not limited to the description of instruments but also engaged with many of the actual procedures of subterranean geometry that have been introduced in the first part of this chapter. Numerous passages dealt with the attribution of concessions and their dimensions, always emphasizing practical details. A 'Christian miner' should always 'let him [the surveyor] measure and set the boundaries of his mine'. Rectangular-shaped concessions should be oriented 'along the direction of the ore vein, from the middle of the winch'. Lengths units were often rephrased to be more tangible for the audience: 'half a *Lachter*' was immediately clarified as: 'as large as a man reaches with his elbows'.[83] In his description, Mathesius was not trying to prove mathematical truths but simply exposed existing methods. In a passage that might even be read as a veiled criticism of his friend Agricola, he made clear that 'the laymen should stay unaware of *Euclid* and of the rigorous geometry'. What mattered for Mathesius was to present how the instruments were used in underground surveying. Interestingly, the word 'layman' was used here to refer to common miners who were neither surveyors nor clergymen. Once again, the knowledge of the Scriptures and of mathematics were equated.[84]

---

[80] Mathesius, *Sarepta oder Bergpostill* (1562), p. 204: 'ein vaß durch … die meßstebe oder visier ruten abgetheilt sein / messen und eichen kan was es halte … also kan man auch durch die vereinigten quadranten erfahren / wie ferne eine stadt von der andern liegt … da man zumal zwo linien des triangels haben kan / darauff die dritte künstlich durch messen / oder rechnen gesucht werde.'

[81] On the early history of triangulation, see Frisius, *Libellus de locorum describendorum* (1533); Haasbroek, *Gemma Frisius, Tycho Brahe and Snellius and Their Triangulations* (1968).

[82] Mathesius, *Sarepta oder Bergpostill* (1562), p. 204: 'Diß alles gehet auß freyer und natürlicher kunst zu / dafür man Gott unnd seinen gaben / dem Magneten und scheiben / quadranten und fleissigen leuten zu dancken hat.'

[83] Ibid., p. 89v: 'Christlicher Bergkman … lest er im die zeche vermessen / und verlochsteinen'; p. 29v: 'nach dem streychen des gangs vom Mittel des Rindbaums anzuhalten'; p. 30r: 'ein halbes lachter'; 'so breit ein Mann mit seim elbogen reichen kan'.

[84] Ibid., p. 204r: 'Es müssen die Leyen so vom *Euclide* und der gründtlichen Geometri unberichtet sein', and later 'Aber der triangel / unnd acht auff die proportion haben / das ist in diesem Fall meyster / wer sich darein schicken kan.'

Lutheran preachers aimed at giving miners a basic understanding of the mathematical principles underlying underground surveying. I do not think that sermons were used to train surveyors, notwithstanding the fact that both Mathesius and Spangenberg had been teachers in mining cities.[85] While they undoubtedly present more details than scholarly books, early-modern subterranean geometry was not reducible to a written discourse, however worldly and precise. More importantly, their descriptions had no normative claims. Unlike Agricola, who would sometimes 'enjoin things which ought to be done', Mathesius was clear: 'I do not intend in any way to reform [*reformiren*] how miners and smelters carry out their work.' Pastors rather wanted to popularize the emerging culture of geometry by integrating it into the authoritative discourse of the reformed religion. In doing so, they demonstrated that, in Spangenberg's words, it was 'impossible to build, maintain, and usefully benefit from mining without measure, numbers, and weight'.[86]

### Obeying the Laws of God and Geometry

Focusing on the immediate experience of their audience allowed pastors to reach a second goal: reinforce the belief of miners in God and in the rationality of geometry. Most miners listening to their sermons would agree that 'mining was a most beautiful page in the book of Nature'.[87] If the relationship between God and the marvellous mining world was obvious to listeners, the underlying role of mathematics was less evident. Miners had to be taught that mathematics had been God's tool to 'order' the world he had created, having 'built his wisdom and skills in people's hearts'.[88] One should trust surveyors, Lutheran pastors claimed, for their art of setting limits was directly based on God's will. When a *Markscheider* used his compass to ascertain the direction of an ore vein, he was not inventing anything but simply revealing an existing truth. Religion was thus able to present subterranean geometry as 'accurate' in a third way. Surveys were not only performed with care, and in agreement with accepted laws and customs, but also according to a higher, divine law.[89] Using

---

[85] Mathesius was first appointed teacher in the Latin school of Saint Joachimsthal, while Spangenberg had taught at the Eisleben's *Gymnasium*. See Boettcher, 'Martin Luthers Leben in Predigten' (2000), pp. 166, 170.

[86] Mathesius, *Sarepta oder Bergpostill* (1562), introduction: 'hab ich mir traun das in keinen weg fürgenommen / Bergkleut und schmelzer inn irer Bergkarbeit zu reformiren'. Spangenberg, *Die XIX. Predigt Von Doctore Martino Luthero* (1574), p. 48: 'ohne Maß / Zal und Gewicht / unmüglich ist Bergwerge zu erbauwen / zuerhalten und nützlich zugebrauchen'.

[87] Eichholtz, *Geistliches Bergwerck* (1655), table of contents: 'was Bergwerck in solchem Natur-Buche für ein schönes Blat sey'.

[88] Mathesius, *Sarepta oder Bergpostill* (1562), p. 204v: 'Denn unser Gott hat seine Weißheit und geschickligkeit ins menschen hertz gebildet.'

[89] According to Witold Kula, this process is characteristic of 'stratified societies': Kula, *Measures and Men* (1986), p. 9: 'in addition to a guarantee from the secular authority, one of a sacred emerges as well'.

mathematics, even when it seemed like a rule of thumb or an efficient trick, was framed as the most legitimate way to understand the Creation and to glorify its creator. Sermons held in mining regions repeatedly presented mathematics and religion as mutually supportive. For instance, Mathesius introduced Lynceus of Argos, 'of which the poets write that he could see through the mountains, through copper and silver', as a metaphor for subterranean geometry. Using art and instruments was becoming a way to manage the invisible. Surveyors would initially have 'imitated' the celestial principles and 'invented an art and a masterpiece, as well as many useful instruments'.[90]

Presenting the rules of mine surveying as divine also carried obvious political benefits for local rulers, since religious rules were respectfully observed by miners. Pastors and mining masters were prone to compare the Ten Commandments of the Bible with a 'Christian mining law' (*Christliche Bergkordnung*). Mining laws had been introduced, pastors argued, to ensure that divine law was respected, and because 'the ancients rightly felt that there are many angles and cross-ways in the hearts of men'. If local mining offices represented the authority of God, their orders suddenly acquired a much higher character. Once again, a geometric vocabulary was used to equate the accuracy of mathematics and the perfection of religion.

Sermons actively supported the idea that rulers used the impartial tools of mathematics to ensure fairness in the extraction processes. Despite being heavily tilted towards political authorities, sermons pretended to simply follow the Scriptures since – as Mathesius was prompt to remind his audience – 'no judge on earth is so high that he won't be judged by a higher one'.[91] When the deacon of Annaberg reminded its auditors that they were made of dust, just as Adam had been, and would turn to dust again, simple miners were encouraged to accept their dangerous fate – although in their case the well-known saying might take a more immediate meaning.[92] Sermons were obviously no passive description of sixteenth-century mines. They actively sought to discipline the unruly mining people and to foster obedience through rationality, ordering miners to respect rules and not to revolt. Mathesius frequently used another popular adage, presenting God as the 'highest mining master' who 'had given concessions and measured everything' as explained in the sixteenth Psalm: 'The lines are fallen

---

[90] Mathesius, *Sarepta oder Bergpostill* (1562), p. 204r: 'Der alte Bergman Linceus darvon die Poeten schreiben / er habe können durch die Berge / und durch kupfferund silber sehen … nachgeammet … eine kunst unnd meisterstück / und vil nützlicher instrument nach einander erfunden.'

[91] Ibid., p. 30r: 'die alten haben wol gespürt / weil ins Menschen hertzen viel winckel und querschleg sein'. 29v: 'Es sol aber ein yeder Richter und Berichter nicht vergessen / das kein Richter so hoch ist auff Erden / er muß von einem höhern gerichtet werden.'

[92] Schreiter, *Decimae metallicae* (1615), p. 107: 'Du bist Erden / und solt zu Erden werden / Gen. 3.'

unto me in pleasant places; yea, I have a goodly heritage.' Complaining about one's own misery or contesting an unfavourable survey was thus presented not only as improper, but as patently unchristian.[93]

Religion and mathematics were also used to criticize the adversaries of local rulers. These enemies were the Catholic Christians in Saxony and in the Harz, or the Muslims in Hungarian kingdoms – in Mathesius's words the 'Papist and Turkish gloominess and blindness'.[94] The devil was often summoned as a bogeyman to frighten miners and presented as a 'fake surveyor'. To illustrate its temptation with a danger well known to the miners, Satan was sometimes conflated with iron: Just as ferruginous ore exerted an influence on the compass and endangered mining operations, the devil used 'magical underground surveying' (*zauberischen marscheiden*) to 'seduce and divert people from God'.[95] To avoid getting fooled, a pious miner had to rely on 'our Christian compass, the law of God'. This, and the Bible, would allow him to 'show the straight line, in the right hour [of the compass] and the right direction, so that we can walk right to the Christ and the eternal life'.[96] Once again, this sermon was written using the *Bergmannsprache* and could only be understood by miners, for this analogy assumed a knowledge of local compasses and their graduations. To follow Luther's example, Spangenberg summed up, a miner should never deviate from the right path:

Concerning this part of subterranean geometry … Doctor Luther has behaved just as wittingly on the mountain of Our Lord God, beginning by giving industrious introduction and instruction to the many, where and what to look for, when one wants to behave rightly in faith as in Christian obedience, to avoid getting either too high or too low, too right or too left.[97]

Mathematics was finally presented as a tool to ensure fairness for both the miners and the shareholders. A system of setting limits was bitterly needed in mining operation, 'for nobody can see through stone, and a *quarter* of favour often

---

[93] Mathesius, *Sarepta oder Bergpostill* (1562), p. 315v: 'Und der oberste Bergmeister hat im darauff verliehen / unnd vermessen / wie der Son Gottes saget. Pfal. 16. Das loß ist mir gefallen auf liebliche / und mir ist eine schöne maß und erbteil worden.' (English translation from the King James Bible.)

[94] Ibid., p. 205r: 'Papistischer und Turckischer finsternuß und blindtheit.'

[95] Ibid., p. 205r: 'die leute von Gott abziehe und verfüre … die zauberischen marscheiden.'

[96] Ibid., pp. 205–206: 'unser Christlicher magnet das Gesetz Gottes … die rechte linie inne / die uns die rechte stunde und weg weiset / darauff wir gestracks zu dem Herrn Christo und dem ewigen leben zufaren können.'

[97] Spangenberg, *Die XIX. Predigt Von Doctore Martino Luthero* (1574), pp. 99–100: 'Was nun dieses stuck deß Marscheidens / nemlich das Abziehen / anlangend / hat sich D. Luther auff unsers Herren Gottes Berge auch gar weißlich verhalten / anfenglich fleissige anleitung unnd unterricht menniglich gegeben / wohin unnd wornach sich zurichten / wenn man im Glauben / und im Christlichen Gehorsam recht fahren wölle / damit man nicht zu hoch noch zu tieff / auch weder zur lincken noch zu rechten zu weit komme.'

weighs more than a *centner* of equity'.[98] Without equity, good investors would be evicted and replaced by scammers. Mining operations were fundamentally built on trust, and surveying played a crucial role in upholding that trust. 'What would you say, dear people', asked Peter Eichholtz in his *Spiritual Mine*, 'if a foreman, well aware of the limits of his pit or concession, crossed it with his workers and went into the neighboring concession? ... Should he get away with it unpunished?' While numerous such cases doubtless happened in early modern mines, the preacher then quoted the Gospel of Matthew to underline the gravity of ignoring boundaries: 'Depart from me, ye that work iniquity.'[99]

The geometry preached by Mathesius, Spangenberg, and their contemporaries was very different from the learned discourse written down by Agricola or Reinhold. The discrepancies, however, do not reflect our modern understanding of labelling the first as 'theological writings' and the second as 'scientific texts'. Mining sermons paradoxically offer a more precise, worldly, and technical view of surveying processes, closer to the descriptions that can be found in contemporary laws and customs. The differences between these two types of text seem to be mostly shaped by their intended audiences. While Agricola wrote a definitive work for scholars, tracking surveying practices back to their Euclidean principles, Lutheran preachers addressed the inhabitants of mining regions: 'I had to speak their language', recounted Spangenberg at one point.[100] The sermons did not offer geometrical demonstrations, as these would hardly have been convincing to an audience of common miners, but rather devoted a great attention to instruments, concrete procedures, and ceremonies. These texts have rarely been used as sources for the history of knowledge because their content is heavily dialectical and culturally located.[101] However, it is primarily by listening to sermons and watching public ceremonies that the mining people encountered, in a very mundane way, a new geometric and numerical culture.

In the discourses of preachers, geometrical methods were offered religious justifications, while mathematics in turn cemented the social order of mining states. Pastors' arguments were more appealing to common men and artisans than to scholars, as the Bible, which was by then available in German,

---

[98] Mathesius, *Sarepta oder Bergpostill* (1562), p. 204v: 'weil niemand durch den stein sehen kan / und offt ein quintet gunst mehr gilt / als ein centner gerechtigkeit'.

[99] Matthew 7:23 (translation from the King James Bible). Eichholtz, *Geistliches Bergwerck* (1655), p. 142: 'Weichet alle von mir ihr Übelthäter ... Was bedüncket dir wohl / lieber Mensche / wenn ein Steiger / der seiner Grube oder Zechen Marscheider wol wüste / würde aber mit seinen Arbeitern übersetzen / in die benachbarte Zechen einfallen ... solte ihm das wol ohngestraffet hingehen?'

[100] Berndorff, '"Und da habe ich müssen mach ihrer sprach reden". Einsichten in die lutherischen Bergmannspredigten des Cyriacus Spengenberg' (2000), pp. 189–203.

[101] A brilliant exception is Sahmland, 'Gesundheitsschädigungen der Bergleute' (1988), pp. 240–276.

emphasized tangible proportions and concrete measurements over abstract proofs and demonstration.[102] Protestantism both supported the development of education and promoted mining geometry as the natural expression of a higher law. If mining sermons were not restricted to mathematics – Mathesius, for instance, described at length the segregation of silver from copper using lead – the emphasis was mostly put on geometry.[103] The frontispiece of Peter Eichholtz's *Spiritual Mine* similarly depicts all mining arts and crafts, but puts the subterranean surveyor with his compass in a privileged position (see Figure 2.5)[104]

Subterranean geometry illustrates how new values associated with practical mathematics gradually spread into early modern societies through multiple channels. This movement was not primarily a descending one, from scholars to practitioners, but reflected a growing use in human affairs, in churches, and of course in the mines. Geometry quite literally pervaded the everyday life of workers, building a mental framework all the more powerful because it was palpable and readily available. A visitor once marvelled at how simply the orientation of underground galleries was made visible to everyone: 'The direction and inclination of the pits are indicated from the outside by the hut erected at the entrance. The inclination of the roof indicates the falling, the [orientation] of the highest wall indicates the direction.'[105] Abstract and complex concepts were here embodied in the design of a hut, in what might be the purest form of constructive geometry. The improving status of mathematics in mining regions was a consequence of its growing use of geometrical and arithmetical methods in daily settings. The new sensibility to abstract values of scientific inquiry came later, likely as an aftermath of these widespread operations. Simply put, more geometry came to be taught as a result of its ubiquitous role in civil life, promoting the value of accuracy and prompting the following remark in Mathesius's *Mining Homilies*:

Thank God, these and other liberal arts are being introduced again in schools, together with the gospel, and many good people know what they serve for, and how one can use the quadrant and the triangle to measure the earth; Mining rulers and mining states should thus be helpful and favorable to the fine minds who are inclined to it, who love

---

[102] Kula, *Measures and Men* (1986), ch. 2, has brilliantly described the concreteness of measures in the Bible and its influence on early modern thought. On manual labour and crafts in the Bible, see also Hooykaas, *Religion and the Rise of Modern Science* (1972), pp. 83–88.

[103] Mathesius, *Sarepta oder Bergpostill* (1562), p. 100r-v.

[104] On Peter Eichholtz, see Lommatzsch, 'Petrus Eichholtz und sein "Geistliches Bergwerk"' (1963), pp. 5–10.

[105] Beck, *Briefe eines Reisenden von *** an seinen guten Freund zu **** (1781), p. 230: 'Das Streichen und Fallen der Gruben wird von aussen durch die bey dem Eingang aufgerichtete Hütte angezeigt. Die Neigung des Daches nämlich zeiget das Fallen; die höhere Wand das Streichen an.' This can be seen on numerous mining maps, for instance on Beyer's rendition of the *Heaven's Crown* mining pit in SächsStA–F, 40040, H3585.

Figure 2.5 The mining arts on the frontispiece of Peter Eichholtz's collection of mining sermons, entitled *Spiritual Mine* (1655). The subterranean surveyor (*Markscheider*) is right above the title, figured with a mining compass and a level. TU Bergakademie Freiberg/Universitätsbibliothek WA XIII 218, detail.

mathematics and the arts, so that these persons embrace subterranean geometry in a proper way, that they also obtain useful and robust instruments, in order to always raise the water and the ore at low costs.[106]

The demand is reminiscent of Melanchthon's and Ramus's well-known defences of mathematics, but instead of asking rulers to hire university professors, Mathesius suggested to train subterranean surveyors and outfit them properly.[107] In the sixteenth century, these endeavours were not competing but perfectly complementary. Both illustrate a new dynamic emerging around mathematics, summed by Spangenberg in one of his sermons: 'there is no craft, duty, or handling that could be practised without counting, measuring, and weighing, or without using or distributing proportions'.[108]

*

In the mining states of the Holy Roman Empire, a popular consensus emerged in the sixteenth century around mathematical accuracy. The development of deep-level mining, combined with the specific framework of mining laws, allowed competing investors to work side by side on the same ore veins. A sense of renewal developed, expressed not only by Agricola, but also by mining officials such as Lazarus Ercker (ca. 1530–1594) or Georg Engelhard Löhneysen (1552–1622).[109] Adapting surveying techniques to one's daily problems was not about rediscovering the perfection of Greek mathematics, but rather copping with new concrete problems. Subterranean geometry gradually emerged as the most reliable method to exploit metal mines, at a time when mathematics was neither the only tool used in technical procedures nor even the most natural one. Traditions and rulings shaped a culture of quantified accuracy by defining procedures and introducing accountability in the social act of measuring. Religious texts in turn cemented these practices by presenting them as worldly reflections

---

[106] Mathesius, *Sarepta oder Bergpostill* (1562), p. 204v: 'Weil aber Gott lob / dise uff andere freie künste / zu diser zeit neben dem Evangelio / wider in die Schulen kommen / unnd viele guter leut wissen warzu sie dienen / und wie man der quadrangel und triangel zur abmessung der Erden brauchen können / Sollen Bergherrn und Bergstet / feinen köpffen die hierzu naturt und geneigt / und lust und lieb zu der Mathematiken und kunst haben / behülfflich und förderlich sein / das sie solch marscheiden auß dem rechten grundt ergreiffen / und auff / nützliche und bestendige instrument trachten / damit man immer von tag zu tag die wasser und berg mit leichter unkost heben könne.' Mathesius then praised the Holy Roman Emperor Maximilian II for his support of crafts and artisans (*Künstler*).

[107] See Hooykaas, *Humanisme, science et réforme* (1958); Hooykaas, *Religion and the Rise of Modern Science* (1972).

[108] Spangenberg, *Die XIX. Predigt Von Doctore Martino Luthero* (1574), p. 46: 'ist zwar kein Handwerck / kein Ampt noch Hantierung / das one Zelen / Messen und Wägen / oder one an unnd außtheilung der Proportionen köndte geübet oder getrieben werden'.

[109] Löhneysen, *Bericht vom Bergwerck* (1650), p. 3, writes for instance that 'current artisans surpass the Ancients by far'.

of higher principles.[110] Looking back at the previous chapter, it appears almost as if scholars and craftsmen were dealing with two different disciplines. While Georgius Agricola and Erasmus Reinhold tried to apply the corpus of *practica geometriæ* to a new material context, actual methods of subterranean geometry belonged to a different order of knowledge. Ceremonies and instruments were rooted in the vibrant material culture of the German-speaking mining states. In the mining pits of the Holy Roman Empire, the values traditionally associated with exact sciences gave way to operational requirements of practicability, efficiency, and above all accuracy. The validity of procedures did not depend on logical consideration nor on their agreement with ancient texts, but rested on consensus and the public display of geometric rules of law.

Subterranean geometry presents interesting characteristics typical of early modern practical mathematics. At the time, a growing number of areas of civil life came to use geometry, following an apparently large pattern. Virtually all of these uses relied on constructive, elementary, and useful methods which largely unrelated to the academic knowledge epitomized by Euclid's *Elements* (this holds, incidentally, even when practitioners themselves claimed to rely on Euclid).[111] Just as wine gauging, mercantile arithmetic or navigation, subterranean geometry was not initially an autonomous profession but gradually specialized from a technical field, in our case from mining. Mathematical principles were here blended with a local knowledge of the structure of ore veins, the legal framework, and a meticulous workmanship grounded on decades of accumulated experiences. Above all, it was from its very beginning oriented towards the solution of technical and legal problems. The richness of craftsmen's procedures, expressed 'in empirical rules and quantitative terms', in the words of Edgard Zilsel, was not recorded in scholarly treatises and therefore overlooked.[112] Still, a new form of rationality was arising outside of learned circles and was written down in jurisprudence, reports, and mining sermons. It was fostered by the rich administrative culture of the mining states, in which written and quantitative information was central.[113] In mining cities, the status of mathematics was raised by its association with the work of miners, artisans, and engineers. It was only a matter of time before early modern rulers realized that the know-how of subterranean surveyors, machinists, and local reckoning masters could be used in much wider settings.

---

[110] The counterexamples of the Kingdom of France and the Spain monarchy, where no comparable legal frame developed, show the richness and efficiency of the mathematical culture that developed in central Europe.

[111] Shelby has shown that medieval master masons, despite referring to Euclid, worked in a largely autonomous tradition: Shelby, 'The Geometrical Knowledge of Mediaeval Master Masons' (1972), p. 420.

[112] Zilsel, 'The Genesis of the Concept of Physical Law' (2000), p. 111.

[113] Ernst Pitz emphasizes the influence of administration in this process and attributes to all surveying activities the rise of a new *Landeskulturtechnik* (technical culture of land). See Pitz, *Landeskulturtechnik, Markscheide- und Vermessungswesen* (1967), p. 8.

# 3    The Mines and the Court

The culture of mathematics that emerged around the extraction of metallic ore found, during the second half of the sixteenth century, a broad resonance within the Holy Roman Empire. Mining regions were densely populated and highly urbanized, attracting printers, school teachers, and more generally some of the finest minds of Central Europe. The cultural and economic significance of mining towns briefly reached great heights and exerted a lasting influence, most crucially on the practical use of numbers and measurements. From there, the geometric and arithmetic skills used for the extraction of ore made their way to the residences of early modern rulers, where they gained an illustrious audience and entered new playing fields. In their capitals the kings, dukes, and electors maintained flourishing court lives, collecting artefacts in cabinets of curiosities as well as books and manuscripts in their libraries. The cultural impact of mining was palpable and it was indeed often celebrated, for instance during parades. In Dresden, the Saxon electors would 'dress as miners', and mining was 'associated with great technical know-how and productive of great and magnificent riches'.[1]

Courts were places of encounters and collaborations, where various kinds of artists and engineers learned to know each other's crafts. Instrument-makers worked with underground surveyors, who in turn cooperated with university professors. Circulations of knowledge between princely courts have been studied, most prominently by Bruce Moran, who described them as 'institutional nodes of technical activities'.[2] No less important, it will be argued here, were the exchanges between mining cities and their respective courts. The preceding chapter has described the birth of a specific mathematical culture in the mining pits of the Holy Roman Empire. In the second part of the sixteenth century, its impact then vastly grew in two different, albeit related, dimensions. First, the impact of subterranean surveyors and mining officials was felt way

---

[1] About the mining parades at the Dresden court, see Watanabe-O'Kelly, *Court Culture in Dresden* (2002), here p. 120.
[2] Moran, 'German Prince-Practitioners' (1981), p. 253. In a brilliant intuition, Moran noted that 'the mining surveyor held a preeminent position at princely courts' (p. 261), but did not elaborate. This chapter hopes to build on his work and present in detail the surveyor's role in sixteenth-century court life.

beyond the mining regions, as the most gifted became courtiers, ensuring the diffusion of their own values of usefulness, versatility, and accuracy. Second, mathematical methods developed in the context of ore extraction were put to use elsewhere, further cementing the status of the discipline and the standing of its practitioners. Surveyors increasingly acted and were recognized as experts, whose extensive experience and 'productive knowledge' was brought into play in the most diverse technical and military settings.[3]

Mining cities were of great cultural significance and, during their brief phases of prosperity, exerted a strong force of attraction.[4] Looking at civil life, it can be seen how the specific art of setting limits, and more generally quantified information, enjoyed wide recognition. Underground surveyors, but also reckoning masters and machinists (*Rechen-* and *Kunstmeister*) were initially given generous benefits, such as stipends and titles, to acknowledge their contribution to the local prosperity. Some mining officials developed engineering and calculating skills that made them valuable beyond the underground world of metal mines. Thanks to the versatility of their expertise, these superior craftsmen moved from mining pits to princely courts, where they were confronted with all kinds of technical issues.[5]

As nascent state administrations faced growing challenges to oversee and improve their dominions, they relied heavily on officials trained in the local mining and forestry administrations.[6] Some became courtiers with official titles and were commissioned to report on existing or projected infrastructures in order to improve lands, canals, and roads. Their skills were also used to design surveying instruments, map realms, or settle contested borders.

Early-modern mathematical skills were often transmitted – orally and directly in the mines or the workshops – from father to sons. This chapter thus focuses on two dynasties of Saxon practitioners whose careers illustrate the interplay between the mines and the court: the Öders were underground surveyors, while the Rieses specialized in arithmetic and accounting. Over generations, their fate was linked to the rise and decline of the mining city of Annaberg, and to the capital of the Electorate, Dresden.[7] Belonging to the local elite of the Ore Mountains, both families won the favour of the Saxon Elector and came to work directly for him as court *arithmetici* and court surveyors. They acquired a significant influence at a time when gifted individuals could develop personal interactions with the powerful rulers of the Empire. Mining

---

[3] Ash, *Expertise: Practical Knowledge and the Early Modern State* (2010).
[4] For a general overview, see Kratzsch, *Bergstädte des Erzgebirges* (1972); Kaufhold and Reininghaus, *Stadt und Bergbau* (2004). More precise references are given over the course of the present chapter.
[5] On the concept of superior craftsmen, see Zilsel, *The Social Origins of Modern Science* (2003).
[6] Recent works on the role of engineers in that period include Vérin, *La gloire des ingénieurs* (1993); Morel, Parolini and Pastorino, *The Making of Useful Knowledge* (2016).
[7] References on the Ries and Öder families are given below. On the history of Annaberg, see the town chronicles in Richter, *Umständliche aus zuverläßigen Nachrichten* (1746); Jenisch, *Chronicon Annaebergense continuatum* (1658).

officials were employed as military or civil engineers and – at the Dresden court – cooperated with instrument-makers to design striking compasses and other magnificent instruments for August of Saxony and his successors. Remarkably, fruitful collaborations between these practitioners and university professors were orchestrated by rulers and mostly took place at court, not in the pits themselves or even in mining cities, allowing us to refine existing analyses about the circulation of practical knowledge.[8] The bond between the court and the mines finally weakened as their respective interests took diverging paths, and was irrevocably broken by the outbreak of the Thirty Years War in 1618.

## The Rise of Mining Cities

Mention of mining regions usually conjures up images of hostile places, mountainous and sparsely populated. Popular visions, and even some historical analyses, depict a wild environment, 'a treasure trove of precious metals and stones, healing springs, and medicinal herbs that are hidden in the deep woods and dark valleys and under forbidding rocks'.[9] Adventurous scholars would venture in these inhospitable environments, harvest useful knowledge and bring it back to the civilized world of towns and universities. However fabled, such descriptions could hardly be further from reality. Early-modern mining regions underwent tremendous changes during the mining rush (*Berggeschrey*) of the fifteenth century. Printed pamphlets promised riches – 'silver is found in numerous prospecting', promised one of them in 1523 – attracting throngs of investors, technicians and miners who dreamed of getting 'rich within the hour, at little cost, and with limited efforts'.[10]

In his *Bermannus*, Agricola vividly described the urban character of the Ore Mountains in the early sixteenth century. Although 'a few years only have passed since the beginning of mining operations here', there were many large towns: 'I could almost take them for one of our great cities, Erfurt or Prague, or among Italian towns, Bologna and Padua.' The landscape also bore the marks of frantic extraction: 'These mountains that you see here deprived of trees were once a dense forest, and these valleys, with thousands of inhabitants, where once full of wild beasts.'[11] Saint Joachimsthal, where silver veins were found in 1516, became

---

[8] A recent collection of studies on 'knowledge formation and the urban context' is De Munck and Romano, *Knowledge and the Early Modern City* (2020), p. 3.

[9] Williams, *Ways of Knowing in Early Modern Germany* (2017), p. 82, and more generally the whole chapter 'Demonology and Topography: Locating Giants and Witches' (pp. 67–110).

[10] Rudhart, *Antzeigung des Nauenn Breyhberuffen Bergwerks Sanct Joachimsthal* (1523), reproduced in Lempe's *Magazin der Bergbaukunde* (1794), vol. 8, p. 161: 'Mancher wird Reich zu einer stundt / Das sylber bricht yn offt Im schurff entkegen / Auf wenig kost und gering darlegen / Wie ich selber hab gesehenn unnd erfahrnn.'

[11] Agricola, Halleux and Yans, *Bermannus le mineur* (1990), p. 20: 'aliquam ex magnis urbibus nostris Erphurdium puto uel Pragam pene uideremihi uideor, aut quantas Italia habet Bononiam et Patauium ... perpaucos esse annos ex quo fodinae hic cultae sunt. ... Quos nunc montes

within two decades the second largest city of Bohemia after Prague.[12] Paraphrasing Agricola, Sebastian Münster described the local atmosphere in his *Cosmographia*: 'for most of the year, there is a fog so large and thick that the hills are covered and the sun's rays cannot pass through it' – a pollution due to smelting operations.[13] Neighbouring forests were gradually cut down to provide combustible material, leaving an industrial landscape surrounded by barren hills (see Figure 3.1).[14]

Far from being secluded, the mining regions of the Ore Mountains, the Harz, or Lower Hungary were at the forefront of urbanization, because a large work-force was needed to extract the ore, drain the mines, as well as in the crushing, smelting, and minting processes.[15] In the Saxon part of the Ore Mountains, two-thirds of the population lived in a dense and well-connected network of thriving towns such as Annaberg, Schneeberg, or Freiberg.[16] These cities dif-fered markedly from their medieval counterparts, as most had mushroomed from a *tabula rasa*, as can be seen from their regular and geometrical town plans.[17] In order to lure workers, rulers had promulgated freedoms, offering liberation from debt as well as rights to prospect and to operate smelting huts. These dynamic regions promoted social mobility and came to exert a major cultural influence within the Empire, sometimes in unexpected ways. In the copper mining city of Eisleben, a mine investor and smelter fathered Martin Luther, who would spark off the Reformation – a movement that in turn proved especially successful in the proto-industrial mining regions.[18]

arboribus spoliatos cernicis, in his densissima sylua erat: et ualles quas iam tot milia hominum habitant, latibula erant ferarum.'

[12] At its height, it counted more than 15,000 inhabitants; see Ernsting, *Georgius Agricola, Bergwelten 1494–1994* (1994), p. 161.

[13] Münster, *Cosmographei* (1550), p. dcccli, 'Von dem Joachims thal': 'Es haben zu etlich zeiten des jars so groß unnd dick nebel diß thal der massen bedeckt daß der sonnen schein gar nit dohin hatt mögen kommen.'

[14] Albinus, *Auszug Der Eltisten und fürnembsten Historien* (1598), vol. 2, p. 49. See a more detailed analysis of this picture in Kratzsch, *Bergstädte des Erzgebirges* (1972), pp. 1–2. Source: Sächsische Landesbibliothek – Staats- und Universitätsbibliothek Dresden (Hist. Sax.A.72-1, http://digital.slub-dresden.de/id301421307).

[15] The 'city-building power' of a mining region marks a noticeable exception to the general trend: most urban economies of the Holy Roman Empire declined in the late fifteenth century. See Strieder, 'Die deutsche Montan- und Metall-Industrie im Zeitalter der Fugger' (1931), pp. 189–226; Suhling, *Aufschließen, Gewinnen und Fördern* (1983), pp. 114–126. Tenfelde, Bartels, and Slotta, *Geschichte des deutschen Bergbaus* (2012), vol. 1, pp. 355–360 on the mining cities of the Harz, and pp. 361–369 on the Ore Mountains. On mining cities in Lower Hungary (now Slovakia), see Štefánik, 'Die Anfänge der slowakischen Bergstädte. Das Beispiel Neusohl' (2004), pp. 295–312.

[16] Vogler, *Le monde Germanique et Helvétique a l'époque des réformes* (1981), p. 222, or more precisely Schirmer, 'Das spätmittelalterlich-frühneuzeitliche Erzgebirge als Wirtschafts- und Sozialregion (1470–1550)' (2013), pp. 61–67.

[17] Among the mining cities, Marienberg and Zellerfeld, but also Scheibenberg, Oberwiesenthal, and Sebastianberg possess perfectly geometrical town plans. See Kratzsch, *Bergstädte des Erzgebirges* (1972), pp. 51–68.

[18] On this, see both my analysis of the relationship between religion and mine surveying above (Chapter 2) and Knape, *Martin Luther und der Bergbau im Mansfelder Land* (2000).

Figure 3.1 Allegorical view of the mining city of Freiberg during the silver rush, by Heinrich Göding in 1597. In the foreground, a team of miners are prospecting, trying to locate a vein with a dowsing rod and ascertaining its direction with a compass. Behind them, several mine workers are hurriedly extracting ore. In the background, smelting huts are spewing out a dense fog while most trees have been cut down. A new city church and walls, symbolizing the prosperity of the town, are being hastily built. Courtesy SLUB Dresden (Hist.Sax.A.72-1).

The meteoric rise of Annaberg illustrates the dynamism of mining towns, and especially the remarkable importance given to geometry and arithmetic there. In the early 1490s, silver ore was found on the 'Mount of Terror' (*Schreckenberg*). The influx of people prompted the Saxon ruler to send the physician and humanist Ulrich Rülein von Calw (1465–1523) – who would later author the first *Bergbüchlein* – to survey the landscape and assess the situation in 1496.[19] City rights were granted in the following year, together with a mining law, and the foundations of a church were immediately laid out. Originally made in wood, it had by 1525 been replaced by the massive stone church of St Anne, home of

---

[19] Ulrich Rülein von Calw would later draw a perfectly geometrical town plan for the city of Marienberg in 1521. A similar process happened in the neighbouring Schneeberg a few years before (1477), when the *Markscheider* Hans Setener was sent by the local count. See SächsStA–D, 10005 Hof- und Zentralverwaltung (Wittenberger Archiv), Nr. Loc. 4322/03, p. 128.

the famous *Bergaltar*, a masterpiece of Renaissance art. In the *Cosmographia*, Münster marvelled at 'the wonderful mines which have bloomed' in Annaberg. Despite being 'such a nice new city', Münster added, it had been 'circled with town walls within five years, and has such a nice church and buildings which can hardly be rivalled'.[20] A Latin school opened in 1498, only one year after the town was officially founded, soon attracting famous teachers such as the humanist Johannes Rivius (1500–1553). The original wood building was later rebuilt in stone with four main *auditoria*, a library, and a herb garden. The school hosted several hundred pupils and was as such bigger than many contemporary universities.[21] The printer Nicolaus Günther settled there 'because of the blooming school', publishing grammars and classics while also acting as deacon.[22] In 1540, a general inspection indicated that the town had six German-speaking schools, as well as one school for girls, a tally that did not include private *Rechenschulen* (arithmetic schools, known as reckoning schools).[23]

This allowed rapid progress in literacy and numeracy, so that by the mid-century most inhabitants were 'able to solve their everyday computing tasks on their own'.[24] 'There are many learned people here', a chronicler aptly noted, and mathematical skills were particularly abundant. The flourishing mining business attracted scholars such as Johannes Widmann (ca. 1460–ca. 1505), now remembered for having introduced the + and – symbols in print. His *Behende und hüpsche Rechenung auff allen Kauffmanschafft* (Mercantile Arithmetic), which appeared in 1489, was one of the first arithmetic textbooks ever published. A well-trained mathematician, Widmann taught the earliest complete course on algebra in the Empire, at the University of Leipzig. And yet, instead of pursuing an academic career, he left his position for the booming city of Annaberg in the 1490s.[25]

---

[20] In the early sixteenth century, the population of Annaberg exceeded 10,000 inhabitants. Münster, *Cosmographei* (1550), p. dxxviii: 'sant Annen Berg / dz ein solich schöne neuwe statt ist / und erst innerthalb fünf jaren gar mit mauren yntgefangen worden / hat auch so ein trefflich schöne kirchen und gebäw / der gleichen man kaum findt.' Münster then makes clear that his discourse concerns most of the mining cities of the time: 'the same [goes for] Joachimsthal, Goslar and Marienberg, where many thousand people are, and uncountable fortunes are made from silver and other mines'.

[21] See Meltzer, *Bergkläufftige Beschreibung der Bergk-Stadt Schneebergk* (1684), p. 226. Bartusch, *Die Annaberger Lateinschule zur Zeit der ersten Blüte der Stadt und ihrer Schule* (1897), pp. 44–48, 51, 151–152; Kratzsch, *Bergstädte des Erzgebirges* (1972), pp. 96–97. On Latin schools in the Ore Mountains, see Uhlig, *Geschichte des sächsischen Schulwesens bis 1600* (1999). On the fact that schools and 'academies' could be bigger and more influential than many universities, see Whaley, *Germany and the Holy Roman Empire* (2013), vol. 1, pp. 504–507.

[22] Kreysig, *Nachlese zum Buchdrucker-Jubilaeo in Ober-Sachsen* (1741), p. 4; Blaschke, *Sachsen im Zeitalter der Reformation* (1970), pp. 96–97. Printers played a crucial role in promoting early modern technical culture; see Eamon, *Science and the Secrets of Nature* (1994), pp. 94–96.

[23] Rochhaus, 'Adam Ries und die Annaberger Rechenmeister zwischen 1500 und 1604' (1996), p. 95.

[24] Rüdiger, 'Adam Ries d. Ä. und dessen Söhne als Rechenmeister und Mathematiker' (2017), p. 152.

[25] Unfortunately, very little is known about how Widmann applied his skills in the mining administration. See Folkerts, 'Die Mathematik im sächsisch-thüringischen Raum im 15. und 16.

The arithmetician Adam Ries (1492–1559) similarly abandoned his prosperous school in Erfurt – a centre of learning at the time – to settle there in 1523, opening a *Rechenschule* to meet the need for skilled calculators. Ries also worked in the mining administration as a mining clerk and accountant. He collected and checked the handwritten reports made by foremen, computing quarterly reports about the quantity of ore and the profits of silver mines. Reckoning masters and 'abacists' were indispensable to the mining administration.[26] Mining regions were among the first in which Indo-Arabic numerals replaced the Latin ones, pioneering double-entry bookkeeping north of the Alps. The transition from using a counting board with tokens to pen computations can be tracked both in mining archives and in Adam Ries's popular textbooks.[27] In this stimulating atmosphere, Ries could, for instance, discuss the *ars magna* – algebra, by then a rare skill – with Hans Conrad, a local smelter and essayer.[28]

Far from being restricted to subterranean geometry, mathematics thus framed in many ways the everyday life of mining towns. A broad class of people relied on quantifying procedures for technical duties; 'geometry ruled the construction of smelting furnaces' just as it governed the attribution of concessions. Besides scaled plans, the so-called *Schmelzbücher* (smelting books) contained numerical lists specifying temperatures and weights for various recipes.[29] With the influx of capital, arithmetic was used for the computation of profits and losses of mining *Kuxen*, financial instruments akin to modern shares, and this information was printed quarterly and widely circulated. Arithmetic also enabled the calculation of loans and inheritances ubiquitous in such dynamic economies.

The city of Annaberg enjoyed the privileges of being a *freie Bergstadt* (free mining town) ruled by a council of twelve *Ratsherren* (aldermen). The council coordinated the development of the city and secured its food supply, ensuring that a fraction of mines' profits went to the construction of hospitals, churches, and schools. Annaberg was important enough to have its own mint,

Jahrhundert' (2011), pp. 11–12. Kaunzner, 'Johannes Widmann, Cossist und Verfasser des ersten großen deutschen Rechenbuches' (1996), p. 48, speculates that Widmann might have then gone to Saint Joachimsthal.

[26] Rochhaus, 'Adam Ries und die Annaberger Rechenmeister zwischen 1500 und 1604' (1996), passim; Rochhaus, 'Zu den Rechenmeistern im sächsischen Erzgebirge während des 17. und 18.' (2002), p. 351: 'The *Rechenmeister* traditionally assumed … in addition to their actual profession, numerous offices in the administration of the mining district.'

[27] Schirmer, 'Das spätmittelalterlich-frühneuzeitliche Erzgebirge als Wirtschafts- und Sozialregion (1470–1550)' (2013), pp. 73–74. There are too many studies on Adam Ries to offer a comprehensive list; the series edited by the Adam-Ries-Bund in Annaberg offers the most thorough collection on his activities. On double-entry bookkeeping in smelting companies, see Kasper, 'Das Rechnungswesen im kursächsischen Hüttenwesen im 16. Jahrhundert' (1985), pp. 67–73.

[28] Jenisch, *Chronicon Annaebergense continuatum* (1658), p. 28; Schellhas, *Der Rechenmeister Adam Ries (1492 bis 1559) und der Bergbau* (1977), p. 6.

[29] Uta Lindgren, 'Maß, Zahl und Gewicht im alpinen Montanwesen um 1500' (2008), pp. 349–350, describes Hans Stöckl's *Schmelzbuch* (Smelting Book).

where highly skilled technicians relied on practical arithmetic on a daily basis for their assaying and alloying works.[30] A whole economy developed around smelting houses and fabrics, giving birth in turn to a cultural life specific to this region, ordered around computation.[31] There were dense exchanges with surrounding regions, for these industrial centres could not be sustained with local agriculture. Metals and coins were shipped to Leipzig and Dresden, while thousands of carts brought back tons of grains for the mining people, as well as coal and lead to feed the furnaces and chemically refine the silver.[32]

Food was one of many areas in which mathematics soon proved useful on a daily basis. The lack of a significant local agriculture made mining cities heavily dependent on imports. In the early years, it led to large fluctuations in the price of bread, which could lead to riots. An obvious answer was to fix the price of bread while adapting its weight to the current price of cereals. Loaves were sold at the same fixed and round prices, each corresponding to one coin so as to avoid giving change (one spoke, for instance, of a *Pfennigbrot* or a *Groschenbrot*). In times of shortage, bakers simply reduced the weight of each bread according to public tables. This rule was meant to avoid both starvation and undue profiteering. In 1533, arithmetician Adam Ries was asked to compute a set of such tables, entitled *Brotordnung*, that were tested with local bakers.[33] The tables were printed in 1536 and their success led many mining cities to adopt similar regulations. At a time when bread was the staple food, the inhabitants of mining cities soon understood that 'the number of loaves of bread to be baked can be easily determined', as it was 'namely twice as much as a bushel costs in *Groschen*'.[34]

It is in the thriving city of Annaberg – where he bought a house in 1506 – that we encounter for the first time Georg Öder, patriarch of a long dynasty of subterranean surveyors.[35] Along with the local mining master, Öder was among the first to measure the local shafts and concessions. His stature was important enough to be appointed alderman of the city from 1521

[30] The mint of Annaberg opened in 1498 and lasted until 1558, when Elector August closed all local mints and centralized coinage activities in its capital Dresden. See Schellhas, *Der Rechenmeister Adam Ries (1492 bis 1559) und der Bergbau* (1977), p. 6.

[31] See Manfred Bachmann, *Der silberne Boden* (1990), pp. 193–194 on mining culture, and pp. 199–200 on industry and fabrics.

[32] Westermann, *Bergbaureviere als Verbraucherzentren im vorindustriellen Europa* (1997), pp. 203–209; Schirmer, 'Das spätmittelalterlich-frühneuzeitliche Erzgebirge als Wirtschafts- und Sozialregion (1470–1550)' (2013), pp. 71–72; Tenfelde, Bartels, and Slotta, *Geschichte des deutschen Bergbaus* (2012), vol. 1, pp. 318–326.

[33] The original text has been re-edited in Gebhardt, *Die Annaberger Brotordnung von Adam Ries* (2004), and analysed in Gebhardt, *Zur Wirkungsgeschichte der Brotordnung von Adam Ries* (2006). See pp. 280–281 for the tests carried out with bakers (*Backproben*) and their influence on other mining cities.

[34] Gebhardt, *Die Annaberger Brotordnung von Adam Ries* (2004), p. 85.

[35] Most of the information in this paragraph comes from Bönisch, *Die erste kursächsische Landesaufnahme* (2002), pp. 6–10; Reichert, 'Die Kurfürstlich-sächsischen Markscheider Georg Öder die Jüngeren sen. und jun.' (2014), pp. 147–188, here pp. 152–153.

onwards.[36] He held various technical positions in mining sites – something common at the time – most of them related to surveying, for example the supervision of drainage tunnels. Öder had nine children, several of whom trained with him or frequented the Annaberg Latin school. Hieronymus became professor of medicine at the University of Greifswald, Jacob became foreman in a mine near Schneeberg, while Hans became subterranean surveyor in the neighbouring Marienberg.

Thanks to his geometric skills, Georg Öder had established himself as an important local figure; more importantly, these abilities ensured the subsequent social ascent of the family. His most famous son was Georg Öder II, born around 1511. He likely studied at the local Latin school and succeeded his father as town surveyor after his death in 1535.[37] He also inherited the position of alderman and stayed in the town council until his death in 1581. Georg II recalled his training with his father in the following way: 'My late father and I were always performing underground geometry, setting the marking stones and bringing the vertical [into the mines].'[38] He enjoyed the reputation of someone 'sufficiently trained and knowledgeable of the work' of a subterranean surveyor, according to a local mining master.[39] His reputation extended beyond the borders of the Electorate. In 1545, when the city of Goslar in the Harz, hundreds of kilometres away, asked the Saxon court to send an expert for the surveying work of a drainage tunnel, Georg II was designated. A decade later, he would be sent to Franconia (now part of Bavaria) for similar work in the gold mine of Goldkronach.[40] Godparents of his children were chosen among the high society of the city, including the master of the mint, various members of the town council, and even the wife of the main pastor.

## An Elector's Geometry of Power

The accumulation of mining wealth by German rulers is symbolized by the legend of Duke Albrecht of Saxony (1443–1500) and the 'silver table'. During the silver rush, Albrecht went to Schneeberg in person in the late 1470s and toured the mines. Having found a huge block made of pure silver, legend has it that he organized a banquet there, boasting that even the *Imperator Germanorum*

---

[36] Öder took his oath as official surveyor in 1518, but was probably active before that time. One could only be an alderman for two years out of three, so he was intermittently member of the council until his death in 1535.

[37] Reichert, 'Die Kurfürstlich-sächsischen Markscheider Georg Öder die Jüngeren sen. und jun.' (2014), p. 154.

[38] SächsStA–F, 40001, Oberbergamt Freiberg, Nr. 3477, f. 2r: 'mein sehliger Vater und ich, alezeit mir das marscheiden gebraucht, den Lochstein yhr und aliger Seiger eingebracht haben'.

[39] Letter from the mining master of Schneeberg to the Elector Johann Friedrich I of Saxony on 31 March 1545, as quoted in Reichert, 'Die Kurfürstlich-sächsischen Markscheider Georg Öder die Jüngeren sen. und jun.' (2014), p. 155.

[40] Ibid., p. 155.

had no comparable table.[41] Local rulers clearly benefited from the new mining dynamic, granting town rights and freedom to prospect in exchange for a tenth of all profits, while investing directly in the most productive concessions. Finally, their monopoly on minting was another source of bullion. The Electorate of Saxony, the Principality of Brunswick-Wolfenbüttel, and most importantly the Austrian crown profited immensely from their metal mines. Rulers used these seemingly inexhaustible resources not only to wage wars on their neighbours but also to maintain luxurious courts. The wiser ones established advanced bureaucracies, reformed their school system, and improved the rest of their dominions.[42] To this aim, they enlisted in many cases the same technicians and engineers who had ensured the success of their mining enterprises in the first place.

The Saxon territories had historically been divided between the Albertine and Ernestine branches. They were reunited by Maurice of Saxony (1521–1553), whose victory at the battle of Mühlberg in 1547 won him the Electoral title. Maurice made Dresden his residential seat, and was succeeded by Elector August (1526–1586), who turned Saxony into a leading power in the Holy Roman Empire.[43] Diplomatically, the Saxon rulers walked a fine line in a complex political environment. Although their realm had been the crucible of the Protestant Reformation, they alternatively sided with the Catholic Hapsburg Emperor or with Protestant rebelling princes to expand their territories. The capital of the Ore Mountains, Freiberg, served as an operating base and became a centre of power. Both Moritz and August were born in this city – in the imposing Freudenstein castle – and were later buried in its cathedral.[44] August was raised in Freiberg and in Innsbruck, there together with the future Emperor Maximilian II. Back in his home town, he was taught in the Latin school of Johannes Rivius, who had by then left Annaberg.[45]

August ruled Saxony for three decades, becoming a powerful patron of mathematics. He created a library in 1556 and opened some years later a *Kunstkammer* (chamber of the arts) to host his ever-growing collection of

[41] The episode is depicted on the frontispiece of Meltzer von Wolckenstein, *Historia Schneebergensis Renovata* (1716). The possibly fictitious event might relate to an actual visit of Duke Albrecht. Georg Agricola mentions it briefly in his *De Natural Fossilium* (1558), lib. VIII, p. 330–331.

[42] Braunstein, 'Les statuts miniers de l'Europe médiévale' (1992), pp. 50–53 states that the mining resources 'ont puissamment aidé à l'affermissement des Etats territoriaux, sans exclure de ce paysage politique la part patrimoniale des dynasties impériales successives'.

[43] On the rulers of Saxony in the sixteenth century, see Rogge, *Die Wettiner* (2009), pp. 185–242.

[44] Their successors would cherish the tradition until their conversion to Catholicism in the early eighteenth century. Moritz visited often as Elector and his personal physician, Johann Neef, lived there (see Long, *Openness, Secrecy, Authorship*, 2001, p. 188).

[45] Moritz stayed more than seventy days in Freiberg during the decade 1541–1551. August lived notably longer in the city, mostly for hunting purposes – see Schattkowsky, *Das Erzgebirge im 16. Jahrhundert* (2013), pp. 42–44.

scientific instruments.[46] Interested in various mathematical fields, from ballistics to astronomy, the Elector himself practised assiduously and ordered masterpieces from the finest instrument-makers. He conversed frequently with other prince-practitioners, for instance Wilhelm IV (1532–1592), Landgrave of Hesse-Cassel, and King Christian of Denmark (1503–1559).[47] The fascination about instrument-making, cartography, and mathematics cannot simply be dismissed as a 'hobby' or a 'monarch's obsession with representation', as some historians postulate.[48] August of Saxony and his contemporaries consciously aimed at improving their land and tightening their grip on ore, wood, and other natural resources.[49] They supported university professors, instrument-makers, and mathematical practitioners, harnessing their talents into a 'geometry of power', in the words of Michael Korey.[50] Among the individuals involved in this rationalization process, subterranean surveyors and mining officials went to play a disproportionate role. In an interesting interplay between mines and courts, the Öder and Ries families were irresistibly attracted to the court of Dresden while maintaining bounds with their home city of Annaberg. Their know-how and experience of administration proved to be perfectly complementary with the knowledge of scholars, as both were frequently exploited in the state's affairs.

The mining boom in the Ore Mountains had made it crucial to ascertain the exact frontier between the Electorate of Saxony in the north, and the Kingdom of Bohemia in the south. Given that some ore veins stretched on both territories, it was a diplomatic conundrum until both sides agreed in 1550 to send 'mathematicians' on a mission to settle the issue. 'Our mountains', wrote a local pastor, 'are the crown and the boundary [*Markscheide*] of the country', purposely using a dialectal mining word as a metaphor: the state was equated to a mining concession whose limits had to be accurately ascertained.[51] In that context, the Saxon Elector sent Georg Öder II from Annaberg and Nicol Patschke (d. 1556), a subterranean surveyor in Freiberg, together with Johannes Humelius (1518–1562), professor of mathematics at the University of Leipzig, all of them 'under special oath' for the occasion.[52] The King of Bohemia sent his own team of surveyors, led by Augustin Hirschvogel (1503–1553), artist

---

[46] The Dresden *Kunstkammer* is featured is countless books and exhibition catalogues, most recently in Koeppe, *Making Marvels* (2019).

[47] See Moran, 'German Prince-Practitioners' (1981), pp. 254–256.

[48] Knothe, 'Die Dresdener Forstzeichenbücher' (1998), p. 17: 'der Besessenheit eines Monarchen für sein Hobby und für sein Repräsentationsbedürfnis'.

[49] This argument is developed in Dolz, *Genau messen, Herrschaft verorten* (2010).

[50] Korey, *The Geometry of Power* (2007).

[51] Lehmann, *Historischer Schauplatz* (1699), p. 85: 'Sind unsere Berge des Landes Cronen und Marckscheiden.'

[52] SächsStA–D, 10024 Geheimer Rat (Geheimes Archiv), Nr. 08342/01, f. 2v: 'So haben wie demmnach die *mathematicos* / als nemlich Augustinum Hirschvogeln / undt magistrum Joannem Humelium / zu solchem werk / sonderlich veraidet.'

and court *mathematicus* known for his etchings and town maps.[53] The mission was meant to be a 'visual inspection' (*Augenschein*), which was the usual procedure at the time to solve legal disputes, but the necessity of a thorough survey and mapping quickly became evident to both parties. It was not exactly natural to send subterranean surveyors in order to 'ascertain with a thorough certitude the height, length, breadth, width, and other characteristics of the bespoken domain'.[54] After all, the operation did not imply a single underground measurement. More importantly, the rules of foreign diplomacy differed markedly from the mining customs they were accustomed to. Öder and Patschke were likely summoned because they were the only persons with sufficient hands-on experience of uneven terrain. A report written in June 1550 detailed that 'some [of the surveyors] had worked with the compass' while the others proceeded to dig furrows and set the actual border. The expedition also gathered from locals the names 'of all hamlets, villages, mountains, mines, fields, and rivers'.[55] We witness here how actual surveying operations blur the categories of cartography as it was then taught in universities. Mathematical geography was mixed with surveying and chorography, the science of describing regions.[56] The result was represented in a scaled map that was sent to the Elector (Figure 3.2).

Looking at the working papers of Georg Öder II preserved in the Dresden archive, it is possible to reconstruct precisely the method that he used. A central point was chosen on the highest local mountain, the *Auersberg*. From this point, surveyors went in straight lines to a dozen prominent points located on the border, measuring lengths with a miner's chain and using a compass to record directions.[57] Numerical data was collected on several sheets using a standardized, almost tabular form, displaying evident skill at managing information. The better part of the survey was performed by Georg Öder II, who also drew the resulting map, although cartography was by then hardly used in subterranean geometry, as will be explained in Chapter 5. The underground surveyor was thus able to adapt to a new context methods that had proven their values in the mines.

---

[53] Augustin Hirschvogel is mainly known for his artistic activities, from glass-making in Nürnberg to his famous landscapes etchings. However, he was a court *mathematicus* who mapped several dominions of King Ferdinand I. See Kühne, 'Augustin Hirschvogel und sein Beitrag zur praktischen Mathematik' (2002), pp. 237–251.

[54] SächsStA–D, 10024 Geheimer Rat (Geheimes Archiv), Nr. 08341/06, f. 296r: 'die Grose, lange, Preite, weite unnd andere gelegenheit gemelter Herrschafft, zu eine grundtlichenn gewissheitt bringen möge'.

[55] SächsStA–D, 10024 Geheimer Rat (Geheimes Archiv), Nr. 08342/01, f. 2v: 'eins thails mit dem Compass zugehen / eins thails aber der grenitz und Rainung der Herschafft erfaren / zugeordnet / welche Personen sie zu abmessung unnd abschreitung / der gebirge unnd ortter / gebrauchen / Unnd alle unnd ide flecken / dorffer / gebirge / Perkwerck / welde / wasserflus / und andere gelegenheit der Herschafft mit Iren nhamen aigentlich benennen.'

[56] On the typology of geographical disciplines in the sixteenth century, see for instance Cormack, *Charting an Empire* (1997), chs. 3–5.

[57] SächsStA–D, 10024 Geheimer Rat (Geheimes Archiv), Nr. Loc. 08342/01.

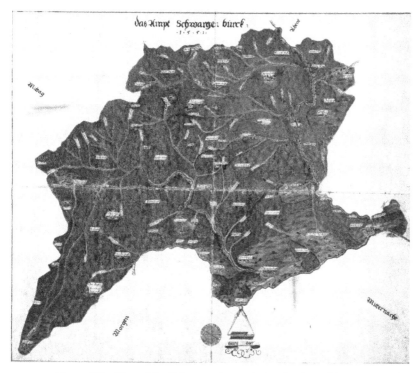

Figure 3.2a  Map of the Schwarzenberg mining district, with the Bohemian
border to the South. Hand-drawn and coloured by Georg Öder II in 1551. The
legend depicts a mining compass with a scale given by a pair of dividers and
the two phrases: '100 cords length' above, and '1 cord has 10 lochter [*fathoms*]' below. Courtesy SächsStA–D, 12884, Karten und Risse, Schr 001, F
002, Nr 010, Makrofiche Nr. 04976.

The success of this first collaboration revealed to August of Saxony the efficiency of underground surveyors.[58] At that time, Georg II had already inherited from his father the position of subterranean surveyor in Annaberg, but
in 1560 the Elector also appointed him as his court surveyor (*Kurfürstlicher
Markscheider*). This new title was created for the occasion and belonged to the
category of 'goldsmith, screw makers, clock-makers, painters, wood turners, and
locksmiths'.[59] Öder thus became a courtly figure, a craftsman who was expected

---

[58] By success, I mean here that the surveying operations were technically adequate. Politically,
the issue would not be settled before October 1556, when both parties signed the Treatise of
Schneeberg. After a short relief, new challenges would emerge in the coming decades, as shown by
the involvement of Matthias Öder in border surveys in 1604, discussed later in the present chapter.

[59] Reichert, 'Die Kurfürstlich-sächsischen Markscheider Georg Öder die Jüngeren sen. und jun.'
(2014), p. 162.

Figure 3.2b A detail from Figure 3.2a showing the mining compass in the legend to the map of the Schwarzenberg mining district.

to live in the capital and received a modest stipend, a horse, as well as some other benefits in kind. His expertise, which was not restricted to mining matters, could be requested at any time. In his first years as court surveyor, Öder seems to have mapped various parts of the Electorate, collaborating with and quickly replacing Professor Humelius, who abruptly died in 1562.[60] He was asked to produce maps of the Elector's forests, as well as of districts in the Ore Mountains. To fulfil his duties, he was usually accompanied by up to four 'chain-bearers', sometimes including his brother Hans, who was also a surveyor.[61]

In 1565, Georg II – who was well in his fifties – was relieved from duty and finally returned to Annaberg. His son Georg III, trained by him in the art of subterranean geometry, inherited his position of court surveyor and continued his activities. Meanwhile, Elector August was spending a substantial part of the mines' revenues to turn Dresden into a leading German court, attracting artists as well as mathematicians seeking patronage. In 1564, the instrument-maker Nicolaus Valerius had sent August a quadrant with a personalized *Geometria* handbook. In 1569, the instrument-maker Christoph Schissler from Augsburg delivered another *quadratum geometricum* with his own

---

[60] On Humelius's surveying activities, see Woitkowitz, 'Der Landvermesser, Kartograph, Astronom und Mechaniker Johannes Humelius (1518–1562)' (2008), pp. 75, 81–87.
[61] Reichert, 'Die Kurfürstlich-sächsischen Markscheider Georg Öder die Jüngeren sen. und jun.' (2014), p. 164.

handwritten *Geometria* booklet to the Elector.[62] When a present was pleasing enough, the Elector rewarded its maker and ordered more items.[63] August also recruited artisans to come to Dresden and work directly for him. Hans Göbe (d. 1574), native of Innsbruck, was hired as clock-maker in 1568 – in the same administrative category as the Öders – and those practitioners indeed collaborated. Göbe was for instance asked by the Elector to make 'many compasses and instruments useful for surveying' for Georg Öder III, following his design instructions.[64] The artisan went on to build several tools under Öder's supervision, either for the surveyor's professional use or to enrich the Elector's chamber of the arts.[65]

Interactions and collaborations at the Dresden court somehow blur modern categorizations between instrument-makers, scholars, and practitioners. One notices that artisans accompanied their instruments with booklets, eager to show that they mastered not only the mechanical and tangible, but understood the underlying theories. Similarly, Georg Öder II and III originally came from a world of mines and practice, yet acquired a courtlier stature. They were not simply carrying out measures any more, but acted as experts, projectors, and designers. The activities of the Öder family in Dresden enlighten us about the standing of subterranean geometry among sixteenth-century prince-practitioners. When Elector August died in 1586, an inventory of his art chamber was soon made, revealing several objects related to mine surveying. The *Kunstkammer* was one of the richest in the Holy Roman Empire, and the inventory listed almost a thousand scientific instruments, including more than a hundred compasses, dozens of quadrants and countless drawing tools.[66] More specifically, it contained 'five wooden compasses with small holes' that had been made 'by the late Elector August himself', together with other pieces relevant for subterranean surveying: 'two gilded mine compasses', 'a silver compass with a silver instrument to do many mining things', 'one round compass with wax rings', to name but a few.[67]

---

[62] Wunderlich, *Kursächsische Feldmesskunst* (1977), pp. 37, 68, 196.

[63] While most of these instrument-makers came from Nuremberg and Augsburg, several lived in the Ore Mountains, attesting an original tradition of precision mechanic. The most famous was clock-maker Andreas Schellhorn from Schneeberg, active in the 1570s. See Bachmann, *Der silberne Boden* (1990), p. 195.

[64] Letter from 19 October 1570, in SächsStA–D Cop. 356a fol. 372A, as presented in Schmidt, *Zur Geschichte der Kartographie unter Kurfürst August von Sachsen* (1899), pp. 155–160, here p. 157.

[65] In that context, Hans Göbe seems to have produced a mining compass in 1571, based on a sea compass brought by the Elector from Denmark, see Schillinger, 'The Development of Saxon Scientific Instrument-Making Skills from the Sixteenth Century to the Thirty Years' War' (1990), p. 278. This compass was suspended in gimbals, in a way prefiguring the suspended mining compasses that appeared in the middle of the following century. An eventual link between the work of H. Göbe and B. Rösler (who lived three generations later) has yet to be investigated.

[66] Ibid., p. 279. For more general context, see Strano, Johnston, Miniati and Morrison-Low, *European Collections of Scientific Instruments, 1550–1750* (2009).

[67] '5 Holtzerne Compaßscheiben mit löchlein zum Marscheiden … hat Churfürst Herzogk Augusten zu Sachßen seliger sellsten gemacht … 2 Mößene vorguldte Gruben-Compaß …

Descriptions are unfortunately often too sketchy to distinguish between ostentatious pieces mostly meant to be displayed and more mundane instruments who would indeed be used in the fields and mountains. It is clear, however, that several of these objects had been used on site, either by the Elector or by the Öders. Intriguingly, the inventory also listed mining maps, manuscript reports and even 'a written and computed instruction on subterranean geometry'. This might indicate a curiosity about this specific art, and means that some practitioners were willing to couch their knowledge on paper. A few years later, when the instrument-maker Tobias Volckmar designed a surveying instrument (*Meßkästchen*) for the new Elector of Saxony Christian I, he followed the ritual of attaching a handwritten instruction. The first part of the booklet displayed his knowledge of the classic *practica geometriæ*, while the last dealt with 'the *Bergschienen*, also known as subterranean geometry'.[68]

The Elector's art chamber illustrates perfectly the influence of underground surveyors on mathematics at court. The Dresden court was noted for its extensive collection of mathematical instruments and books, which reflected a profound interest in questions of measure, survey, and quantification. It also acted as a crucible. By opening a library and collecting instruments, August and his curators brought classics such as Petrus Apianus' *Cosmographia* or Ptolemaeus' *Geographia* – works that figured prominently in many European libraries – together with local surveying instruments and their practitioners. Mining officials were asked to collaborate efficiently with foreign artisans. In that context, the descendants of Georg Öder and Adam Ries, moving from Annaberg to Dresden, were able to share their eminently practical know-how with the specific skills of university professors and instrument-makers. One should, however, reflect a moment on an unexpected consequence of this situation: these contacts between scholars and miners took place at court rather than in the mines. The very local knowledge of underground surveyors, odd as it might be, met the more classical geometry of its time in Dresden, rather than in the Ore Mountains of Saxony. In other words, the careers of Öder, Ries and their descendants reveal a surprising social mobility.

Historical studies largely see practical knowledge as highly localized and tied to the conditions in which it emerged. While it is true that this type of

---

1 Silbern Compaß mit 1 Silbern Instrument zu mancherley sachen Bergkwergk ... 2 Mößinge runde Stunden Compaß mit waxenen Ringen', in Dolz and Schillinger, 'Markscheideinstrumente in den sächsischen kurfürstlichen Sammlungen im Spiegel handschriftlicher Inventare und Kataloge' (1996), p. 94. The inventory of 1587 has been published in Syndram and Minning, *Die kurfürstlich-sächsische Kunstkammer in Dresden. Das Inventar von 1587* (2010).

[68] Wunderlich, *Kursächsische Feldmesskunst* (1977), pp. 108–109, 114–115: 'Vom Bergschienen, welches man auch Markscheiden nennet.' It should be noted that Volckmar was not a miner and this manuscript (destroyed during World War II) probably did not go into much detail, for this part of the manuscript was only four pages long.

mathematics indeed developed in the very specific conditions of ore shafts, tunnels, and galleries, it later took a new dimension at the Dresden court by interacting with other forms of knowledge. This often-overlooked dimension is crucial, for it demonstrates the versatility of this mathematical culture. Subterranean surveyors have sometimes been presented as mere labourers, whose contribution to court life was limited to mapping the Elector's hunting forests, or other idiosyncratic behaviours.[69] On the contrary, what defines them here as local experts was their ability to apply their surveying and counting skills in areas apparently unrelated to mining. Presented and analysed together, the careers of the Öders and the Rieses illustrate how a culture of mathematics born in the mines spread within the Holy Roman Empire. Practitioners were not confined to remote mines where scholars came to observe them; their expertise would soon be employed in the most diverse settings around Saxony.

## Engineering the Mining States

Dresden was a major meeting place, where the Elector's power acted as a catalyst between miners and scholars. It was not, however, the sole point of contact between the culture of mining cities and the civil life of the Electorate. Strong economic and cultural ties existed between the Ore Mountains and the university towns of Wittenberg, Leipzig, and the neighbouring Erfurt, in mathematics as well as religion, law, and medicine. Prosperous mining cities possessed some of the best Latin schools and sent gifted students to nearby universities. Their economic dynamism opened pathways and careers, attracting in return numerous graduates as physicians, pastors, and administrators to fill important positions in the mining offices. Georg Sturtius (1490–1548), for example, was the son of a rich miner in Annaberg. He studied and later taught medicine in Erfurt, before heading back to the Ore Mountains. Sturtius became town physician in Annaberg, where Adam Ries dedicated to him his most famous mathematical work, the *Coß*. He then headed to Saint Joachimsthal, where he was later replaced by the then young Georgius Agricola. Leaving the Ore Mountains, he went back to teach in Erfurt and eventually cared for the elderly Martin Luther there in 1537.[70]

In the larger European context, the sixteenth century marks the heyday of the versatile *ingénieurs* and *mechanicii*. In the second half of the century, their status singularly improved as their expertise was required in many domains of civil life. They possessed useful knowledge that was certainly localized, but more importantly, very dynamic. Studying the migrations of German workers all over Central Europe, in Scandinavia or – as Eric Ash has beautifully

[69] Schmidt, *Kurfürst August von Sachsen als Geograph* (1898), p. 7.
[70] Müller, 'Sturtz, Georg', in *Allgemeine Deutsche Biographie* (1894), vol. 37, pp. 54–56.

shown – in England, reveals that local knowledge can travel very far, as long as the context it similar enough to ensure its efficient application.[71] In the Netherlands, Simon Stevin (1548–1620) published on anything from navigation to compound interest, while filing patents to improve the design of windmills. In France, Jean Errard (1554–1610) wrote an influential treatise on fortification and was interested in surveying instruments. The *Theatrum Instrumentorum et Machinarum*, published by Jacques Besson in 1578, features windmills, cranes, and winches with their proportions.[72]

German princes took an active role in that development, from Landgrave Wilhelm IV of Hesse-Cassel to the Emperor Rudolf II (1552–1612).[73] Lesser rulers such as Henry Julius of Brunswick-Lüneburg (1564–1613) – whose coffers were filled with silver extracted in the Harz – similarly hired several engineers to improve his duchy. Julius recruited Dutch technicians to design and supervise the digging of canals, while surveyor Franz Algermans mapped his provinces. Most of these experts were hired precisely for their wide range of skills. When Julius hired a certain Johann Tiele in 1586, it was as 'wine-gauger, surveyor, and timber measurer'. Moreover, his contract specified that he was also required 'to assume duties in all other geometrical, mathematical, and architectural arts, to the best of his knowledge'.[74]

In the mining states of the Holy Roman Empire, many of these functions came to be fulfilled by underground surveyors. For a short time span in the sixteenth century, there had been 'no clear separation' and there were regular exchanges between university professors and mathematical practitioners, as the natural collaboration of Georg Öder II with Johannes Humelius to map the Bohemian border illustrates.[75] Practitioners gradually established a foothold in the learned and courtly world, with Adam Ries obtaining the title of court *arithmeticus* and Georg Öder II becoming subterranean surveyor for the Elector.[76] Thanks to his new status, Adam Ries was able to send one of his sons, Abraham, to the University of Leipzig where he learned Latin and Greek. Abraham subsequently translated classical texts of Archimedes and Euclid into German, although these works were never published.[77] Mining regions once again appear as dynamic

[71] See Ash, *Power, Knowledge, and Expertise in Elizabethan England* (2004). On the general status of mathematics in relation to technology and civil uses, see Morel, 'Mathematics and Technological Change' (2023).

[72] Van Berkel, 'Stevin and the Mathematical Practitioners, 1580–1620' (1999), pp. 13–36; Métin, *La fortification géométrique de Jean Errard et l'école française de fortification (1550–1650)* (2016); Ravier, *Voir et concevoir* (2013).

[73] See Moran, 'Princes, Machines and the Valuation of Precision in the 16th Century' (1977), pp. 209–228; Trunz, *Wissenschaft und Kunst im Kreise Kaiser Rudolfs II* (1992).

[74] On Henry Julius of Brunswick and his mathematical practitioners, see Pitz, *Landeskulturtechnik, Markscheide- und Vermessungswesen* (1967), pp. 84–86, 106–108.

[75] Folkerts, 'Die Mathematik im sächsisch-thüringischen Raum im 15. und 16. Jahrhundert' (2011), p. 19.

[76] Ries and Deschauer, *Das 1. Rechenbuch von Adam Ries* (1992), p. 3.

[77] Rüdiger, 'Abraham Ries (1533–1604) und sein Werk' (2011), p. 25.

intellectual centres: Annaberg and other cities provided a large share of the professors and ruling circles in Central Europe. Mining regions not only supplied their rulers with precious commodities such as metals, wood, and fur, but less apparent – but no less important – were the experienced officials and skilful craftsmen whose expertise could be applied to various endeavours.

As sixteenth-century rulers tried to improve their dominions using land reclaiming and canal digging, hydraulics was an obvious field to focus on. Subterranean surveyors had an experience comparable to the expensive Dutch experts working for many European courts.[78] In the Ore Mountains of Saxony, for instance, the cheapest way to supply the mining cities with wood – needed to power the furnaces and consolidate the galleries – was to float trunks from neighbouring forests. In 1564 Georg Öder II designed and supervised the construction of a new ditch to supply his hometown of Annaberg, which opened two years later. The skills of underground surveyors were well adapted to these issues: they had a wealth of experience in using water levels to dig drainage tunnels and knew how to carefully control slopes over long distances. Öder managed to plan an 11-kilometre ditch that crossed the watershed with a carefully controlled drop of about 8 *Lachter* (ca. 16 m).[79] This operation showed that mine surveyors could apply their skills to related problems outside of mining sites. August of Saxony, who had contributed to the funding of this project, could once again observe first-hand the efficiency of the Öders.

In 1578, August of Saxony wished to establish saltworks in his Electorate in order to avoid relying on its northern neighbour, Prussia.[80] There was a promising brine source close to the city of Leipzig, but it was located far from any forest, although substantial quantities of wood were needed in the evaporation process to produce salt. A new canal, the *Elsterfloßgraben*, was thus planed, whose main goal was to float combustible to the future saline. The project was supervised by the mining master and machinist Martin Planer and the forest master Paul Gröbel, illustrating how the administration of natural resources purveyed the nascent states with a more general capacity of expertise. As was usual at the time, the work was subcontracted to projectors, in that case Georg Öder III and another expert from the Ore Mountains, Nicol Lippold. While the extensive levelling work seems to have been correctly made, the two men had underestimated both the cost and the difficulties of the earthworks, and could not complete the project on time. After this failure, Georg Öder III fell out of favour, and the contract was simply allocated to another mining official.

---

[78] A map of major European projects led by Dutch experts can be found in Van Veen, *Dredge, Drain, Reclaim* (1955), pp. 52–53.

[79] For 11 km, a height difference of 16 m means that the slope was about 1.5 ‰. See Reichert, 'Die Kurfürstlich-sächsischen Markscheider Georg Öder die Jüngeren sen. und jun.' (2014), pp. 170–171.

[80] On the Elster canal, see Andronov, Baum, *et al. Der Elsterfloßgraben* (2005).

However, the situation escalated when one of Georg's brothers, Hieronymus, was caught on the construction site, performing measurements 'with a compass, a water level, and its accessories.'[81] Hieronymus Öder argued that he was trying to remedy the defective canal, hoping to rehabilitate his brother. In any case, he was charged with displacing levelling stakes, unauthorized measuring and note-taking. Moving marks was severely punished: Hieronymus was tortured but ultimately released, while Georg III was exiled. Yet technical experts were highly sought after at the time, and Georg III soon entered the service of Rudolf II at the imperial court. A few years later, he is found supervising the digging of an underground canal supplying water to the *Letna* royal gardens in Prague.[82]

Subterranean surveyors could also be used as military engineers, as happened during the three-and-a-half month siege of Gotha in 1567. August of Saxony, backed by the Emperor and the Imperial Diet, declared war on his cousin Johann Friedrich II, who contested him for the Electoral title. Two subterranean surveyors from Schneeberg, Hanns Voigtlander and Melchior Pfeilschmidt, helped in diverting the Leina River that supplied the city with fresh water.[83] Miners were similarly involved when Elector August decided to build a formidable fortress on a cliff in Königstein. *Bergmeister* Planer travelled from Marienberg, and accompanied by local technicians dug a well to a depth of 152 m, ensuring the castle's water supply. The technical prowess required six years of work, involving not only a main shaft, but galleries and water-wheels adapted from the mining context; the experience was replicated in the following years in other fortresses in Saxony.[84] The Elector did not rely solely on mining officials and also employed foreign military engineers, the most famous being Roch Guerini Linari (1525–1596), a Tuscan specialist of modern fortification.[85] Underground surveyors were nevertheless ubiquitous in all duties related to practical geometry, from hydraulic to military engineering. Surprisingly, I found no contact with craftsmen involved in military mining, as the methods of destroying fortification seem to have substantially differed from the knowledge of the *Markscheider*.[86] Moreover, unlike foreigners who came and went, local practitioners had a total obedience to the Elector and a more situated expertise.

---

[81]  Reichert, 'Die Kurfürstlich-sächsischen Markscheider Georg Öder die Jüngeren sen. und jun.' (2014), p. 182: 'Wasserwaage sampt derselben Zugehörungen und Compast.'
[82]  Stein, 'Zur Tätigkeit der Familie Öder in Böhmen' (1988), p. 18.
[83]  Rudolphi, *Gotha* (1717), p. 126, and Meltzer von Wolckenstein, *Historia Schneebergensis Renovata* (1716), pp. 470, 635: 'in der Belagerung Gotha … als Markscheider / die Wasser abzuwägen / beschrieben worden'.
[84]  Gleue, *Wie kam das Wasser auf die Burg?* (2008), pp. 53, 63–69, 92, 268–274. Planer also supervised the digging of wells in the fortresses of Stolpen and Augustusburg.
[85]  On the relationship between surveying and military sciences in Saxony, see Wunderlich, *Kursächsische Feldmesskunst* (1977).
[86]  On military mining, see Ageron, 'Mathématiques de la guerre souterraine' (2018), pp. 211–224. Pierre Ageron confirmed to me that he knew of no such collaboration.

Besides their missions for the Elector, practitioners from the Ore Mountains also disseminated throughout Saxony. Their versatility led them to various professions in which their mathematical skills could prove useful. The pathways of later Öder and Ries family members illustrate this gradual, low-key mathematization of civil life. The Öders mostly stayed involved in mining operations, in which their geometrical knowledge could immediately be useful: Jacob and Hans, two brothers of Georg II, are known to have been respectively foreman and underground surveyor, and their sons took over in the following decades.[87] In addition to fulfilling his duties for the Elector, Georg III opened an inn – 'At the Three Lilies' – in Dresden.[88]

The case is even clearer with the descendants of Adam Ries, since all of his sons took up professions related to mathematics in the broadest sense. The most gifted, Abraham, studied at the University of Leipzig and became mining clerk and accountant. He succeeded his father as court *arithmeticus* of the Elector August, implemented a coinage reform and acted occasionally as land surveyor. Jacob Ries stayed in Annaberg as *Rechenmeister* and gauger, computing new tables for the price of bread. A third son, Isaac, moved to Leipzig where he opened a school for mercantile arithmetic. He published sets of tables as ready-reckoners (*Rechenknecht*) and was also appointed official wine gauger. Although it is difficult to follow the numerous grandsons of Adam Ries, several of them seem to have pursued similar careers well into the seventeenth century.[89]

By the end of the sixteenth century, the quantifying culture born in the mining cities of Saxony had experienced a dramatic expansion. Beyond the specific context and technicalities of metal mining, mathematical practitioners developed a workmanlike expertise. Spurred by the versatile demands of the Electors, they figured how to survey or compute their way out of all kinds of challenges. Within two generations, experts from the Ore Mountains extended their field of operation to the whole Electorate, often encouraged by August himself. The Öders and the Rieses dealt with canal digging and military operations in rural areas; they opened reckoning schools, gauged barrels, and practised accounting in the cities. Most of these tasks did not belong to the standard duties of mining officials. What they had in common was a culture of quantification used for technical and economic purposes. The stark contrast between the numerous opportunities available in early modern Saxony and the shortage of technical skills made such engineers invaluable. In a world in which mathematics

---

[87] Hans Öder was *Markscheider* in Marienberg, where his sons Hans II (d. 1594) and Jacob (d. 1603) succeeded him. See Bogsch, *Der Marienberger Bergbau seit der zweiten Hälfte des 16. Jahrhunderts* (1966), p. 242.

[88] Reichert, 'Die Kurfürstlich-sächsischen Markscheider Georg Öder die Jüngeren sen. und jun.' (2014), p. 179.

[89] Rüdiger, 'Adam Ries d. Ä. und dessen Söhne als Rechenmeister und Mathematiker' (2017), pp. 160–162, 167.

was increasingly perceived as a powerful general method, skilled practitioners were a scarce and valuable commodity. People knowledgeable about geometry and arithmetic often came from mining regions, where specific conditions had spurred an advanced culture of practical mathematics. Rulers of Central Europe happily employed such engineers to improve their lands, seizing the opportunity to assert their nascent territorial power.[90] In this framework, they crucially contributed – in Saxony as elsewhere – to the development of cartography.

### Charting Kingdoms

Mapping enterprises enjoyed a great development in the early modern Holy Roman Empire. The goals and tools of cartography could vastly differ, from the survey of a local monastery to the construction of terrestrial globes. Artists or local engineers were regularly asked to describe and draw city views, works that were subsequently gathered by editors and culminated in great geographical works such as Münster's *Cosmographia* (1544), or Braun and Hogenberg's *Civitates orbis terrarum* (1617). In this section, I focus on an intermediary level undertaken by many rulers: mapping one's possessions.[91] In a context when realms were rarely contiguous and mostly entwined, territorial maps were powerful tools to monitor one's borders and possessions. Frequent wars and the ensuing negotiations made it necessary to understand the relative positions of villages, roads, and natural resources. Rulers thus increasingly had their forests, mines, or saltworks surveyed. Portions of the Swiss Confederacy, Bohemia, and Bavaria had already been mapped in the early sixteenth century – the latter in collaboration with the famous humanist Petrus Apianus (1495–1552). Such attempts, surprising at it may be, rarely implied extensive surveying operations, which by their very nature were expensive. More often than not, map-making relied on existing knowledge – found for example in Ptolemaeus' *Geographia* – completed with some astronomical observations and brought up to date using travel memories as well as 'oral investigations' with local inhabitants.[92]

---

[90] Chandra Mukerji makes a detailed case about seventeenth-century France in Mukerji, *Impossible Engineering* (2009), pp. 15–35, ch. 2, 'Territorial Politics'.

[91] This is precisely what Augustin Hirschvogel (encountered earlier in this chapter while discussing the border dispute with Bohemia) did for the city of Vienna in 1547. For an overview of the beginning of modern cartography in Central Europe, see Stams, 'Die Anfänge der neuzeitlichen Kartographie in Mitteleuropa' (1990), pp. 37–105; Lindgren, *Astronomische und geodätische Instrumente zur Zeit Peter und Philipp Apians* (1989); Lindgren, 'Land Surveys, Instruments, and Practitioners in the Renaissance' (2007), pp. 477–508.

[92] Pitz, *Landeskulturtechnik, Markscheide- und Vermessungswesen* (1967), p. 78: 'auf Grund mündlicher Erkundigungen'. On the various surveying techniques, from the more scholarly to the more practical ones, see Lindgren, 'Land Surveys, Instruments, and Practitioners in the Renaissance' (2007); Bourguet, Licoppe and Sibum, *Instruments, Travel and Science* (2002), pp. 8–9.

When Petrus Apianus offered Georg the Bearded – at the time Duke of Albertine Saxony – to map his territories in 1532, his nephew and rival Johann Friedrich – Elector of Ernestine Saxony – flatly refused, fearing that such an instrument would 'reveal the landscape of both our cities and domains as a register' and open their realms to invasions.[93] A generation later, when Elector August rose to power in a then-united Saxony, the atmosphere had changed: one of his first actions was precisely to map portions of the country.[94] After settling the contested border with Bohemia, as has been described above, Georg Öder II was frequently asked to survey Electoral forests. This cartography was initially meant to enhance the administration's knowledge of its rights and possessions, not to fuel the general growth of knowledge. German princes rarely considered printing their patiently hand-drawn maps, and often explicitly forbid it. When Duke Julius of Brunswick decided in 1572 to have his domains surveyed, he contacted Gottfried Mascopius (d. ca. 1603), professor of mathematics at the Pädagogium of Gandersheim. Mascopius was instructed to produce 'a map or land table' of 'all mines, smelting huts, and saltworks' with the help of the mining administration, without disclosing its result to 'anybody, be it from low or high extraction'.[95]

During the three decades of August's reign, numerous projectors contacted the Dresden court, offering their services in producing maps of the country. Hiob Magdeburg, a native of the Ore Mountains and teacher in a prestigious Latin school, sent to the Elector a huge hand-drawn map in 1566, whose scale also featured illustrations depicting surveying processes.[96] Cords are being pulled in straight lines and marked using wooden stakes, while a surveyor aims at or assesses a direction using a compass. In most cases, however, these projectors had little concrete improvements to offer and essentially collated existing sources. Johann Criginger (1521–1571), another native of the Ore Mountains, was pastor of Marienberg when he contacted the Elector in 1567, asking him to finance the printing of a map of Saxony. Criginger had – by his own account – 'brought together' his map 'at home, without all the hiking and inspections' that came with extensive surveying operations.[97] Criginger

---

[93] SächsStA–D, 10024 Geheimer Rat (Geheimes Archiv), Nr. 09762/03, *Mathematica, Mechanica et Geographica 1532–1726*, f. 53r: 'unser beyderseits Landschaft an Steten und Ritterschaften als ein Manregister offenbart'.

[94] Within a generation, it had become clear that suspicion against cartography was not only unfounded, but also prejudicial. In 1553, Humelius would say: 'ich bedauere es, daß eine solche nicht schon früher von Apianus herausgegeben wurde', in Woitkowitz, 'Der Landvermesser, Kartograph, Astronom und Mechaniker Johannes Humelius' (2008), p. 81.

[95] NLA HA BaCl Hann, Historische Nachrichten über die Harzverhältnisse 84a, Nr. 7/5, Letter from 16 January 1572: 'Mappe oder Lantaffel ... Niemandt es sey hohes oder Nidrigs standts'; Nr. 7/10, 24 January: 'Bergwerke, Hütten, Salzwerke und anderes mehr.'

[96] Dresden, SLUB, Kartensammlung, A13534.

[97] Kirchhoff, 'Matthias Öder grosses Kartenwerk über Kursachsen aus der Zeit um 1600' (1890), p. 322: 'daheim ohn alles wandern und besichtigen zusammen bracht'. Other projectors who produced maps of Saxony were Matthaus Nefe and Scultetus Bartholomäus.

eventually obtained 50 Gulden for his efforts, a sum too modest to justify extensive cartographic operations and likely meant to reward the goodwill of a loyal subject.[98]

In order to carry out an accurate survey, merely reading about mathematical geography was not enough. Cartographers who only relied on existing works or book knowledge could hardly be trusted with real-life instruments. The scholarly tradition of *practica geometriæ*, and indeed most of the existing literature, was to compile problems about isolated measurements when surveying of fields, properties, and – more rarely – small pieces of land. These works offered little advice to deal with hills, cliffs, and all the practicalities and legal questions encountered in large-scale mapping campaigns.[99] On the other hand, August of Saxony had experienced the skills of mine surveyors, and knew of their ability to map whole districts. When he finally decided in 1586 that more precise maps of his possessions were needed, underground surveyors were natural candidates for the task. Georg Öder II and III had proved their skills on many occasions, most recently in surveying the hunting forests belonging to the Elector, before Georg III was exiled following the fiasco of the Elster Canal. When he was stripped of his title of court surveyor to the Elector in 1579, the distinction was passed on to his brother Matthias Öder, who was logically appointed for the task.

Matthias Öder was asked by the Elector 'to produce a map of our land, and where our hunting [domains] extend, and feature on it all of our forests, together with neighboring cities, villages, and rivers'. The Elector would defray all the costs of the surveying campaign and pay for 'four persons to constantly hold the cords'.[100] Unlike previous attempts, this survey was conceived as a collective enterprise, in which several chain-bearers worked under the supervision of an underground surveyor. Matthias Öder did not complete it before his death in 1614, and the significance of the expense receipts (at least 6000 Gulden) attest of the large scale of the operation.[101] Once again, the Elector decided to pair a seasoned practitioner with a scholar, Melchior Jöstel (1559–1611). Born in Dresden, Jöstel had studied at the University of Wittenberg and had been assistant of Abraham Ries in Annaberg. Once again, the diffuse influence of the mathematical culture

---

[98] See Wunderlich, *Kursächsische Feldmesskunst* (1977), pp. 20, 48–50. Another map was produced in 1568 by Bartholomäus Scultetus (1540–1614), who taught mathematics at the *Gymnasium Augustus* in Görtliz. Prepared 'at his desk with the help of messages' from collaborators, it was already pretty accurate and was adapted by Ortelius in his *Theatrum Orbis Terrarum* (1573).

[99] Lindgren, 'Land Surveys, Instruments, and Practitioners in the Renaissance' (2007), pp. 481–487, 500–507.

[100] Letter from the Elector Christian to Matthias Öder, 6 July 1586. Reproduced in Ruge, *Geschichte der sächsischen Kartographie im 16. Jahrhundert* (1881), pp. 231–232: 'ein mappa vnsers ganzen lanndesumkreiss, wiefern sich itzunter vnsere Jagten erstrecken zuuerfertigen, vnd darein alle vnsere Holtzer, sambt den vmliegenden Stedten, Dorffern vnd wässern zubringen … hierzu vier Personen, so stets auff die schnure warten'.

[101] See Bönisch, *Die erste kursächsische Landesaufnahme* (2002), p. 13.

born in the Ore Mountains is noticeable. From 1595 on, Matthias Öder was also assisted by his nephew Balthasar Zimmermann (ca. 1570–1634), who came from a family of mining clerks in Annaberg. After Öder's death, Zimmerman inherited his position of court surveyor and carried on the land survey.

The *modus operandi* of the survey illustrates how an early modern cartographic enterprise was actually conducted. Moreover, contemporary archival material and printed sources make it possible to reconstruct the specific methods that were used. Matthias Öder's material included standard mining instruments, that is measuring chains and compasses, together with more elaborate tools such as quadrants and theodolites to record inclinations. For instance, when a forest parcel close to the Bohemian border was surveyed in 1608, 'one began on the big Hem Mountain, where there is a flat space, in the middle of which a 12-sided pillar is located'. From this central point, 'the forest is divided like a spiderweb in 10 circles and 12 main blades', as presented in Figure 3.3.[102] Matthias Öder used this web as a round grid on which all villages, trees, mines, and noticeable places could be plotted. Measuring with a chain how far they were from the central point, or the closest circle, places were accurately represented on the map.

This surveying method is remarkable for two related reasons. First, using a spider web grid is original and exemplary of the kind of mathematics developed by early modern practitioners. It does not correspond to standard ways in which printed books of the time presented surveying operations. Neither Erasmus Reinhold – who had written about underground surveying – nor Augustin Hirschvogel – working for the Bohemian crown during the litigious border survey of 1551 – seem to have known about it, or found it useful to describe it in their published works.[103] Such methods shows little connection with the contemporaneous development of geometry. One can thus speak of 'underground mathematics' in a broader sense, meaning mundane methods widely used but ignored by contemporary scholars, and thus ignored in the classical historiography. Matthias Öder's mapping of Saxony does not rely on astronomy to ascertain the coordinates of major cities or significant points. Neither does he use triangulation, theoretical geodesy, or any kind of projection.[104] Second, it continues the tradition developed six decades before in order to map the Bohemian border by radiating from a main summit.[105] Looking at

---

[102] Lehmann, *Historischer Schauplatz* (1699), p. 128: 'Man machte den Anfang auf dem grossen Hemberg, alwo oben ein ebener Platz … auf demselben stehet mitten ein 12 eckige Säule … der gantze Wald als ein Spinnen-Rad in 10 Rundungen und 12 Haupt-Flügel abgetheilt / also daß iede Rundung 1200 Doppel-Schritt von der andern entfernet ist.'

[103] Even Martin Grosgebauer, who presented himself as *Forstmeister* (Master of the Forests), did not introduce it in his Grosgebauer, *Vom Feldmessen* (1596).

[104] This point is nicely made by Kirchhoff, 'Matthias Öder grosses Kartenwerk über Kursachsen aus der Zeit um 1600' (1890), p. 323; Bönisch, *Die erste kursächsische Landesaufnahme* (2002), p. 31.

[105] Georg II Öder had used a central point from which the *Flügel* were drawn and measured using a chain and a mining compass, as has been shown above in the determination of the border with Bohemia.

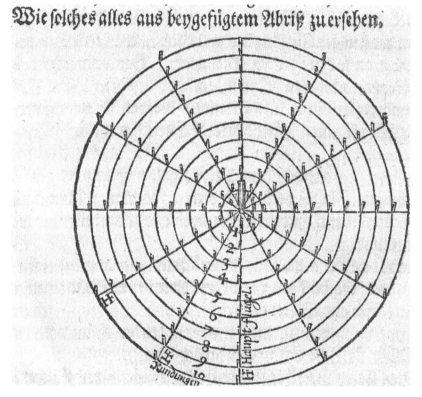

Figure 3.3a Design of the *Spinnen-Rad* (spiderweb) method used by the subterranean surveyor Matthias Öder in 1608 relying on the method employed
by his father. Principle illustrated by C. Lehmann, *Historischer Schauplatz*
(1699), p. 128–132, Universitäts- und Landesbibliothek Sachsen-Anhalt.
From a pillar located at the top of a hill, straight lines named *Flügel* (wings)
were traced, probably using a chain and a compass. The *Rundungen* (concentric circles) indicate that the surveying team set marks at regular intervals.
Courtesy of the Universitäts- und Landesbibliothek Sachsen-Anhalt Halle,
Sign. Pon Vk 106.

the successive surveys of Saxony from the point of view of practitioners, one
notices that knowledge and know-how circulated from fathers to sons within
the small circle of subterranean surveyors. These intrinsically local methods
were well adapted to the mountainous topography, and the forestry administration seem to have used comparable tools.[106] In the Ore Mountains, it was

[106] This is all the less surprising that mining and forestry officials were working closely in the Ore
Mountains. See Wilsdorf, Herrmann and Löffler, *Bergbau – Wald – Flösse* (1960).

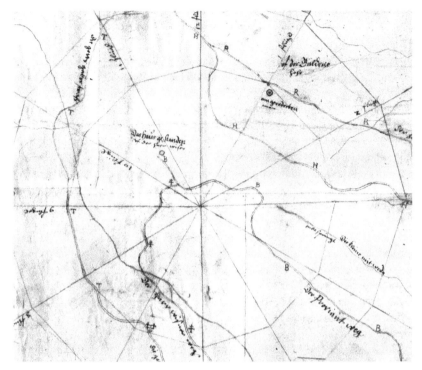

Figure 3.3b Corresponding actual survey made by Matthias Öder in 1608, using the 12-sided *Spinnen-Rad* (this place is today known as *Conrasdswiese*), Courtesy SächsStA–D, 12884 Karten und Risse, Schr R, F 003, Bl 2 + 5 (Makrofiche Nr. 00704), detail.

fitting to locate a higher summit and operate from this central point. Not only did Matthias Öder take over a method that his father Georg II had used half a century before, but evidence shows how customary these operations were. Describing the operation, a chronicler of the Ore Mountains complained that the 'wings and paths have been broken, overgrown and marshy' in recent years, recounting how the same roads were usually used both for hunting, for forestry, as well as for cartographic purposes.[107] Before the actual survey, workers usually performed a 'wood cleaning', trying to 'find again the old numbers' carved on prominent trunks by previous generations of surveyors and then 'so overgrown by the trees' as to be hardly noticeable.[108]

---

[107] Lehmann, *Historischer Schauplatz* (1699), p. 128: 'sind die Flügel und Wege sehr verbrochen / verwachsen und vermorschet'.
[108] Ibid., p. 128: 'da man etliche mahl die alte Zahl / so in den Bäumen sehr verwachsen war / wieder fande'.

This method, apparently well known to local specialists, was naturally seen as belonging to subterranean geometry, although neither the Öder family nor other practitioners had published about it at the time.[109] In spite of this absence of printed instructions, there is ample evidence that it was applied, on a large scale, by several generations of subterranean surveyors. The spiderweb method, moreover, was implemented in regions that were notoriously hard to survey, with rough and hilly terrains and hunting forests. While it was not described by contemporary scholars, it seems that at least Humelius used it from time to time, although his maps 'cannot compete with the accuracy and the richness of details of the Öder-maps'.[110]

The results of such mapping campaigns were indeed astonishing: modern analysis of the grid distortion shows that 'Öder's achievement is unique in Germany' for its precision, and probably 'in the whole world' at the time.[111] One seems here to encounter a general law of practical mathematics. The efficiency of a method rests less on its sophistication than on the skills and experience of the practitioner, as well as the time spent on it. In other words, sixteenth-century surveyors skilfully using a robust procedure outperformed newer scholarly methods, such as triangulation, in spite of the latter's genuine theoretical superiority. Moreover, following customary procedures was the surest way to ensure that results would be perceived as accurate, and thus be socially acceptable.

Matthias Öder had been raised in the culture of geometry characteristic of mining cities, yet he also assimilated the Dresden courtly culture. Having benefited from a hands-on training from an early age, the Öders were uniquely fitted for mapping mountainous and forested regions. Their superiority can be noticed in observing minute material details rarely brought up by historians. For instance, an underrated difficulty of large-scale surveying campaigns was to manage a large quantity of data and find one's way through a labyrinthine terrain. Given their experience in sinuous mining pits, underground surveyors were well aware of such problems and had already devised a solution. Trained to carve a wealth of symbols in stones, they knew how to deal with large amounts of information, a skill they would seamlessly transfer to land surveying.[112]

---

[109] Ibid., p. 128, recounts how a local pastor casually spoke about the 'surveying according to the *Marckscheider Kunst*', as if it was a well-known thing in the Ore Mountains.

[110] Wunderlich, *Kursächsische Feldmesskunst* (1977), pp. 51–53, elaborates on how this method was used to map Saxony.

[111] Bönisch, 'The Geometrical Accuracy of 16th and 17th Century Topographical Surveys' (1967), pp. 62, 69.

[112] On the signs used in underground surveying, see Adlung, *Markscheiderische Tafeln und Inschriften im sächsischen Erzbergbau* (1999). Similar signs were also used in forestry administrations, see Knothe, 'Die Dresdener Forstzeichenbücher' (1998), pp. 6–17.

## Courtly Instruments in a Practical Context

The depth of relationships between the mines and the court can be illustrated by a final example. In 1604, a new survey of the ever-contested border between Saxony and Bohemia was carried out.[113] This work was undertaken by Matthias Öder and Melchior Jöstel during their survey of Saxony, although it was a distinct political issue. In a letter directly written to Öder, Elector Christian II of Saxony spoke of existing 'errors' (*Irrungen*) and asked to 'survey comprehensively the contested border territories [*Kriegsstücke*] and bring them on a plan'. To this aim, the underground surveyor would have to work not only with Professor Jöstel, as he was accustomed to, but also to hire 'a notary and a painter' to draw the map.[114] A first series of measurements was made by the two mathematicians, each relying on his own preferred method, but Jöstel stayed on site and decided to repeat it on his own.[115] In a letter to Öder, he explained his decision: 'my good friend, I shall not conceal to you that I have decided to survey the contested border anew, with my method'. Jöstel, who was by then professor of mathematics at the University of Wittenberg, went on to explain that he used an 'astronomical measuring device'. This is a rare mention of a learned instrument being used in actual surveying works. Jöstel specifically mentioned it because his task would 'take its proper time', and therefore instructed the painter to wait before drawing the map.[116]

What was this instrument, exactly? Interestingly, the painter Hans Richter ultimately represented the above-mentioned tool, which combines a semicircle and a compass, on the resulting map (Figure 3.4). It closely matches a piece belonging to the Elector's August *Kunstkammer* and made by the instrument-maker Christoph Trechsler (1546–1624).[117] Son of a Dresden gunsmith, Trechsler had

---

[113] From the moment Bohemia was attached to the Hapsburg dominions in 1526 until the final settlement of 1635, the Saxon–Bohemian border was contested, and therefore surveyed and marked, every few years. See Wetzel, 'Der Erzgebirgische Kreis im Ausgestaltungsprozess des frühen albertinischen Territorialstaates' (2013), pp. 42–43.

[114] SächsStA–D, 10024 Geheimer Rat (Geheimes Archiv), Nr. Loc. 09762/05, f. 4v: 'die streitigen Grenzörter durch dieses werkes verstandige abgemessen unnd in einen Abriss zu gebracht werden solten … ein *Notarius* und ein Mahler'.

[115] The protocol drafted by the *notarius* is an exceptionally detailed source of information. A partial transcription is given in Reichert, 'Zur Geschichte der Feststellung und Kennzeichnung von Eigentums– und Herrschaftsgrenzen in Sachsen' (1999), p. 82.

[116] SächsStA–D, 10024 Geheimer Rat (Geheimes Archiv), Nr. Loc. 09762/05, f. 2r, letter from 25 June 1604: 'Matthes Oder, günstiger gutter freundt, euch sol Ich nicht verhalten, das Ich die stritige Landtgrenz wiederumb vom neuen auf meine arth bezogen, und ezliche nothwendige *circumstantias*, einer und ausser derselben, dazu genommen, und solches mit dem Astronomischen Massstab anfragen wil, welches alles seine gebührliche Zeith haben muß.'

[117] Mathematisch-Physikalischer Salon, Staatliche Kunstsammlungen Dresden inventory number C III f 3 (destroyed during World War II, picture by Ernst Zinner). The instrument was photographed and extensively described before its destruction during World War II. Another hint that the drawing depicts the actual instrument is that the 1589 instruments contains an engraving (under the main plates) with a quadrant that matches the drawing.

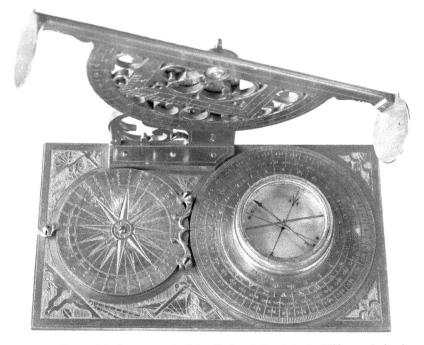

Figure 3.4a Instrument built by Christoph Trechsler in 1589 on a design by Abraham Ries and Melchior Jöster. Courtesy Mathematisch-Physikalischer Salon Dresden (Nr. C III f 3).

begun his career in the Dresden arsenal before turning to the production of surveying instruments. In that context, he collaborated for a long time both with Abraham Ries and with his student, Melchior Jöstel. Asked by the Elector, the pair designed a specific instrument built by Trechsler in 1589 (also represented in Figure 3.4). It works as a proto-theodolite recording both elevation angles using a semi-circle and directions using a compass. The history of this instrument illustrates how the know-how of scholars and surveyors blended at the Dresden court. For instance, the instrument included a compass featuring both the hours units commonly used in the mines and the regular degrees used in universities.[118] Interestingly, Trechsler's instrument was originally kept by Abraham Ries himself, who only sold it to the Elector in 1594, when he was getting older. One can assume that Jöstel, when he was asked ten years later to survey the Bohemian

[118] On this instrument, see Schillinger, 'The Development of Saxon Scientific Instrument-Making Skills from the Sixteenth Century to the Thirty Years War' (1990), 282, 289; Weißflog, 'Die Bergwaage des Abraham Ries' (2014).

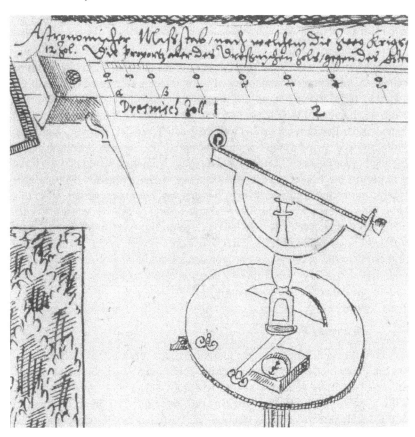

Figure 3.4b Drawing of the 'astronomical measuring device' used by Melchior Jöstel, in order to survey with Matthias Öder the border between Saxony and Bohemia. The semi-circle is used to measure inclinations, while the compass records directions. Courtesy SächsStA–D, 12884 Karten und Risse, Schr 001, F 013, Nr 001 Makrofiche Nr. 04986.

border, seized the opportunity to test his instrument in real conditions, and subsequently asked the painter to represent it on the map.[119]

The methods and instruments employed for the survey of the Bohemian border exemplifies once again how the Saxon Elector brought together a variety of

---

[119] Egon Weissflog has analysed the Trechsler instrument in detail, but does not mention the 1604 map by Jöstel. See Weißflog, 'Die Bergwaage des Abraham Ries' (2014). This analogy between the map and the instrument is hinted at in Schillinger, 'Zur Entwicklung der Vermessungsinstrumente im 16. Jahrhundert' (1988), p. 49, footnote 15.

experts to strengthen his own power.[120] The central collaboration between M. Jöstel and M. Öder, a university professor and a court surveyor, illustrates the larger trend described in this chapter. The numerous ramifications of this example, however, show the scale of these operations and the variety of involved technicians. A painter was summoned to produce an illustrated map, while a lawyer recorded the operation. The instrument that Jöstel used likely belonged to the art chamber of the Elector and was the product of another collaboration shaped at the Dresden court, this time between Trechsler and Abraham Ries. Furthermore, drawing a map was only one part of the complex operation of setting the border between Bohemia and Saxony. Numerous workers were hired to mark out the border once it had been set.[121] Border stones (*Reinsteinen*) were positioned at regular intervals, but this was not nearly sufficient: in this dense forest, stones would be barely visible after a couple of decades. Border ditches were dug, trunks were felled on large portions and trees planted anew in straight lines. Some prominent trees were carved with specific symbols, and all these material operations were represented on the map itself. In the early modern period, a large panel of skills were necessary to process, represent, and embody the data collected by surveyors.[122]

## Diverging Paths

For a couple of generations, strong relations existed between the Ore Mountains and the court of Dresden. Mining cities could have remained purveyors of bullion – and more generally natural resources – carefully controlled by the political power of the Saxon state. Instead, the culture of geometry spurred by the developments of ore extraction was co-opted by the ruling elite. The Öder and Ries families were outstanding members of a larger group of superior craftsmen. Having proven their usefulness, first in the mining administration of Annaberg and then for missions at the behest of the Elector, they received official positions and entered the Dresden court. Under the influence of August, mining officials came to be seen as versatile experts and interacted with scholars and instrument-makers. Surveying practices were essentially

---

[120] The exact method used by Öder for this map is not precisely known, but a technique akin to his 'spiderweb' method is likely, at least judging by his field book (see SächsStA–D, 10024 Geheimer Rat (Geheimes Archiv), Nr. Loc. 09762/05, unpaginated, '*Anno 1604. Das Buchlein von der außmeßung das Streitiges holzes.*'

[121] On the material aspect of surveying and border setting, see Reichert, 'Zur Geschichte der Feststellung und Kennzeichnung von Eigentums– und Herrschaftsgrenzen in Sachsen' (1999), pp. 3–30.

[122] See Ziegenbald, 'An Interdisciplinary Cooperation' (1993), pp. 313–324. On instrument-makers from the mining regions themselves, see Weißflog, 'Uhrmacher und Mechaniker im Umfeld von Adam und Abraham Ries' (2008), pp. 341–356.

based on their experience and transmitted from fathers to sons, orally or in manuscripts. While their methods had – more often than not – little in common with the published works and scholarly reflections of the time, this did not prevent genuine exchanges with university professors from Leipzig or Wittenberg.

Mathematical practitioners had one foot in the mines and one at the court. While this chapter has focused on the Electorate of Saxony, similar interplays happened with the rulers of Brunswick, Austria and Hesse-Cassel. Underground surveyors, who had learned to deal with levelling and technical issues, and above all knew how to adapt to an ever-changing environment, were uniquely set for the task. A combination of favourable conditions created a space in which in-depth exchanges between subterranean surveyors, scholars, and instrument-makers took place until the outbreak of the Thirty Years War. Collaborations were spurred by a handful of German prince-practitioners, who used court positions and university chairs to bring together and reward their most useful subjects.

Along with military captains, navigators, architects, and other superior craftsmen, underground surveyors contributed to the mathematization of nature in early modern Europe. In addition, their brief involvement in court life – where they often had close contact with rulers – amplified the cultural impact of mining and the discourses on the efficiency of engineering.[123] In her study of sixteenth-century 'trading zones', Pamela Long has highlighted the importance of these exchanges and the catalystic role played by court culture. In such settings, 'the skilled acquired some learning, and the unskilled learned acquired some skill. The two groups came closer together.'[124] This description applies to practical geometry in the mining states with two significant developments. First, such exchanges were less spontaneous than spurred, mediated, and paid for by rulers. Consequently, they took place not only in the mines, but more often in a courtly setting. Second, the processes were in no way irreversible, and collaborations indeed stopped in the early seventeenth century, when favourable conditions ceased to be met.

It is fruitful to see knowledge exchanges as essentially dynamic and ephemeral, remembering that such cooperation in early modern Europe was the exception rather than the norm. The overlap of interest was only partial, and yet this time span produced wonders. Court mathematicians combined their immersion in learned culture with the practical applications of mathematics. The accuracy of underground surveying had been proven by its daily successes in mining towns, most prominently in Annaberg. This culture was replicated and adapted for canal digging, border setting, and mercantile arithmetic. Practitioners' knowledge 'gained general currency', as Long puts it, due to its association

---

[123] Hilaire-Pérez, Blond and Virol, *Les ingénieurs, des intermédiaires?* (2022).
[124] Long, *Artisan / Practitioners and the Rise of the New Sciences, 1400–1600* (2011), p. 126.

with the power of the court.[125] Acute technical problems, such as the enduring border dispute between Bohemia and Saxony, truly brought together not only practitioners and scholars, but all kind of experts from neighbouring countries, forcing them to expose and confront their methods.

The interplay between mining regions and their respective courts gradually came apart at the beginning of the seventeenth century. Mining regions were hard hit by the influx of gold and silver from the New World's mines, just as the raging Thirty Years War disrupted the complex economy of ore extraction. The slowing of these once dynamic regions lessened the need for mathematics, gradually reducing the pool of available practitioners in mining cities. In Annaberg, the great fire of 1604 exacerbated the decline of extraction, and the city never recovered its past glory.[126]

More broadly, the court culture took its own path and became less immediately relatable to practical concerns, as the descendants of Elector August had less personal interest for mathematical pursuits. As time passed, the heirs of great practitioners successfully entered the learned world of the courts. As a result, they often left the mountains for the capitals, giving up hands-on work. As they were recognized as scholars, they also became authors, but did not write about the crafts that had made their forefathers indispensable. Abraham Ries, despite being the son of a practitioner and working for the mint, is a transitional figure in which the courtier sometimes outweighs the mining official. One of his students and a native of Annaberg, Lucas Brunn (d. 1628), went on to study in Leipzig and Altdorf. He was eventually appointed inspector of the Saxon Elector's art chamber.[127] In his German version of Euclid's *Elements*, published in 1625, Brunn boasted to 'extract all theorems from Euclid and bring them to the light of day'.[128] The unmistakable mining metaphor alluded to the extraction of ore, and more specifically to a central duty of subterranean surveyors. A reader might expect a useful work bridging the gap between the Greek corpus and the idiosyncratic methods of the Öder family. The book, however, was written purely in a patronage context and scarcely addressed concrete issues. It offered neither applications to mining nor actual surveying procedures. Brunn's rhetoric was still heavily tilted towards usefulness – entitling

---

[125] Ibid., p. 128: 'This proximity to centers of power meant that the empirical values promoted in these trading zones gained general currency.'

[126] On the decline of reckoning masters in the Ore Mountains during the seventeenth century, see Rochhaus, 'Zu den Rechenmeistern im sächsischen Erzgebirge während des 17. und 18. Jahrhunderts' (2002), pp. 343–352.

[127] Brunn describes his training in the introduction of *Euclidis elementa practica* (1625).

[128] Ibid., An den Leser: 'Theoremata / auß dem Euclide zu *vertiren* und an Tag zu bringen'. In the dedication, he explicitly linked the publication of this work to his appointment to the *Kunstkammer*. On this book, see also Folkerts, Knobloch and Reich, *Maß, Zahl und Gewicht* (1989), p. 69.

his chapters 'handwork of the first book of Euclid' – and yet not a single use for human affairs was presented.[129]

Perusing through the manuscripts of court mathematicians such as Lucas Brunn or Abraham Ries, one marvels at the depth of their interests. They wrote on music, harmonies, and astronomy; they were able to discuss both the Copernican hypothesis and puzzles related to number theory.[130] Unlike their forefathers, this generation had learned to read Latin and Greek and possessed an extensive knowledge of theoretical mathematics. Brunn and Ries did not despise technical issues and occasionally wrote about them. Still, they had come to perceive practical geometry in the much broader realm of the courtly culture of their time, and as such their relationship to mathematics greatly differed from those of mining practitioners. In Annaberg, Clausthal, and other mining cities, underground surveyors were still busy with a limited set of methods, tirelessly applying them in diverse settings. Courtly culture continued to flourish in different directions, while mine surveyors receded in the background. In the seventeenth century, subterranean geometry was barely mediated by printing and developed into a very dynamic but strictly handwritten tradition.

---

[129] Brunn, *Euclidis elementa practica* (1625), p. 1: 'Handarbeit deß ersten Buchs *Euclidis*'. The assertion was correct inasmuch as he dealt with the construction of specific figures with straightedge and compass, but it was far removed from any practical problems to be dealt with in civil life.

[130] Abraham Ries wrote mostly unpublished works on *rithmomachia* (SLUB Mscr. C.433), astronomy (SLUB Mscr. C.1) and many collected works and translations (SLUB Mscr. C.3, C5, C81). See also Rüdiger, 'Abraham Ries (1533–1604) und sein Werk' (2011), pp. 23–36.

# 4 Writing It Down

## Innovation, Secrecy, and Print

Practical conditions for mining, and hence for subterranean geometry, drastically changed in the first decades of the seventeenth century. The political context was volatile: the precarious Peace of Augsburg (1555) between the Catholic Holy Roman Emperor and the Lutheran princes suddenly shattered in 1618, as Europe spiralled into the Thirty Years War. Mining regions were highly contested in a bid to control flows of bullion.[1]

Freiberg, capital of the Saxon Ore Mountains, was besieged three times, its smelting huts burnt to the ground and neighbouring factories destroyed.[2] Contemporaries complained that 'because of war, pestilence, dearth, and turmoil, many glorious and profitable [mine] workings must remain idle'.[3] More disruptive than armed soldiers, Spanish gold and silver from the American colonies flooded the European market from the latter part of the sixteenth century. The ensuing monetary crisis and inflation promptly made many mines unprofitable.[4] In the meantime, the introduction of gunpowder blasting transformed the very way in which extraction was conducted. Virtually unknown before 1625, it had replaced the usual hammer and chisel mining, or the dangerous use of fire, by the middle of the seventeenth century. Since ore veins are softer than surrounding rocks, miners used to follow their sinuous paths. Using gunpowder, miners were now able to dig deeper and quicker. More importantly, tunnels could advance in straight lines through hard stone and take shortcuts instead of following the softer veins. Subterranean surveyors,

---

[1] On mining regions as strategic targets, see Fessner and Bartels, 'Von der Krise am Ende des 16. Jahrhunderts zum deutschen Bergbau im Zeitalter des Merkantilismus' (2012), pp. 453–455.

[2] The city was besieged in 1632 by imperial troops and – after the Elector changed sides – in 1639 by the Swedish army. The third, and ultimately unsuccessful, siege lasted for most of the winter 1642–1643. See Breithaupt, *Die Bergstadt Freiberg im Königreich Sachsen* (1847), pp. 15–24.

[3] Kirchmaier, *Hoffnung besserer Zeiten* (1698), p. 77: 'durch Krieg / Pestilenz / theure Zeit / und Auffruhr / viele herrliche und bauwürdige Wercke liegen bleiben müssen'.

[4] See De Vries, *Economy of Europe in an age of crisis, 1600–1750* (1976), pp. 113–120; Fessner and Bartels, 'Von der Krise am Ende des 16. Jahrhunderts zum deutschen Bergbau im Zeitalter des Merkantilismus' (2012). On mining in early modern Latin America, see Brown, *A History of Mining in Latin America* (2012), pp. 1–45.

adapting to these new conditions, acquired more flexibility to plan ever larger operations.[5]

Geometry played an essential role in meeting these new challenges, although the lack of sources still hinders our understanding of this turbulent period. While the rich literature produced during the mining and metallurgical boom has been abundantly studied, the corresponding seventeenth-century writings are generally perceived as 'lacking new ideas' or even 'apathetic'.[6] This chapter attempts a reappraisal, showing the lively development of subterranean geometry in spite of unfavourable circumstances, on the basis of a set of hitherto unstudied manuscripts. The textual production of surveyors provides a unique window on the evolution of the mathematical culture of early modern mines. Its development and ambivalent relationship to the printing press offer rich material to observe how the use of various written media affected an idiosyncratic set of practices.

In the aftermath of the Thirty Years War, subterranean surveyors came to write down their know-how and circulate it in a highly original scribal tradition, the *Geometria subterranea* or *New Subterranean Geometry*.[7] The handwritten tradition of mine surveying was allegedly created by an elusive mining official, Balthasar Rösler (1605–1673) and disseminated by his students. This corpus illustrates the ways in which practical mathematics gradually developed in the seventeenth century. Moreover, it provides us with a missing link between the frantic activity of sixteenth-century mines and the analytic approach that developed in the mining academies of the late eighteenth century. In this chapter, we will witness the introduction of specific instruments, a new approach of data collection, and crucially the resourceful appropriation of theoretical tools in a world of practices. More broadly, the mines show how practitioners were able to assimilate results and ideas from the booming mathematical sciences of the time.

Writing down practical knowledge opened a rich horizon for improvements, but manuscripts were only truly useful when used in a specific technical and cultural setting. Building on administrative documents, one can reconstruct precisely how these texts were used to train surveyors. Early modern administrations played an active role in the preservation and circulation of skills, establishing a well-structured companionship system that ensured an efficient circulation of knowledge. The paradoxical paucity of printed documents in that

---

[5] Bartels, *Vom frühneuzeitlichen Montangewerbe zur* Bergbauindustrie (1992), pp. 170–186. analyses in detail the introduction of gunpowder blasting (*Sprengen mit Schwarzpulver*).

[6] Koch, *Geschichte und Entwicklung des bergmännischen Schrifttums* (1963), p. 60: 'teilnahmlos ... arm an neuen Gedanken'. The comparatively richer mining literature of the sixteenth century has been studied, besides Koch, in Darmstaedter, *Berg-, Probir-und Kunstbüchlein* (1926); Long, *Openness, Secrecy, Authorship* (2001), pp. 175–209.

[7] A first description of this corpus can be found in Morel, 'Le microcosme de la géométrie souterraine' (2015), pp. 17–36.

period has to be accounted for, given the public curiosity of contemporaries and its importance for the European economy. I argue that mining manuscripts were designed to be used in the microcosm of the mining states, and could not function independently, as printed books might.

In 1686, Nicolaus Voigtel made some of this knowledge broadly available when he published his *Geometria subterranea*, whose frontispiece has been presented in the general introduction. The positive reception of the book indicates a lasting impact on the German intellectual spheres. Surprisingly, this publication did not end the manuscript tradition and had at best an ambiguous impact on actual practices. Encouraged by the commercial success of Voigtel, several scholars authored mathematical textbooks on mine surveying in the early eighteenth century. Such works could not, however, replace the original literature of surveyors, whose authority rested on skills and know-how more than literary reputation.

## An Elusive Tradition

During the Thirty Years War (1618–1648), sieges and destruction not only slowed extractive activities but also disrupted the transmission of mining knowledge. Important skills in smelting processes, mineralogy, and machine building had to be transmitted face-to-face, in the mines or the workshops. Bankruptcies and deaths meant that this know-how risked disappearing as the hostilities dragged on for decades.[8] Little is known about the cultural history of subterranean geometry during these troubled times. Mining archives mostly focus on technical, administrative, and economic issues, and it is therefore unsurprising that they only offer a sketchy account of the renewal of the subterranean geometry tradition. In this regard, practical mathematics generally stands in stark contrast with scholarly mathematics, a domain for which discoveries and priority disputes are extensively documented. Such circumstances complicate the task of modern historians, for this is precisely when a new corpus of surveying practices developed around the figure of a Saxon surveyor, Balthasar Rösler (1605–1673), and an instrument, the *Hängekompass* (suspended compass).

Rösler was born in 1605 in a mining village of the southern Ore Mountains, then belonging to Catholic Bohemia.[9] He was raised in a Protestant family of miners, which fled religious persecutions and emigrated to nearby Saxony at

---

[8] One could argue that it is precisely this risk of loss that led surveyors to put their knowledge in writing. While interesting, I have found little evidence to support this hypothesis so far. More generally, one should be cautious in not assigning the logical or natural impulses of modern scholarship to early modern practitioners.

[9] The scarce information available about the personal life of Rösler is summed up in the excellent Meixner, Schellhas and Schmidt, *Balthasar Rösler* (1980).

the outset of the Thirty Years War. Nothing is known about his youth, although he does not seem to have been university trained and likely had little formal education. Sons of mining officials usually received on-site training in the mines, either with their family or with local officials. One way or another, Rösler learned to use surveying instruments and to solve the typical technical problems. He subsequently found a place as subterranean surveyor in Marienberg in the early 1630s, but complained that 'the high threat of war has caused decline in all neighbouring mines'. Surveyors being at the time mostly paid by the job, Rösler complained that he 'could only earn very little performing subterranean geometry.'[10] It turned out that things could still get worse, for the city of Marienberg was soon besieged and plundered, leading to an outbreak of the plague in 1633. When his patron died the following year, Rösler decided to try his luck somewhere else and came back to Bohemia, where he worked as mine foreman in the free mining city of Graslitz (now Kraslice in the Czech Republic). In the aftermath of the war, he was among the Bohemian Protestants who once again fled to Saxony. Now considered as a 'good miner and skilled surveyor', he was recruited in Freiberg in 1649. Later in his career, he was appointed mining master in Altenberg, another major mining city, where he died in 1673. Chronicles and archives unanimously attest that he trained a large number of surveyors from various mining regions: Saxony, the Harz, and obviously Bohemia, where he had worked for seventeen years.[11]

An important disciple of Balthasar Rösler was Adam Schneider (1634–1707). Schneider was born in the Harz, a mining region then belonging to the House of Brunswick. Nothing is known about his life until 1669. At that point, he had recently moved to Altenberg and began to work on a *New Book on Subterranean Geometry*, whose title page is presented in Figure 4.1.[12] The timing suggests that Schneider moved there precisely to learn subterranean geometry from Balthasar Rösler, and that his book was indeed a copy of his master's. Rösler's original manuscript, which has since been lost, was the first handwritten document dedicated to the actual practices happening in the mines, and described the contemporary innovations. The *New Book on Subterranean Geometry* written by Schneider is likely the oldest dated version still extant. Its author subsequently became subterranean surveyor in the neighbouring city

---

[10] Letter to the *Berghauptmann*, 10 October 1632, reproduced in Meixner et al., *Balthasar Rösler* (1980), p. 11: 'wegen der großen Kriegsgefahr die Bergkwerge allerorthen in großes abnehmen gerathen, das ich auch das wenigste mit Marckscheidten verdienen können'.

[11] SächsStA–F, 40001, Oberbergamt Freiberg, Nr. 3477, f. 30r-v and Meixner et al., *Balthasar Rösler*, p. 22. Rösler replaced Elias Morgenstern, who had recently died. At least three subterranean surveyors from the Harz were then trained in Altenberg by Rösler in the early 1660s. See Henning, *Acta Historico-Chronologico-Mechanica* (1763), p. 5; and Freiesleben, *Bergmännische Bemerkungen* (1795), p. 366.

[12] On Adam Schneider and his son Johann Adam, see Riedel, 'Adam und Johann Adam Schneider' (2006), pp. 201–209.

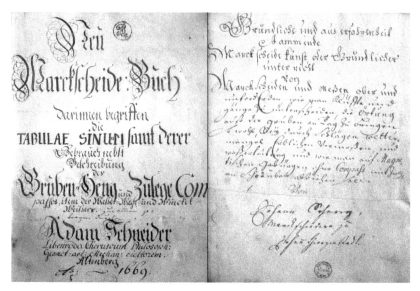

Figure 4.1    Title pages of two subterranean geometry manuscripts written in the Saxon Ore Mountains in the second part of the seventeenth century. TU Bergakademie Freiberg/Universitätsbibliothek XVII 18 and 15.
**Left**: *New Book on Subterranean Geometry* by Adam Schneider in Altenberg (ca. 1669).
**Right**: *Systematic and Experience-based Subterranean Geometry* by Johann Scherez in Johanngeorgenstadt (ca. 1693).

of Marienberg. His son Johann Adam Schneider (1674–1723), after being his assistant, succeeded him and introduced in turn several students to the art of setting limits.

The biographies of Rösler and Schneider are fairly well known, at least compared to mining officials of similar status, and yet they appear as elusive figures. The exact contribution of Rösler to the development of subterranean geometry is still unclear. Dozens of mining maps by his hand show both his immense workload and the improvement of his skills over several decades. These documents, however, hardly tell us why and how exactly his new approach was introduced.[13] While Balthasar Rösler was the first subterranean surveyor to extensively write down an organized corpus of methods, he had been trained in wartime, in a tradition of which we have little direct traces. Neither Rösler nor Schneider published anything in print, and such practitioners' manuscripts do not possess introductions or presentations about their context. This was a

---

[13] Jobst and Schellhas, *Abraham von Schönberg, Leben und Werk* (1994), D 198, p. 129, indicates that his main manuscript was written 'around 1652', a plausible but unsubstantiated intuition.

standard behaviour for seventeenth-century mining officials, which neverthe-less hinders our comprehension of the early development of this new tech-nical surveying. The reception of new mathematical methods by the mining offices, and their integration within existing legal systems, is equally poorly documented.

This approach quickly diffused in the aftermath of the Thirty Years War and had become widespread by 1700. The teaching of Balthasar Rösler spread among Central Europe through an elaborated system of manuscripts, even-tually reaching the mining regions of Scandinavia and Mexico in the early eighteenth century. The individual texts were hitherto scattered over sev-eral archives and known only to local archivists. These documents are often undated and unsigned, although most copies mention the name of the surveyor who authored the first drafts. Some are fancy exemplars with elaborate title pages and a handwritten table of contents, whereas others are working copies full of corrections and addenda. This might explain why the coherence and sig-nificance of this corpus had not been recognized. All were written by the sur-veyors themselves and their circulation reflects the travels of mining officials disseminating their useful knowledge. Most of the copyists, Balthasar Rösler included, possessed little formal training in mathematics. The *Geometria sub-terranea* manuscripts, as will be detailed below, were transmitted from master to apprentices. The few dated copies suggest that they were written during a training phase, a subtitle indicating, for instance, that the scribe had been 'taught from the attached knowledge'.[14] The following analysis focuses mainly on Adam Schneider's manuscript, since it is authored, dated, and has a direct connection to Balthasar Rösler (Figure 4.1). All the manuscripts share a simi-lar structure, although many offer variations and original sections. The first pages are dedicated to the 'fundamentals', presenting notions of arithmetic and geometry. The following sections deal with the suspended compass and the other surveying instruments. The bulk of the text consists of a long list of propositions, presenting dozens of mining tasks that are then detailed and solved, often with real data and maps as examples.

### Actions, Words, and Numbers

The renewal of subterranean geometry is closely associated with the use of the suspended compass (*Hängekompass*). This instrument has its base plate, on which the magnetic needle is set, suspended in metallic circles, also known as gimbal or cardanic suspension, as depicted on Voigtel's frontispiece in the introduction to the present book (see Figure I.1). While the mining lit-erature often credits Rösler with the invention of this new compass, it would

[14] Gotha Landesbibliothek, Chart A 972, title page: 'aus diese mit beygefügte Wißen dociert'.

be more accurate here to talk about an innovation because there is historical evidence of earlier use of suspension. In the early modern period, navigators' compasses were sometimes suspended to allow for precise measurements on a moving ship.[15] In the sixteenth century, the Elector of Saxony had already received a suspended sea compass, which had been used by courtly surveyors to map hunting forests. The object was stored in his chamber of the arts, and its influence is certainly conceivable, although no concrete evidence corroborates this.[16]

In any case, adapting suspensions to the existing mining compass allowed a surveyor to significantly improve his observations.[17] Adam Schneider instructed that it was necessary to have a measuring chain, a semicircle, and a suspended compass. A survey began by attaching the cord and 'plumbing from the middle of the winch' set atop a shaft.[18] Each time the chain, pulled in straight line, reached a wall, a segment was physically defined. In the end, the surveyor had produced a broken line going from the entry of the mine to its very end, and this object was called a *Zug* (a pull). If recorded accurately, these observations allowed in theory to precisely represent any mine working, either on a map or on the surface. The main difficulty was that chains were neither pulled horizontally or vertically, but obliquely, making it difficult to assess the exact direction and inclination of the gallery using a regular compass. Previous surveyors seemed to rely on rules of thumbs and approximations to deduce the vertical and horizontal components of the trajectory. Balthasar Rösler was apparently the first to systematically use a suspended compass around the middle of the seventeenth century. The new instrument made it easier to 'break down' the course of the inclined chain: the suspended compass would record the horizontal angle, and a semicircle (*Gradbogen*) indicated the inclination (see Figure 4.2).

'The true foundation of this art', continued Schneider in his manuscript 'rests on a *triangulo rectangulo* which is named *master of mathematics*.' The special status given to right-angled triangles was a standard consideration, which can already be found in medieval manuscripts on practical mathematics. At this point, the text turned immediately to the mining context: 'this triangle is always given to me by the semicircle suspended [to the chain] …

---

[15]  Taylor, 'The South-pointing Needle' (1951), pp. 1–7.

[16]  The subterranean surveyor who mapped the Elector's Forest was Georg Öder III, working in the 1570s. These maps were 'directed with the sea compass' ('nach dem Seekompaß gerichtet'), and incidentally used the spiderweb approach described in the preceding chapter.

[17]  The simple idea of attaching an instrument to the surveying cord is older. Already in 1574, C. Spangenberg described a water level and a goniometer 'that has two ears, so that man can suspend it in the middle of a cord freshly stretched between two rods'. The difference is that previous instruments were used to deal with vertical angles, while the suspended compass measures horizontal angles. Spangenberg, *Die XIX. Predigt Von Doctore Martino Luthero* (1574), p. 57.

[18]  TU BAF – UB XVII 18, f. 36r: 'von der Mitte des Rundbaums hinein seigern'.

Figure 4.2   The right triangle, also known as 'master of mathematics', drawn by Adam Schneider. The hypotenuse AC represents the surveying chain stretched in a narrow gallery. The suspended semicircle gives the angle of inclination. Together with the length of the chain, it allowed the miner to compute the two other sides of the triangle, inaccessible to direct observation. TU Bergakademie Freiberg/Universitätsbibliothek, XVII 18, f. 4r.

whose degrees are cut off by the thread', as presented in Figure 4.2.[19] This triangle epitomizes the evolution of subterranean geometry. Each measured segment of the *Zug* could abstractly be considered as the hypotenuse of a triangle. Its sides were labelled with the Latin names used in the academic world: *hypotenusa*, *basis*, and *cathetus*. Each side was then associated with its corresponding names in the dialect of miners. Since the chain was always pulled along the *hypotenusa*, this side was routinely called the 'inclined' (*Fläche* or *Donlege*). The base of the triangle represents the 'sole' (*Sohle*) of mining galleries, while its *cathetus* corresponds to the change in 'perpendicular depths' (*Seigerteuffe*) in the shaft. Presenting Latin terms along a vocabulary known to miners ensured communicability between the daily practice of the mines and the then burgeoning trigonometry, the science of solving triangles.

The writing style, shared by all the manuscripts, is remarkable. Most of the text is made up of geometrical instructions which are lively, informal, and heavily dialectal. The succession of measuring operations reads precisely like a recipe:

Suspend your *compass*, a[nd] align *sept*[entrion] in the direction you want to survey; and look once again below, in your writing table, if you have correctly inscribed the

---

[19]  TU BAF – UB XVII 18, f. 3v-4r: 'Der wahre Grund dieser Kunst beruhet in einem *Triangulo rectangulo*, welcher *Magister Matheseos* genannet wird, solchen giebt mir allezeit die angehangte Waßerwaage, oder *Quadrante* mit seinen *Perpendiculo*, von welchen derselben *Gradus* … mit dem Fädlein abgeschnitten werden.'

*graduation* of the semi-circle; In the meantime the *magnet* needle will have taken a rest, and show you the hour and the location of [the direction] where you have surveyed; But in order to be more certain, suspend it once again at another place [on the cord], enliven the needle and check if it has met the previous hour or its parts; If this is correct, then writes the cardinal point in the right *column*, and the hour with its parts also in its place. However, the most important thing is that you should not forget, just as you have noted the [graduation of the] semicircle, the [number of] *lachter* with their parts.[20]

The reader receives advice and orders such as *mußtu nicht vergeßen* ('you should not forget') or *ermuntre das Zünglein* ('enliven the needle'). The structure reflects a succession of actions, while the rationale of the whole operation is never articulated; needless to say, no proofs were offered.

Such a text was clearly not meant to be used alone, but had to be elaborated on by a teacher who would both make the meaning of the operations clear and show how the instruments had to be concretely placed on the chain. Alternatively, these instructions could serve as *vade mecum* for the surveyor himself, while the tradition of written instruction, considered as a whole, 'standardizes and widens the repertoire of the practitioner'.[21] The *geometria subterranea* tradition focuses on the concreteness of measuring practices. For instance, taking into account that a magnetic needle needs a moment to settle, it suggests to use this time to check if the previous results have been correctly entered in the survey book. This kind of particulars, which tell us about the exact *modus operandi* of practitioners, are all the more fascinating because they rarely find their way into scholarly discourse. Instead of the rational reconstruction offered by scholarly treatises, even when seemingly dealing with concrete practices, these texts collect step-by-step procedures based on individual experiences.

Manuscripts further instructed how to use and process the data. Subterranean surveyors came to use a standardized table as repository, the *Grubenzug*, as can be seen in Figure 4.3. In a metonymic transfer indicative of the down-to-earth approach of surveyors, the action of pulling a cord (*Zug*) gave its name to the entire broken line representing the survey (*Zug*), and by extension to the data table in which it was stored (*Grubenzug*). These tables thus embodied the very act of surveying, but were nevertheless complex information systems, undoubtedly challenging for people without formal training. Each line

---

[20] TU BAF – UB XVII 18, f. 36v – 37r: 'Nachdeme hange deinem Compass an, u. kehre Sept: aldahin, wohin du ziehen wilst; und siehe, unterdeßen noch einmahl in deine Schreib-Tafel, ob du beÿ der Waage die Grad recht eingeschrieben, mittlerweile wird das Magnet Zünglein sich zur Ruhe begeben, und dir die Stunde und deßen Orths, wohin du gezogen hast, weisen; Damit du aber desto sicherer gehast, so hange ihn nocht einmahl an einen andern Orthe an, ermuntre das Zünglein, und nun wahr, ob es auch die vorige Stunde oder deren Theil eines wie vor, genau getroffen; Triffs zu, so schreibe den Orth der Welt in seine gehörige Columnan, und die Stunde mit ihren Theilen auch an seinen Orth. Vor allen Dingen aber mußtu die Lachter mit ihren Theilen, zugleich, wenn du die Waage eingeschrieben, nicht vergeßen.'

[21] Eamon, *Science and the Secrets of Nature* (1994), p. 132.

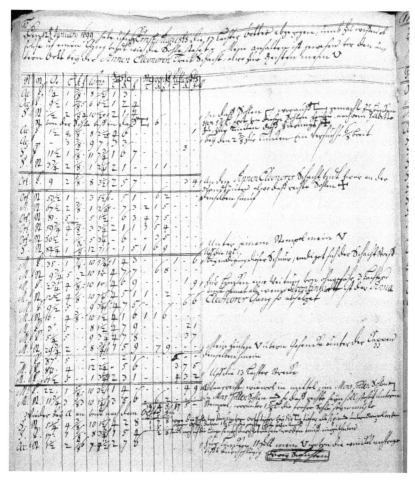

Figure 4.3    Mine survey (*Grubenzug*) recorded on 12 January 1699, by surveyor Jobst Henning Tolle in Clausthal. Each line presents the data corresponding to one pull of the surveying chain. Commentaries on the right explain how
to locate the symbols regularly carved in stone by the surveyor. A line that
reads, for instance, 'S | f | 11 | ½ | 1 | 7/8 | 11 | 7 ¾' corresponds to a chain length
of 1 *Lachter* and 7 *eighths*. The inclination given by the semicircle is 11.5° and
the direction given by the suspended compass indicates that the gallery was
heading north (*Septentrion*), in the eleventh hour, with 7¾ *eighths*. Courtesy
Bergarchiv Clausthal. NLA HA BaCl Nds 6692/1. Photo: Thomas Morel.

recorded one set of measurements, that is, the move from one point in the mine
to the next. The full table therefore corresponded to a given path underground.
With the accompanying remarks, it was sufficient to fully describe, represent,
and eventually reproduce a mine working.

The tabular depiction of data was a stark departure from the methods used in the previous centuries.[22] Instead of directly reproducing a survey with cords – on the surface or in another gallery – data was stored in a table and could be used whenever a surveyor needed. Trigonometry became indispensable to this practice of subterranean geometry in order to process these measurements and compute the *sole* and the *perpendicular depth*. Trigonometric tables, however, are not something an average surveyor could easily understand, let alone deduce from his experience in the mines. We are here confronted with an instance in which practitioners unavoidably had to reach out to scholarly mathematics. Adam Schneider's manuscripts introduced '*tabulae sinuum rectorum et versorum*' as 'an extract from Simon Stevin tables', dutifully specifying that they 'have been introduced by Mr. Balthasar Rößler, mining master and *Markscheider* in Altenberg, in 1664'.[23] One can infer that Rösler, around the middle of the seventeenth century, formally introduced trigonometry in the mines by abridging and adapting to the use of his peers tables that had been published by the famous Dutch engineer.[24] His tables were no exact copies of Stevin's, for he had painstakingly adapted their content to fit the non-decimal length units used in the mines, allowing for faster computations and purveying instructions on how to compute the *Grubenzug* from raw measurements.[25]

To sum up, this forgotten manuscript tradition allows us to reconstruct not only how subterranean surveyors worked, but also the theoretical knowledge they possessed. Using a standardized set of instruments, they recorded and processed data in a specific tabular form.[26] Since most observations are dated to the day, it is even possible to know how long a given task required. In the mid-seventeenth century, surveyors developed generic methods to tackle a wide range of professional duties and technical problems, either by performing calculations on the raw data, or by drawing mining maps on which graphical operations were performed. Some of these tasks were not new and had indeed existed for centuries; the real innovation was the introduction of a unified method of collecting and using data. If investors asked, for example, for the depth reached by a shaft, a surveyor would simply reach to his survey book, perform a few calculations and give a precise result. Being asked the

---

[22] For a general introduction to the historical use of tables and diagrams, see Flood, *The History of Mathematical Tables (2003)*, introduction; Bigg, 'Diagrams' (2016), pp. 557–571; Nasifoglu, 'Reading by Drawing' (2020).

[23] TU BAF – UB XVII 18, f. 8r: '*tabulae sinuum rectorum et versorum … wie solche von Herrn Balthasar Rößlern* Berg-Meister und Marck-Scheider zum Altenberg geführt worden sind anno: 1664'; f. 20r 'Extractum Tabularum Sinuum Simonis Stevini'.

[24] A plausible source would be the German translation *Simonis Stevini Kurtzer doch gründlicher Bericht* (1628), of which an example is still stored in the library of the TU Freiberg.

[25] There seem to have been multiple pathways and later updates, for some mine surveyors used tables computed by other Dutch mathematicians such as Ludolph van Ceulen (1540–1610) and Adriaan Vlacq (1600–1667). See TU BAF – UB XVII 14, f. 105v–109r and Beyer, *Gründlicher Unterricht von Berg-Bau* (1749), p. 136.

[26] NLA HA BaCl Nds 6692/1, *Markscheiderobservationsbücher*, 12 January 1699.

same question several months later, for instance to check the progress of digging operations, the surveyor did not have to survey everything once again. He would simply go back to the mine and locate the last sign he had previously carved in stone. Once cross-referenced to a particular line of his survey book, it was possible to resume operations from there. In short, a computational approach was replacing a mostly instrumental one, providing mining districts with a much-needed aide-mémoire in the form of data tables.

## Learning the *Geometria subterranea*

Subterranean geometry was not an art taught in schools, which – even in the thriving mining cities – were mostly dedicated to Latin and religion. Neither was it lectured on in universities, where only mining law and mineralogy were occasionally presented. At a time when most crafts and skills were learned 'on the shop floor', mine surveying was taught directly in the pits and galleries.[27] It was thus impossible to introduce the relevant mathematical knowledge *in abstracto*: apprentices had to be presented with actual problems and shown how measurements interacted with existing customs, local topography, and mineralogy. The transmission of geometrical knowledge among practitioners had always been a crucial issue. Early modern rulers knew well that know-how was a coveted resource, and that mathematical skills were perishable. In the sixteenth century, it was a common trope to deplore – usually in the prefaces of geometry textbooks – that 'many a fine and useful art … has in our time fallen into oblivion due to envy and carelessness'.[28]

Duke Julius of Brunswick, for instance, devoted a lot of attention to the mines of the Rammelsberg and Zellerfeld, which provided a sizeable portion of his revenues. He was thus receptive when he received a letter from his surveyor Wolf Seidel, on 1 June 1577. Seidel was getting older and sought permission to teach his nephew Zacharias the 'noble art' (*edle Kunst*) of surveying.[29] The duke answered four days later, encouraging Wolf to do so 'for the sake of the mines' so that 'the noble art of subterranean geometry does not disappear with him'. Zacharias was subsequently granted access to all mines and began to follow his uncle in drainage galleries 'once a week', in order to 'acquire this art'.[30] This decision triggered, a couple of months later, an angry letter sent by foreman Michael Pullemann. He had worked as mine supervisor

---

[27] See the introductory essay in De Munck, Kaplan and Soly, *Learning on the Shop Floor* (2007).
[28] Schmid, *Das erst buch der Geometria* (1539): 'zu unsrn zeiten / vil un mancherley schöner nutzlicher künst … aus neyd und farlessigkeit … in vergessen un undertruckt / gelegen seind'.
[29] NLA HA BaCl Hann. 84a Nr. 12, f. 78r: 'die edle kunst meynen vettern Zacharias Schneider zu lernen'.
[30] NLA HA BaCl Hann. 84a Nr. 12, f. 82r: 'gemeinen Bergkwercke zu gute … die edle kunst der Marscheider damit dieselb nicht mit ihm zugleich unterginge zu unterweisen … alle wochen einmahl … er diese kunst alda an die hand nehmen'.

for 'eighteen years', and had regularly 'helped the local officials and Wolf Seidel'. Pullemann has 'not only helped subterranean surveyors to sink [shafts] on deep tunnels, but also to diligently find [the position of] places in mountains, in all places and tunnels, when necessary'. Despite his lack of formal training, he had learned by assisting the official surveyor to 'inspect, level, and survey' and had hoped to secure Seidel's succession.[31]

In the sixteenth century, mining offices were already actively intervening in the training process. Every new surveyor had to take an oath, and rulers rewarded loyal subjects with positions of official *Markscheider*, which came with a monopoly on surveying activities in a given district. Restricting the circulation of knowledge was a risky strategy, as underlined by a father trying to get his son an apprenticeship: 'I cannot think of more than three or four persons in the region … knowledgeable and competent in underground surveying; should one or two of these people decease, there would be a shortage of experienced and ingenious surveyors.'[32] Administrations tried to avoid a deficit of experts, while at the same time exerting a tight control on the diffusion of knowledge. It was equally important to limit the number of skilled people: once they were knowledgeable in the art of setting limits, able foremen such as Pullemann inevitably asked for wage increases, which he got, and might leave the country if offered a better position, which he did not.

The circulation of knowledge and training of subterranean surveyors gradually reached a stable form relying on a *Lehr-Contract* (teaching contract). By the early seventeenth century, we see instances of agreements formalizing both the content and the form of the apprenticeship.[33] On 1 May 1617, Captain-general Caspar Rudolph von Schönberg signed a tripartite contract with David Köhler, a surveyor who was by then 'getting pretty old', and the father of a then-underage Johann Nicolaus Kunert.[34] The Saxon administration agreed to pay the surveyor the large sum of 100 Gulden, a sum roughly equal to two years of a common miner's salary, if he correctly trained his successor. The young Kunert, for his part, had to commit 'not to work for foreign rulers', 'stay

---

[31] NLA HA BaCl Hann. 84a Nr. 12, f. 102v: '18. Jharlang Inn dem Rammelsberge auß und ein gefahren'; 103r: 'Gesenke auf tieffen stoln, nicht allein helffen Markscheiden, sondern auch ingleichen die orttung im berge, ahn allen orten und enden, der es nottig gewesen, bestes vleisses werden helffen … zubesichtigen, abzuwegende unnd abzumessende.'

[32] SächsStA–F, 40001 Oberbergamt Freiberg, Nr. 2310, f. 12r: 'Nuhn weiß ich mich nicht zuerinnern, das solcher Personen, die vnnd die nechste Markscheider Kunst grundlichen Bericht vnnd wissenschafft hetten, inn … Landen … außer drey ader viere, vnnd so einer ader zweien mit todte abgehen solten, wurde es anerfahrenen vnnd kunstlichen Markscheidern inn deroselben landen … mangeln.'

[33] On teaching contracts in early modern guilds, see Reith, 'Know-How, Technologietransfer und die *Arcana Artis* im Mitteleuropa der Frühen Neuzeit' (2005), pp. 349–377. Conditions for subterranean geometry seem to resemble the ones in dyeing, weaving, and cloth shearing.

[34] SächsStA–F, 40001 Oberbergamt Freiberg, Nr. 3477, f. 6r: 'sich der nun auch ziemlich alt machet'; f. 7r: 'weil ich noch nicht mündige Jahre erreichet'.

in the homeland', and enter the service of the Saxon Elector.[35] The training was contracted to last for four years, during which the apprentice was not allowed to practise alone or without his master's consent. The money was paid from the tithes paid on all revenues by mining investors.[36] In this frame, surveyors could teach their know-how without fear of unjust competition, while rulers ensured the preservation of knowledge and the coherence of practices in their mining offices. Using a tax on mining to finance the training of future generations was a marked progress from previous circumstances, in which the widow of a late surveyor was sometimes pushed to sell instruments, maps, or other documents to survive. The duration of the training, which could last from eighteen months up to four years, was in line with the trade education of the time and similar to the apprenticeship of artisans or reckoning masters.[37]

Teaching contracts grew more precise as the new manuscript tradition developed in the second part of the century. In 1682, a mining juror who had learned the discipline with his father tried to convince the administration to hire him. He began by highlighting his 'experience in mines', but claimed above all to 'be able to handle this art *ex professo*'. It was clearly not enough anymore to have learned some kind of surveying: one had to know the accepted way of doing it. Only in this way could a surveyor produce measurement that would be legally accurate, and later teach one's successor the proper *Geometria subterranea*.[38] The mining administrations came to systematically ask aspirants two questions: Did the candidate know the '*fundamenta*' – the elementary knowledge presented in the first sections of the manuscripts? Moreover, did he master the 'propositions', containing the know-how and practical methods? In the seventeenth century, trainees also began to receive certificates at the end of their apprenticeship, in which the master confirmed that their apprentices were 'well able to indicate and discuss the principal and most important *propositiones* of the aforesaid subterranean geometry' (see Figure 4.4).[39] These documents might go as far as to detail what had been taught: 'survey a mining structure and draw it on paper, indicate the clefts and veins, bring the principal points to the day'. Both these tasks and the ensuing procedures correspond precisely with the 'propositions' elucidated in the manuscripts of Balthasar Rösler, hinting at a global training system based on written material.

---

[35] SächsStA–F, 40001 Oberbergamt Freiberg, Nr. 3477, f. 6v: 'in keines fremden Dienste or Bestellung … in Lande bleiben … keiner Arbeit mich hierrinnen unterstehen, viel weniger dieselbige anzunehmen, und vollbringen'.

[36] Incidentally, the chemical training of ore essayers was financed in a similar fashion.

[37] Schneider, 'Ausbildung und fachliche Kontrolle der deutschen Rechenmeister' (2002), pp. 6–7.

[38] SächsStA–F, 40001, Oberbergamt Freiberg, Nr. 2310, f. 29r: 'unterschiedlichen Zechen und Bergkgebäude inn und aus[er]halb Landes privatim gebraucht … Solcher kunst *ex professo* bedienen möchte'.

[39] UAF – OBA 5, f. 122r: 'Die in gedachter Marckscheider Kunst vornehmsten und Wichtigsten *Propositiones* zu erörthern auf- um anzugeben vorgelegt.'

Figure 4.4    Subterranean geometry certificate of Gottfried Klemm (1685–1745), obtained in the Ore Mountains of Saxony and signed by surveyor August Beyer in 1707. Beyer attests that his student knows the 'information', describes all the 'propositions' that he can solve, and specifies that he can do it 'by the arithmetic and by the geometry', that is both by computation or operating with mining maps. With a document attesting his skills, Klemm quickly found a position in Kongsberg, Norway, where this document is now stored. Courtesy Norwegian Mining Museum.[40]

At this point, the apprenticeship fees seem to have already included the right for the trainee to realize a copy of his master's *Handschrift* (manuscript). This training tradition, which emerged out of specific technical constraints, was thus embedded in a social and administrative framework. Teaching contracts and certificates ensured that surveyors shared a common set of methods and used similar instruments. The validity of subterranean geometry was not merely guaranteed by its inner coherence or a set of postulates, but by its integration into the mining administration. Why, then, was this system only sporadically mentioned in administrative sources, and not even hinted at in printed laws? I would argue that this knowledge was

---

[40] I thank Björn Ivar Berg for sending me this image. On Klemm and the early history of mine teaching in Norway, see Björn Ivar Berg, 'Das Bergseminar in Kongsberg in Norwegen (1757–1814)'. Klemm was by no means an exception, as numerous German specialists worked in Scandinavian mines; see Hillegeist, 'Auswanderungen Oberharzer Bergleute nach Kongsberg/Norwegen im 17. und 18. Jahrhundert'.

not hidden or secret, but simply esoteric, meaning literally that it was aimed at a small group of professionals.[41] We have seen that rulers and surveyors shared a common interest in restricting its diffusion while ensuring its preservation. Far from being actively dissimulated, the geometry of mine surveyors was widely circulated, albeit within specific circles and following precise rules.

## Manuscripts, Secrecy, and Knowledge Circulation

In the second half of the seventeenth century, the skills and methods of subterranean surveyors underwent important transformations. The profession increasingly relied on elaborate mathematical methods that were taught within a codified – and contractual – apprenticeship system. Measurements had once been mostly performed in public during ceremonies, following procedures codified in mining laws, as described above (Chapter 2). With the introduction of the suspended compass and the growing use of trigonometry, a good part of surveyors' duties became invisible. Their work now included lots of calculations, data processing and, as will be presented in the following chapter, mapmaking. These scientific developments are intrinsically linked with the birth of a specific technical literature. The next logical step, one might understandably imagine, would be to put this knowledge into print. True enough, the use of manuscripts was heavily criticized by contemporaries.

Scholars raged against a system that was alien to their concept of open knowledge. Professor Leonhard Christoph Sturm (1669–1719) acknowledged that subterranean surveyors might be 'otherwise pretty honest people', but his discussions with them always left him with the impression 'that they are somewhat secretive with the art'. Christian Wolff summed up the general impression in his authoritative *Mathematical Lexicon*: 'This art has always been kept secret by the subterranean surveyors.'[42] Even Nicolaus Voigtel, himself a mining official and surveyor, deplored that 'such an art is kept very secret by most *Markscheider* in the sense that, even when they have already promised someone to teach it, they still don't do it completely but keep the last step secret and for themselves'.[43]

---

[41] On secrecy and esoteric knowledge in the early modern period, see Long, *Openness, Secrecy, Authorship* (2001).

[42] Sturm, *Vier kurtze Abhandlungen* (1710), p. 46: 'mit Marckscheidern / die sonst gar *honnête* Leute waren / gesprochen / weil mir aber dünckete / daß sie etwas geheim mit der Kunst haben'. Wolff, *Mathematisches Lexicon* (1716), p. 672: 'Diese Kunst ist von den Marckscheidern jederzeit geheim gehalten worden.'

[43] Voigtel, *Perfectionirte Geometria subterranea* (1692), Introduction ('An den geneigten und Christlich gesinnten Leser'): 'solche Kunst von denen meisten Markscheidern dermaßen sehr geheim gehalten worden / daß / wenn sie schon iemanden selbige zu lernen versprochen / solches doch nicht gäntzlich ins Werk gerichtet; sondern immer noch einen Sprung vor sich und in geheim behalten haben.'

Voigtel completed his criticism by deploring the bad 'example of others, who often keep their *arcana* only for themselves' in a clear allusion to the enduring manuscript system.[44] Still, this tradition would continue until the very end of the eighteenth century.

The issue of secrecy is by no means specific to subterranean geometry, for a general debate about the openness of knowledge was reaching a high point in the seventeenth century. William Eamon has skilfully shown how 'natural philosophers condemned esotericism in alchemy and the crafts as obstacles to the growth of knowledge natural philosophy'.[45] Mathematics represented for many scholars the paragon of rationality, a model they were trying to emulate in the other sciences. In that context, the inner logic of a scribal system of practical geometry and its slow evolution was for them inconceivable. They mistook the dialect of miners for mystery-mongering (*Geheimtuerei*) and attributed the long training that was needed to collusion. Without print, they assumed that any craft or science would necessarily by imperfect or at least, as Voigtel beautifully phrased it, stagnate in a state of 'complete incompleteness' (*vollkommene Unvollkommenheit*).

The secrecy of this art was, however, all relative, as the first part of this chapter has shown. Most manuscripts, despite being handwritten, strove for clarity. These writings possessed a title page with a table of contents, shared a similar structure, and a coherent system of numbered 'propositions'. They were standardized precisely to be circulated and were so, albeit within a carefully controlled social setting. Such handwritten instructions were not prepared for bookshops, scholars, or curious amateurs, but for apprentices learning under the auspices of the mining offices. The original instruction quickly disseminated in the Ore Mountains and beyond. Contemporaries testified that Rösler's 'manuscript of subterranean geometry had come to the hand of various learned people's hands, and had been paid quite dearly'.[46] Another surveyor, August Beyer, was a successful teacher with an exceptionally long career. 'In this long time', he recalled at the end of his life, 'it could not fail that I collected many things … and in this way this treatise went through the

---

[44] Voigtel, *Vermehrte Geometria subterranea* (1713), An den geneigten Leser: 'wenn ich dem Exempel anderer hätte folgen wollen, die *Arcana* offtmahls vor sich allein behalten'.

[45] Eamon, *Science and the Secrets of Nature* (1994), p. 319, adding that 'the debate over secrecy versus openness in science reached a climax in the seventeenth century, when esoterist practices came under sustained attack'.

[46] Henning, *Acta Historico-Chronologico-Mechanica* (1763), p. 5: 'desselben Manuscript vom Markscheiden nach seinem Tode in verschiedener Gelehrten Hände kommen, und ziemlich theuer bezahlt worden'. In publishing Rösler's *Bergbau-Spiegel*, Golberg himself acknowledged in the preface that copies existed and had been sold, but described them as 'sehr vitios', while he offered to present the 'Autoris Original-Concepten'. We know from at least three surveyors from the Harz who visited him in the 1660s and taught adapted versions to their own students: Peter Heinrich Tolle (?–?), Valentin Decker (?–1705) and Daniel Flach (?–1694). See Meixner et al., *Balthasar Rösler* (1980), pp. 78–80.

hands of many hundreds of people, and always in a different form.' Many of these students copied his manuscript for their own purpose and at least three versions have survived to this day.[47]

If these texts were meant to circulate, why were they not printed? There are two pitfalls to avoid in considering the issue. First, print was not the natural evolution for early modern knowledge, mathematics or otherwise. While it is certainly true that the printing press immensely facilitated the circulation of information, it also modified the nature and content of what was circulated.[48] Most of the published technical books were not merely clean versions of the manuscripts that had preceded them, but new works 'deliberately composed with the press in mind'. Second, scribal knowledge continued to be relevant in the context of mine engineering and practical geometry. The coexistence of printed and manuscript culture was similar among artists-engineers and superior craftsmen.[49] Technicians had long used the handwritten medium: civil architects carried their commonplace book with them and kept sketches of machines, proportions of famous buildings, and even excerpts from useful (printed) books.[50] In commercial arithmetic, countless published textbooks promised to teach the latest techniques, but did not replace the informal notes during the sixteenth and seventeenth century. In their correspondence, unmistakably, merchants rarely mentioned printed material and referred mostly to the genre of handwritten *pratica*.[51] In most professions, printed and scribal documents thus exerted complementary functions. In that context, understanding the circulation of a text and its limits helps us better appreciate its content.

The issue of publication can be tackled from another angle by remembering that knowledge about the underground was particularly coveted. In the

---

[47] Beyer, *Gründlicher Unterricht von Berg-Bau* (1749), introduction: 'Es kunte nicht fehlen, durch die Länge der Zeit kame mir manches für die Hand, so ich ändern oder hinzusezen muste, auf welche Art dieser Aufsaz viel hundert Personen und immer einem anders als dem andern in die Hände gekommen ist.' Surviving copies are dated 1718 (Landesbibliothek Gotha, Chart A 72), 1727 (TU BAF – UB XVII 12), and 1739 (Berbaumuseum Bochum, Sign. 875). On the origin and evolution of this text, see Morel, 'Five Lives of a *Geometria subterranea* (1708–1785)' (2018).

[48] See Eisenstein, *The Printing Press as an Agent of Change* (1980); Blair, *Too Much to Know* (2010), pp. 53–60, 230–264.

[49] Hall, 'Der meister sol auch kennen schreiben und lessen' (1979), p. 49. For an example in eighteenth-century natural history, see Yale, 'Marginalia, Commonplaces, and Correspondence' (2011), pp. 193–202.

[50] On the culture of commonplace books, see Marr, 'Copying, Commonplaces, and Technical Knowledge' (2013), pp. 421–446. On architects' notebooks, a recent study is Merrill, 'Pocket-Size Architectural Notebooks and the Codification of Practical Knowledge' (2017).

[51] Pierre Jeannin has written numerous papers on mercantile arithmetic and the circulation of knowledge, compiled in Jeannin, *Marchands d'Europe* (2002). See also Denzel, 'Die Bedeutung der Rechenmeister für die Professionalisierung in der oberdeutschen Kaufmannschaft des 15. /16. Jahrhunderts' (2002), pp. 23–30.

sixteenth century, rulers and mining cities had publicly displayed geometry to ensure its social acceptance. After the Thirty Years War, however, surveyors and their art came to be used more strategically. Rulers of the Holy Roman Empire and Central Europe were prompt to use this useful knowledge as leverage in their foreign policy. The Saxon Elector provided his ever-changing political allies with subterranean surveyors from the Ore Mountains, or alternatively delivered *laissez-passer* allowing selected foreigners to visit mining cities. This was the 'open sesame' to learn the arts of smelting, assaying, and subterranean geometry.[52] The circulation of mine geometry was at times much wider than expected. For privileged foreigners, the usual procedure was to find a mining official ready to teach his knowledge and pay the heavy sum that the administration would not forget to ask.

The early modern period saw no complete ban on strategic knowledge: countless books were for instance published on castrametation and poliorcetics – respectively the arts of fortifying places and conducting sieges.[53] Just as was the case for subterranean geometry, however, the most useful and recent developments of these disciplines were rarely available in print. Vauban (1633–1707), the famous engineer of the French Sun King, did not publish his treatise on the art of conducting siege and yet, 'despite the interdiction, the work circulated in the country and abroad in a manuscript form'.[54] The military manuscripts of Bernard Forest de Bélidor (1698–1761), dealing with explosive mines employed during sieges, were only published well after his death.[55] A similar phenomenon can be observed with mathematical geography, when it possessed a military relevance. Sturm, whom we encountered above complaining about the arcane character of subterranean surveyors, similarly deplored the secrecy surrounding maps and military geography: 'Yet since those who produce them need to be very experienced and skilful, there are few such maps, and where they are, they are generally kept secrets.'[56] Closely regulated, subterranean geometry became an asset that could be traded, a regalian matter akin to the geometry of war. The Elector of Saxony used subterranean geometry in a similarly strategic fashion way when he authorized the Count of

---

[52] Mining administrations held registers of the foreigners allowed to enter a mining apprenticeship. For Saxony, see UAF – OBA 181 and OBA 182.

[53] On the early modern military engineers and their science of fortification, see Métin, *La fortification géométrique de Jean Errard et l'école française de fortification (1550–1650)* (2016).

[54] On the history of Vauban's *Traité de l'attaque des places* (1704), see the corresponding section in Virol, *Vauban* (2013).

[55] See Glatigny and Vérin, *Réduire en art* (2008); Ageron, 'Mathématiques de la guerre souterraine' (2018), pp. 211–224.

[56] Sturm, *Geographia mathematica* (1705), pp. 78–79: 'Allein weil große Kunst und Erfahrenheit auf seiten derer so sie machen … erfordert werden / so giebts solcher Carten gar wenig / und wo sie sind / hält man sie doch insgemein sehr geheim.'

Robilant, native of Sardinia, to come to Freiberg with four men and 'learn the noble art of subterranean geometry' for the sum of 200 thalers.[57] All things considered, the *Geometria subterranea* manuscripts played a well-defined role in mining administrations. Written and circulated within a professional group, these texts were copied by practitioners for their own daily practices. A young surveyor obtained access to his master's text during his apprenticeship, completing it over the years with new observations, methods, and data. Adolph Beyer, for instance, began copying the *Geometria subterranea* of his uncle August on 8 November 1727, during his apprenticeship, and finished this first part five months later. He then continued to add to his manuscript new cases (*propositions*) as well as many surveys and observations over fifteen years.[58] This technical genre was thus both highly specialized and in constant evolution. Surveyors transmitted their manuscripts – the work of a lifetime – to the following generation. In the absence of specialized periodicals, customs ensured both the preservation of knowledge and its growth in a fluid fashion hardly compatible with the printed medium of the time. These texts belong to a larger category of mathematical practitioners' handbooks that were routinely used as a 'working tool' and could be 'redesigned according to one's need, improving it with new details'.[59]

## How to Print a *Vade Mecum*?

I assure everyone, and especially those eager of the art, that there will be enough to learn from this modest book of mine – with God's blessing – for one to be acknowledged and praised as a skilled surveyor. For I know of several subterranean surveyors who, despite having been appointed and sworn in, do not seem to me to have understood and exhausted all that is contained in this modest work.[60]

In completing the preface of his *Geometria subterranea*, Nicolaus Voigtel used strong words to convince his prospective readers that the book they held

---

[57] SächsStA–F, 40001 Oberbergamt Freiberg, Nr. 3477, f. 11r: 'Erlernung der edlen Marckscheider-Kunst.' Saxony was at the time allied with Austria-Hungary, itself bound to Sardinia, to counter the rising influence of Prussia. Providing a foreign power with an access to this coveted knowledge was a powerful negotiating tool.

[58] TU BAF – UB XVII 12, the first part (copied from August Beyer) was completed on 5 March 1727 (see f. 141v). The last observation of the book is dated 5 April 1742 (f. 179r).

[59] Jeannin, *Les marchands au XVIe siècle* (1957), p. 106: 'On remanie l'instrument de travail selon les besoins, en le perfectionnant avec des précisions nouvelles', and more generally pp. 104–108.

[60] Voigtel, *Geometria subterranea* (1686), introduction: 'Unterdeße versichere einen ieden / sonderlich die Kunstbegierigen / daß aus diesem meinem Wercklein so viel / durch Göttlichen Seegen / zuerlernen sein wird / womit er vor einen fertigen Marckscheider erkant und gepriesen werden möge. Wie ich denn unterschiedliche Marckscheider kenne / welche / ob sie wohl in Bestallung und Pflichten sitzen; dennoch nicht alles / was in diesem Wercklein begrieffen / *exhauriret* zu haben mir vorkommen wollen.'

in their hands would teach them the esoteric art of underground surveying. Voigtel, a mining official, emphasized his familiarity with practitioners, while claiming superiority over most of them, an undoubtedly correct assertion. For the first time, a surveyor with extensive experience put into print the texts that were circulating in manuscript form. And yet, even this well-written work was hardly enough to become a proficient surveyor. In fact, and despite its impressive commercial success, it barely changed the inner dynamic of the discipline, its development, or its teaching.

Voigtel was born in 1658 near Freiberg. Son of a mining official, he learned the art of underground surveying under Valentin Decker and Adam Schneider, both of them former students of Balthasar Rösler.[61] After being appointed mining master in a little Thuringian mining district in the early 1680s, he published his sole book in 1686. While the circumstances of the publication are not fully explained, it is clear that this happened with the consent – and probably at the behest – of the Saxon higher administration, for the book was published 'with the merciful privilege of the Saxon Elector'.[62]

Voigtel likely did not see himself as the author of the book in the narrow, scholarly sense of the word. In the beautiful frontispiece of the book, he merely presented himself as its 'publisher' (*Herausgeber*, see Figure I.1). Where its content came from was an open secret for contemporaries, one of them writing: 'one believes that he had made this book helped by a manuscript of underground surveying, and this would have been Balthasar Rösler's'.[63] The structure is indeed strikingly similar to the manuscripts I have described in the first part of this chapter. In that sense, Voigtel put into print a 'collectively gained, shared resource', a typical endeavour in the early modern period.[64] From now on, anyone entering a bookshop could read about an art whose esoteric teaching took years to complete.

Could anyone be interested in reading about the obscure art of setting limits? The answer is resoundingly positive, for the first edition – self-published by Voigtel – turned out to be tremendously successful. It sold out so quickly that it was reprinted three times in 1688, 1689, and 1693.[65] There were two subsequent editions and at least nine printings by 1714. Some copies of the first edition are even undated, something one usually only sees on works so popular

---

[61] Voigtel's youth and training in Mansfeld and Marienberg are only known through Calvör Henning's description in Henning, *Acta Historico-Chronologico-Mechanica* (1763), pp. 5–6.

[62] Voigtel, *Geometria subterranea* (1686), title page: 'Mit Churfl. Sächs. Gnädigsten *Privilegio*'.

[63] Henning, *Acta Historico-Chronologico-Mechanica* (1763), p. 5: 'man glaubt, daß ihm ein Manuscript vom Markscheiden zur Verfertigung solches Buch geholfen, und dieses wird Balthasar Rößlers Schrift gewesen seyn'. It is more likely that Voigtel offered to his audience a modified version of Schneider's manuscript, but since his master had himself learned with Rösler, this opinion is essentially correct.

[64] Smith, 'Why Write a Book?' (2010), p. 38.

[65] Voigtel, *Geometria subterranea* (1686), title page: 'In Verlegung des *Autoris* selbsten'. It was printed in Eisleben, a mining city, by Johann Dietzel.

as to be reprinted without pause.[66] Very definitely, it means that thousands of copies of the *Geometria subterranea* were disseminated all over Europe. Judging the significance of a work solely by its sales numbers is hardly a recommendable historical method, but this commercial success is nevertheless worth mentioning. Practical mathematics tends to be viewed as a niche topic. Scholarly works, having been disproportionately better preserved in today's libraries, give a distorted impression of the public conversation and issues of their time. The fact that modern historiography has forgotten both Voigtel and his best-seller should not lead us to belittle its cultural importance; it simply highlights a biased vision of the history of knowledge.

The year 1686 thus marks a break in the public perception of subterranean geometry. The success of Voigtel's textbook demonstrates a large public interest about the discipline. The beautiful frontispiece of the work, on which two surveyors present surveying instruments as the best way to find underground riches, resonated with its intended audience. After that time, we can assume that any well-to-do amateur of mining (*Bergwercks-Liebender*) could easily purchase a copy.[67] Prospective buyers probably included cameralists, projectors, or anyone interested in mining in the administrations of the numerous German states. Many a scholar must have rushed for the opportunity to discover this esoteric knowledge. The *Geometria subterranea* was read in detail by no less than Gottfried Wilhelm Leibniz (1646–1716), who immediately secured a copy and worked on it in the summer of 1686. Leibniz copied and annotated numerous passages for his personal use as mining projector in the Harz mines.[68]

For many readers, however, reading Voigtel's prose likely proved to be a disappointing experience. Written by a practitioner, the book followed the typical problem-solving structure of technical texts. Moreover, it was written using the *Bergmannsprache*, a technical dialect hardly intelligible for outsiders. 'A proper explication of the terms would be necessary', deplored a puzzled contemporary.[69] When the philosopher Christian Wolff published his acclaimed *Mathematisches Lexicon* a few years later, he highlighted on the title page his presentation of 'the dialect and expressions of subterranean surveyors'. One

---

[66] The first edition (*Geometria subterranea*) was printed in 1686, 1688, 1689, and 1693. The second edition (*Perfectionirte Geometria subterranea*) was printed in 1691, 1692, and 1714. A third edition (*Vermehrte Geometria subterranea*) was printed in 1713, shortly before Voigtel's death, and reprinted in 1714.

[67] The book was seen and reviewed as useful for the general *oeconomy* of the mines. See Seyffert, *Bibliotheca Metallica* (1728), p. 61: 'Er darrinen zur Ehre Gottes und dem Bono Publico zum Besten, solche Arcana entdecket hat, welche alle Berrgwercks-Liebenden zu wissen höchst nöthig sind.'

[68] Koser, *Über eine Sammlung von Leibniz-Handschriften im Staatsarchiv zu Hannover* (1902), p. 565.

[69] Sturm, *Vier kurtze Abhandlungen* (1710), p. 46: 'vor die / welche dessen nicht kündig sind / eine ordentliche *Explication* der *terminorum* nöthig wäre'.

should thus imagine projectors and scholars at their desks, working their way through Voigtel's *Geometria subterranea* with the help of a mining lexicon, trying to penetrate its arcane procedures.[70] The impact of this publication on the perception of subterranean geometry in society at large was therefore meaningful, if ambiguous. Its influence on mining practices is much more difficult to evaluate. Some practitioners might arguably have developed acrimony against Voigtel for breaking a tacit agreement and devaluing their knowledge. The author incidentally took the precaution to have the local priest write a poem glorifying him for presenting underground surveying 'in open print, very clearly, to the world ... even though one might encounter lots of resentment'.[71] Some aspects were more positive: printing resulted in the circulation of thousands of books, in comparison with dozens, or at most a few hundred handwritten copies. More importantly, Voigtel offered a unified text largely free of errors, whereas some surveyors, having a below-average understanding of the discipline, would otherwise pass 'very vitiated' copies to their successors.[72] There are strong clues that this book, unlike the humanist works of Agricola and Reinhold, was indeed widely read and used by practitioners themselves. Adam Schneider, for instance, who had been Voigtel's teacher, referred to his student's book in a latter part of his manuscript. Printed sources and manuscripts not only coexisted, but interacted fruitfully.[73]

## Authority, Authorship, and Practices

At the turn of the eighteenth century, the most emblematic manuscripts had been made available as a printed book. In standard accounts of early modern science, the transition from manuscript to print is often presented as a one-way evolution geared towards progress, ending generations of obscurity and stagnation. Writing

---

[70] Wolff, *Vollständiges Mathematisches Lexicon* (1734), title page: 'die Mund- und Redens-Arten derer Marckscheider'. In the article 'Marckscheide-Kunst', Wolff himself underlined that the *Geometria Subterranea* was hard to understand ('zu verstehen allerdings beschwerlich fallen müssen') and presented his own lexicon as a helping tool.

[71] Voigtel, *Geometria subterranea* (1686), poem by Johann Casper Franck, preacher of the Saint Nicolas Church in Eisleben: 'Der Welt im offnem Druck gantz deutlich vorzulegen / Es sey der Mißgunst auch so sehr es woll entgegen.' Voigtel, remarking that he worked in Eisleben, 'the homeland of Luther' (*Luthers Vaterland*), felt compelled by God to improve mankind by improving its knowledge of nature.

[72] This was argued by Golberg to defend the publication of Rösler's *Mirror of Mining* (1700). Existing copies would have been 'sehr *vitios*' while Golberg offered to present the 'Autoris Original-Concepten'. See Meixner et al., *Balthasar Rösler* (1980), pp. 78–80.

[73] See TU BAF – UB XVII 18, f. 61v: 'Hiervon lehret Herr Voigtel in seiner *Geometria Subterranea* folgendes'. Another example of a surveyor whose manuscript interacts with Voigtel's printed text is August Beyer. His 1708 manuscript includes references to Voigtel in ch. 18, part 6, dealing with the difficulties of using a compass in ferruginous mines. Beyer even took pains to update the reference when the third edition of Voigtel was published in 1713, adding in the margins 'Editionis Nov: fol: 32 § 34' (TU BAF – UB XVII 12, f. 18v).

specifically about mining arts and the related manuscripts, a modern analysis presented such textbooks as having 'weakened the power of the old mystery-mongering', embracing uncritically the accusations of early modern scholars.[74] First of all, it is a shibboleth that technicians are by nature secretive and resistant to progress. Within two centuries, subterranean surveyors had already managed to introduce tremendous changes in their measuring practices, including new instruments as well as a distinct push towards calculations. This prejudice comes from a fundamental misunderstanding of how practices evolve. In this section, I argue that the driving forces behind the evolution of practical mathematics were experience and authority. The authority of seasoned engineers, respected by their peers for their ranking and their material achievements, usually superseded the prestige associated with authorship in scholarly circle.

Subterranean geometry offers a case in point. Voigtel's book mostly described the best existing practices its author knew of, and the success of his book likely helped homogenize procedures and get rid of some gross mistakes. The work is not purely descriptive, however, as it also contained proposals for a few rational changes, chiefly by introducing a decimal length system. Voigtel noticed correctly that an 'irregular measure ... has made it difficult to learn to the many'.[75] Since the standard length unit of around two metres, named *Lachter*, was divided into eight *Achtel*, even basic multiplications or the rule of three could be difficult to perform. Writing at a time when subterranean geometry had come to rely heavily on computation, Voigtel emphasized how the hundreds of operations would be easier to perform in a decimal system. He took pains to underline the advantages of his proposal, explaining how to convert the ancient data into decimals, and even printing new trigonometric tables in which both systems coexisted.[76]

The editorial success of his book, however, proved ineffective against established standards. In a 1708 composed manuscript, August Beyer acknowledged that the current system, used 'here in Freiberg and in most mining cities', was prone to errors 'if one does not pay strict attention and proceeds with the 8[ths] as with the inches'.[77] Beyer nevertheless dismissed the proposal of a new decimal system for precise reasons. A decimal system was hardly compatible with existing customs, for '[digging] contracts are

[74] Mihalovits, 'Die Gründung der ersten Lehranstalt für technische Bergbeamte in Ungarn' (1938), p. 10: 'Die Fachbücher haben zwar die Macht der alten Geheimthuerei geschwächt, aber die radikale ausrottung der letzteren können wir noch mehr der systematischen staatlichen Fachausbildung verdanken.'
[75] Voigtel, *Geometria subterranea* (1686), introduction: 'einen *irregularen* Maaße ... vielen dadurch schwer zu lernen gemacht worden'.
[76] Voigtel describes the operations on pp. 1–16 and the trigonometric tables on pp. 34–43 (the tables are then presented on pp. 44–66).
[77] TU BAF – UB XVII 12, f. 137v: 'Allhier in Freyberg und in denen meisten Berg städten'; 138r: 'da dann leichters versehen werden kann wenn er nicht genau acht giebet, daß er mit den 8tel gleichwie mit den Zollen *procedi*ert'.

usually awarded using quarter and half-quarters or eighth' of a *Lachter*. The half of the half of a half-*Lachter* – one eighth – was very simple to express in the old system. In the new one, however, this widespread measure would amount to 1.25 tenths, a quantity that would 'not be easily understood by common miners'. Beyer added a more theoretical argument by stating that ten was generally 'not as divisible' (*teilhafftig*) as eight, making widespread computing tricks ineffective.[78] This example pinpoints perfectly how abstract and social considerations were of equal weight when mathematical practitioners assessed a new method. It was not enough for the decimal system to be theoretically better. It also had to be practicable and compatible with customs, or 'the miner would scratch their heads and ask what the gentlemen might mean' with such 'rubbish subtleties' (*Subtilitätenkrämereyen*).[79]

In order to be accepted, radical suggestions – however rational and reasonable they were – could not be justified on purely theoretical grounds. Practitioners did not yet form a homogeneous and compliant community.[80] They were spread over several regions, obeyed different rulers, and bought their tools from different instrument-makers. Changing the subdivisions of a *Lachter* would have implied sweeping changes in maps, instruments, and customs. Voigtel's proposal was ultimately enforced in his mining district, but officials noted that 'apart from Eisleben, this arrangement, however useful, has not been adopted anywhere'. Local surveyors ended up in the weird situation of having to learn the two systems, or they could not have worked anywhere else.[81]

The age-old subdivision of a *Lachter* in eight parts ultimately prevailed due to inertia and tradition. Labelling this reluctance of practitioners as backwardness would be missing the point, and tradition should be here understood in a positive way. It simply meant that measurements had to be accurate according to the legal system in order to preserve the consensus that had slowly developed around the use of mathematics in mining. Once a *Markscheider* such as Beyer had decided to stick to the old system, he possessed the authority to teach it to his numerous students. Printed books, no matter how brilliant their authors were, hardly competed against an integrated training system. In

---

[78] TU BAF – UB XVII 12, f. 138r: 'die Verdingung nach Viertel und halben Viertel od. 8Theils insgemein geschieht … von denen gemeinen Berg-Leuten nicht recht zu begriffen … nicht so teilhafftig'.

[79] Von Trebra, *Erfahrungen vom Innern der Gebirge* (1785), p. 52: 'So würden wohl die Bergleute ihre Köpfe ziemlich zusammen stecken, und sich fragen, was die Herren wohl meynen möchten? … Subtilitätenkrämereyen.' Interestingly, Von Trebra still mentioned Voigtel's attempt three generations later.

[80] For an overview of the attitudes of practitioners towards the decimalization of measures, see Goldstein, 'Les fractions décimales: un art d'ingénieur?' (2017), pp. 185–203.

[81] Von Trebra, *Erfahrungen vom Innern der Gebirge* (1785), p. 52: 'Ausser Eisleben hat es nirgends gefallen wollen, diese allerdings nützliche Einrichtung nachzumachen … in anderen Gegenden in Diensten gut fortkommen konnten.'

the short term, the printed medium was probably bound to fail in its efforts to upend practices. It nevertheless succeeded in making actual procedures known to a general audience, a long-term and diffuse cultural impact that should not be understated either. While the description of geometrical methods could not replace a training *in situ*, it improved the basic understanding of subterranean geometry for the public. In a sense, it promoted for mine surveying what Eamon has labelled 'technical literacy'.[82] Printed textbooks on subterranean geometry might have paved the way for – and in any case accompanied – the nascent system of training institutions that emerged during the eighteenth century. This literature belonged to the category of 'semi-popular' textbooks, which according to Hall was gaining popularity at the time:

By 1600, technical information was being transmitted through three main channels: the older, oral medium of the craft group; the manuscript that circulated among interested parties; and the printed work intended for a 'semi-popular', moderately learned, lay public. With minor modifications, this situation persisted until the late eighteenth century and the rise of schools of engineering.[83]

In the eighteenth century, there was no head-to-head competition between published textbooks and manuscripts. A new market simply developed for a printed introduction to subterranean geometry. This explains, for instance, why someone like Johann Friedrich Weidler (1691–1755) decided in 1726 to write a Latin compendium on the subject entitled *Institutiones geometriae subterraneae* (Principles of Subterranean Geometry). Such printed booklets generally accompanied general courses in elementary mathematics and were devised for university students. The author was professor at the University of Wittenberg and had previously published a successful textbook, the *Institutiones mathematicae*, that included 'sixteen disciplines of pure and mixed mathematics', ranging from optics to hydraulics. In all likelihood, Weidler thought that subterranean geometry would be a topic worthy of a Latin textbook. Two centuries after the *De re metallica*, one observes that the structure of scholarly books on mining was still closely modelled on the classics used in universities. The genre of German university textbooks was highly standardized, because such works had to be short and elementary enough to be read and discussed within an academic semester.[84]

---

[82] Eamon, *Science and the Secrets of Nature* (1994), p. 130. Eamon even presents the printing of technical works as 'a precondition for science'. I would not go that far in this case, and I generally think that the impact was less immediate and more diffuse, but the cultural influence in learned and ruling circles was certainly consequential.

[83] Hall, 'Der meister sol auch kennen schreiben und lessen' (1979), p. 49.

[84] As such, it began with definitions and *scholia*, before turning to theorems and problems. See Weidler, *Institutiones mathematicae* (1718), title page: 'decem et sex purae mixtaeque matheseos disciplinae complexae'. On textbooks and university teaching in eighteenth-century German universities, see Kühn, *Die Mathematik im deutschen Hochschulwesen des 18. Jahrhunderts* (1988), pp. 63–88.

Unlike manuscripts, such works were not embedded in the specific microcosm of the mines. Their authors might never have entered a shaft or used a suspended compass more than a few times. Another eminent university professor gushed in his preface about the 'sojourn' he had once made in a mining city, twenty-five years before writing his textbook on subterranean geometry, considering that it gave him the authority for writing on it.[85] Scholars were keen to offer their views, suggestions, and judgements, but their textbooks were not focused on describing or improving actual practices. They were written for the same reason scholars also wrote on wine gauging, ship building, and other useful subjects: practical mathematics had become worthy of a learned audience. These books were written with an audience of *amateurs* and students in mind, and could generate a substantial profit. It was clear that presenting the tedious typologies of ore veins, or using too much mining dialect, might turn potential readers off. In subterranean geometry as elsewhere, 'early publishers found that, to be a commercial success, a treatise should only contain what the majority of its readers already knew'.[86]

Both in practical mathematics and in mining sciences, the practical authority of experience still mattered more than authorship. Despite being spread all over Europe, the milieu of subterranean surveyors was a closed one. In such circles, everyone knew that Balthasar Rösler was the one who had formalized the new subterranean geometry. His influence rested on his technical accomplishments, his position of mining master, and his numerous students. The fact that none of his working documents were published until long after his death – and even then in a much abridged form – did not diminish his authority. The imbalance between authorship and authority extended even further. Many mining officials placed little emphasis on originality, being mostly focused on their daily tasks and the long-term development of their district. In a logic of administration, this was by far the surest path to fortune and recognition. When Adolph Beyer, the nephew of surveyor Beyer, reluctantly published his *Metallic Leisures* in 1748, the introduction illustrated how uncomfortable the status of author was in a community that valued cooperation and discretion above all: 'First of all, and concerning me, is the following: I found it unnecessary for the moment to put my name on the title [page]', and indeed the work was published without his name.[87] He could have sought pride 'or used the pen for the money', as he phrased it, but decided to imitate his peers and do 'as most mining officials'. The book was a collection of old texts dealing with mining laws, finance, and

---

[85] Kästner, *Anmerkungen über die Markscheidekunst* (1775), introduction.
[86] Briggs, 'The Development of Mine Surveying Methods from Early Times to 1850' (1925), p. 120.
[87] Beyer, *Otia Metallica* (1748), introduction: 'Das erste und was mich selbst belanget, ist dieses: Wie daß es mir unnöthig geschienen, zur Zeit meinen Nahmen auf den Titel zu sezen.'

sometimes geometry, to which Beyer added his own commentaries. Just as Rösler's manuscript, it seems to be the product of a collective elaboration or, in Adolph's own words, a 'collection of all kinds of written and printed mining matters, documents and so on', accumulated over 'twelve years', reworked and improved.[88] Beyer, who had studied law at the prestigious University of Leipzig, nevertheless looked down upon the cameralists and learned projectors who flooded German courts with their reform proposals:

Concerning the mining subjects dealt with [in this book], no one should expect self-excogitated *Projects* and wind machines, or empty shells of various *reasoning*, let alone mining and smelting books cast in a new mold ... Concerning the subjects related to natural history and *mathematics*, they are based on experiments made by me or by others.[89]

Court projectors were seen as ambitious or greedy people who knew nothing of the concrete realities of mining. The polymath Gottfried Wilhelm Leibniz, for instance, had famously failed with his proposal to introduce wind machines to drain the mines.[90] Cameralists were routinely presented as plagiarists who appropriated the hard-won experience of practitioners gathered in old mining and smelting documents. For mining officials, authorship and originality were not associated with authority, because anyone could publish a book that would likely prove useless in practice. Manuscripts, on the other hand, came to symbolize the authority of those who had truly contributed to the discipline, the *Bergverständiger*.

This explains why manuscripts remained the dominant medium for circulating mining knowledge well after the publication of detailed technical books. These patiently collected leaflets served as a repository for ongoing works and as *vade mecum* for engineers. In the eighteenth century, they still embodied the authority of mining experts and were associated with the figure of the travelling engineer. Until the foundation of mining academies, and even a generation later, knowledge was mostly circulated by the experts themselves, who used their certificates and manuscripts as recommendation. Georg Zacharias Angerstein (?–1757), born in Sweden, was for instance successively hired by Hanover and in several mining regions of Austria-Hungary. He carried with him a personal manuscript library that counted sixty manuscripts, as detailed in a report of 1761, several of them dealing with mathematics – one of the in-quartos in particular entitled *Fundamenta*

---

[88] Ibid., introduction: 'eine Sammlung von allerhand geschriebenen und gedruckten Berg-Sachen, Urkunden und dergleichen'.

[89] Ibid., introduction: 'Was nun die abgehandelten Berg-Sachen betrifft, darf sich niemand selbst ausgesonnene *Projecte* und Wind-Machinen, oder leere Schaalen von allerhand *Raisonnements*, vielweniger in andere Forme gegossene Berg- und Probier-Bücher allhier versprechen ... Was die zur Natur-Lehre und *Mathematic* gehörige Sachen betrifft, sind solche auf eigene und anderer gemachte Versuche gegründet.'

[90] Details and original sources about Leibniz's attempts are gathered in Horst, 'Leibniz und der Bergbau' (1966), pp. 36–51; Stiegler, 'Leibnizens Versuche mit der Horizontalwindkunst auf dem Harz' (1968), pp. 265–292.

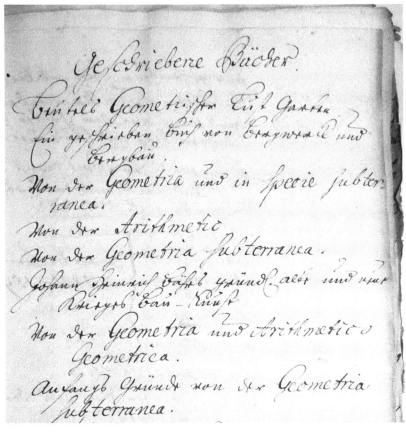

Figure 4.5    Register of the 'written books', meaning the collection of manuscripts, of the late surveyor Sartorius, including numerous German works on
arithmetic, geometry, and underground geometry (excerpted from an administrative report of 28 November 1739). Courtesy Bergarchiv Clausthal (NLA
HA BaCl Nds 6699, unpaginated (details). Photo: Thomas Morel).

*geometriae subterraneae.*[91] Far from being antiquated, such documents were
deemed precious and, although they belonged in principle to the surveyor himself, were tightly monitored by the administration. When surveyor Sartorius passed
away in November 1739, his professional documents were kept separate from his
personal affairs. Among the maps and reports, all carefully 'inventoried', a list of
his 'written books' presented his manuscripts (see Figure 4.5). In the list, one finds

[91] SUBA, HKG I, Report dated from 17 November 1761. The report also lists 132 mining maps
from various regions and numerous mining instruments, among them a suspended compass, a
*transporteur*, a 'magnet' (likely a compass) and a 'triangl' (likely a level).

several exemplars (now lost) of the *Geometria subterranea* genre described in the present chapter. One also sees other manuscripts meant to be used with them, such as elementary geometry and arithmetic, or a 'table of sines necessary for subterranean geometry, 1735'.[92]

\*

Spurred by a handful of practitioners, most prominently Balthasar Rösler, underground geometry underwent an impressive transition during the seventeenth century. Subterranean surveyors assimilated theoretical results and new ideas, for instance trigonometry, but in a piecemeal fashion and on their own terms. Moreover, they elaborated a tabular way to gather data and approached technical issues in a more systematic manner. This transformation and standardization of practices paved the way for the subsequent theoretical development that would eventually culminate in the foundations of mining schools and academies. Remarkably, most of these innovations occurred within a closed community and relied on a manuscript tradition, making this sweeping change hard to notice. Meanwhile, rulers encouraged the signing of teaching contracts, including a copy of one master's *Geometria subterranea*, sanctioned by a system of certificates. These instructions were abundantly amended, annotated, and widely circulated between mining regions. Examples from Norway, Mexico, and Sardinia show how this system was globally recognized by the beginning of the eighteenth century, ensuring an efficient circulation of knowledge while avoiding the printed press.[93]

Nicolaus Voigtel, a surveyor and tax collector, subsequently put this knowledge into print. The textbook proved very popular but its success did not trigger a wave of publishing. Indeed, the next practitioner to author a textbook would be August Beyer in 1749, almost three generations later! Although anyone could now buy a copy of Voigtel's *Geometria subterranea* – and indeed thousands of people did so – a textual instruction by itself was still woefully inadequate to train as subterranean surveyor. As most early modern crafts, mine surveying could not be learned in books at a time when, as Liliane Hilaire-Pérez has shown, 'the impact of reading was still modest' and 'everything rested on the position of seasoned technicians'.[94] The influence of the

[92] NLA HA, BaCl Hann. 84a, Nr. 6699, Markscheiderregistraturen und -inventarien – St. Andreasberger Revier 1739 – 1819, EP dated 28 November 1739, unpaginated, 'registriert ... Tabulae Sinuum ad Geometriam Subterraneam Necessaria. 1735'.
[93] On the circulation of underground surveying on the American continent see Nuria, 'Underground Knowledge' (2016).
[94] Hilaire-Pérez, 'Transferts technologiques, droit et territoire' (1997), pp. 551–552: 'Les effets de la lecture restent limités. Tout repose sur le rôle des techniciens expérimentés.' This is a general point about eighteenth-century technology, see also for Scandinavia Lindqvist, *Technology on Trial* (1984), p. 277.

printed medium was more subtle and indirect. Textbooks changed the way the discipline was considered, triggering a gradual evolution in the relationships between abstract principles and concrete actions. A book market developed in which scholars presented their own vision of subterranean geometry. Most of these works, however, still separated the mathematical theory from its mining context, failing to communicate the uncanny know-how that only years of practices could provide.

The valuation of authority over authorship explains why subterranean surveyors rarely published their manuscripts. It was less an active defence of secrecy than a matter of priority. A successful career might lead them to become mining master or counsellor for local rulers, and subterranean surveyors indeed often reached high administrative positions.[95] Literary activity, especially geared towards a more general and learned audience, hardly fitted into that scheme. As counter-intuitive as it may seem, practitioners abstained from publishing their knowledge of geometry precisely because it was valued within their professional community. In the minds of mining experts, the use of textbooks was too anonymous, not versatile enough to function as working tools and *vade mecum*. What mattered most was the authority given by experience and the ability to solve problems in practice. While the tradition of subterranean geometry continued its evolution, most of it took place silently within this close community, in the mining pits and the administrators' offices. Printed knowledge did not replace manuscripts as these were only one, albeit crucial, component of the mining microcosm.

---

[95] Balthasar Rösler became mining master in Altenberg (Saxony) in 1664. In the following decades, one can name Daniel Flach (d. 1694), Valentin Decker (d. 1705), Johann Christoph Buchholz (d. 1703) in districts of the Harz region; Rösler's grandson Golberg, Johann Conrad John and Friedrich Richter in Saxony; Johann Martin Schweiger in Austria–Hungary, all of whom became mining masters.

# 5 'So Fair a Subterraneous City'
## Mapping the Underground

> After I had seen many of the most remarkable places in the Mine, I returned to the Administrator, and put my cloaths on again in the Stove: where we were afterwards very kindly entertained. He shewed me a Map of that Mine wherein we had spent most of that day; and the delineations of all those places we had been at, with a Scale to measure the lengths and distances of all Passages and Places in the Mine; and it was very delightful to see so large a Draught or Picture of so fair a Subterraneous City.
>
> Edward Brown, *A Brief Account of Some Travels* (1673) p. 110

In 1681, when he commissioned a portrait of himself from the Dresden court painter, Abraham von Schönberg (1640–1711) was at the height of his power. Born in the lower branch of a powerful Saxon family, he had quickly climbed the ranks to become *Oberberghauptmann*, that is, Captain-general of the Saxon mining administration. Schönberg, however, was frustrated in his hopes to revive the local mining districts struggling in the aftermath of the Thirty Years War. Engaged in an important political battle to obtain a new mining law and significant investments to modernize the ageing mine workings, he was about to lose it. The Saxon Elector, Johann Georg III, needed money to entertain a permanent army and did not want to risk alienating the estates by changing the mining laws.[1] The portrait depicts Schönberg wearing an opulent outfit that symbolizes his power as Captain-general (see Figure 5.1). The artist underlined his expertise by discretely featuring miners' tools, such as a mining knife and waist-bag.[2] Yet the most obvious artefact on the portrait is rather incongruous. In place of the traditional silver hatchet – symbolizing his authority on all miners – the Captain-general holds a mining map. The map is realistically drawn

---

[1] Johann Georg III felt he needed a standing army to rival Brandenburg-Prussia, its increasingly menacing neighbour. On Abraham von Schönberg's reform agenda, see Jobst and Schellhas, *Abraham von Schönberg, Leben und Werk* (1994), pp. 74–133.

[2] This portrait has been analysed in Fritzsch, 'Die Schönbergporträts der Bergakademie Freiberg' (1962), pp. 311–317; Felten, 'Mining Culture, Labor, and the State in Early Modern Saxony' (2020), pp. 127–128.

Figure 5.1    Abraham von Schönberg (1640–1711), head of the Saxon mining
administration, represented at work with a pair of compasses and a mining map
(a plan or *Grundriß*), 1681, presumably by Dresden court painter A. Bottschild.
Courtesy Schönbergsche Stiftung (Photo: Ingo Ladleif von Schönberg).[3]

[3] The attribution of the painting to Andreas Bottschild is not certain, see Schmidt, 'Die Familie
von Schönberg und das sächsische Oberbergamt' (2004), p. 38. While Andreas Warnitz, a
Freiberg artist, has sometimes been credited, he rather seems to be the creator of a second, later
portrait painted in 1711 (see n. 100 at the end of this chapter).

with great details, representing the ore veins, galleries, and whims; it even includes the scale and a compass card.[4] During a career that spanned over half a century, Schönberg increasingly relied on technology to offset the unsteady commitment and investments of Saxon rulers. Having his portrait drawn with a mining map was all the more remarkable given that this tool, which later came to characterize underground surveying, was then a recent development. Schönberg's commitment to scaled maps, which were absent from the older mining laws and virtually unknown in the previous century, calls for an explanation.[5] I argue here that the history of technical maps – both plans and vertical sections – should be understood in the context of broader management reforms in the mining states of Central Europe. While political support was at best intermittent, embracing cartographic enterprises offered a straightforward and enduring opportunity to strengthen one's administration from within. The most spectacular example of this technical shift is the *Freiberga subterranea*, a gigantic cartography of the Ore Mountains running continuously over several hundred sheets. Ordered by Schönberg, it was patiently realized by a surveyor and mine inspector, Johann Berger (1649–1695).

Mining plans themselves, be they small formats describing a specific section or several metre-long rolls representing a whole district, were elaborate artefacts. Their figurative part, like the one held by Schönberg in his left hand, was usually surrounded by lengthy explanations, technicalities, and occasionally the data used to draw them. Map-making was based on careful measurements to ensure its accuracy. These plans were no ornamental objects meant to be hung or sold, and were indeed not printed: they were first and foremost used as decision tools. Schönberg and Berger, the former as Captain-general and the latter as surveyor, both tried to answer a crucial question: How to make a geometrically acquired knowledge visible and understandable to their early modern fellows? There had always been a need to represent mountains and concessions, but the new technical depiction of mining pits nevertheless faced many hurdles. Such highly elaborate drawings, combining a plan and the corresponding vertical section, were by no means the natural evolution of the reports, sketches, and models that had preceded them. This type of cartography indeed derived directly from the new subterranean geometry introduced by Balthasar Rösler around the middle of the century.

Mining culture had long valued direct observation (*Augenschein*) and the immediate re-enacting of measurements, which were – in spite of their

---

[4] The captain-general (*Oberberghauptmann*) was traditionally depicted with his silver hatchet (*Bergbarte*), while the mining map belongs to the subterranean surveyor. See Fehling, *Die Kleidungen derer hohen und niedren Berg Officiers* (1719).
[5] This discrepancy has been noticed; see, for example, the pioneering work of Alberti, 'Entwickelung des bergmännischen Rißwesens' (1927), p. 21.

shortcomings – pillars of its legal system. Maps finally answered the age-old dream of seeing through stone, provided one knew what to look for and how to use them. As such, these artefacts had to be actively promoted before they could be trusted. Only then could they offer an alternative way to monitor and manage the mines. With a precise representation, the administration could investigate the direction of veins at a distance, or foresee the costs of future works. This leads us to a last feature of the portrait: together with the map, Abraham von Schönberg holds a pair of compasses. The Captain-general is depicted at work, using his compass and the map scale in order to obtain quantitative information about the underground from his office. Visualization could be a powerful tool to convince a thrifty ruler of the need for major investments. Fitted with graphical representations and, most crucially, their underlying data, administrators advanced the reforms of their subterraneous cities.

## The Mine as Subterranean City

From the Middle Ages on, graphical representations had been used first and foremost for legal purposes, hence earning the label 'forensic cartography'. In the sixteenth century, such 'figured views' – as they were often referred to – were commonly produced as evidence in property disputes, despite being 'conspicuously absent from scholarly legal treatises'.[6] Using the testimonies of the senses, be it through direct observation or drawings, was an innovation of medieval jurists that quickly rose to prominence. This explains why mining laws frequently mentioned jurors, whose testimonies made under oath lent legal value to visitations and could later be used as evidence.[7] The idea of asking painters and artists to draw contested borders or properties arose in this context. Such depictions were meant to reproduce the experience of vision as closely as possible, and were made accordingly. These *Augenscheinkarten* – literally 'maps by visual inspection' – were illustrative plans. They contained figurative elements and buildings in a bird's-eye view, depicted as naturally as possible, to help locate the marking stones separating the concessions set along ore veins. Legal considerations similarly played an important role in the development of graphical depictions of the underground. Producing a visualization, however, was considerably more difficult in that context. A defining characteristic of mine workings was their depth, which is inherently hard to transcribe on paper. For courtly artists, used to giving spectators a sense of harmony and

---

[6] Horst, 'Kartographie und Grundstückseigentum in der Frühen Neuzeit' (2014), p. 370: 'Forensische Kartographie'; Dumasy-Rabineau, 'La vue, la preuve et le droit' (2013), p. 805: 'vues figurées … restent singulièrement absentes des traités de droit savant'.

[7] On the legal concept of *notorium facti* ('obvious facts') in relation to maps, see Dumasy-Rabineau, 'La vue, la preuve et le droit' (2013), pp. 813–815.

proportions, the poor visibility and sinuosity of galleries were a conundrum. Bird's-eye view had been well suited for representing concession limits at the surface, but failed to offer satisfying depictions of shafts and tunnels.

In addition, profound economic developments made mine depictions necessary, not only as legal documents, but as tools of supervision. In the aftermath of the Thirty Years War, local authorities became increasingly prompt to intervene when investors failed to provide capital or when major disasters occurred. As Europe entered the age of absolutism, rulers of the Holy Roman Empire gradually enforced a 'principle of direction', according to which the states' administrations took charge of all technical matters, while investors simply prospected and supplied capital.[8] With this *Direktionsprinzip*, mining offices increasingly resembled town halls.

By the mid-seventeenth century, the once scattered mining operations had mostly been grouped into large districts supervised by the state. There, subterranean surveyors and machinists worked tirelessly to link formerly separate galleries and optimize the flows of men, ore, and water. These new developments gave the traveller Edward Browne the impression of being in a 'Subterraneous City': 'Nor can I term it less, in which there is more building than in many', he added: 'The extent surpasseth most, and the number of the Inhabitants are considerable, their Order admirable, their Watches exact, their Rest undisturbed.'[9] Indeed, mine development and machine building were routinely labelled *architectura subterranea* in learned treatises.[10] In that context, the main purpose of maps became technical. While it is true that sketches had long been used by mine technicians early, the imprecision necessarily limited their use in engineering.[11] Planning a series of coordinated water wheels or connecting a shaft with a gallery was no simple matter of proportion: it implied precise measurements and an accurate use of reduced scales. The Freiberg mining administration considered it its responsibility to administrate and regulate the underground districts, and came to promote maps as its main tool for this purpose. The administration had its own jurisdiction and fiercely defended its prerogatives against general Saxon

---

[8] On the introduction of the *Direktionsprinzip*, see Tenfelde, Bartels and Slotta, *Geschichte des deutschen Bergbaus* (2012), vol. 1, pp. 476–505. Recent studies have put the notion of 'absolutism' in the Holy Roman Empire in perspective. See Whaley, *Germany and the Holy Roman Empire* (2013), vol. 2, pp. 187–191.

[9] Brown, *A Brief Account of Some Travels in Divers Parts of Europe* (1673), p. 110.

[10] Besides *architectura subterranea*, there was also a *geographia subterranea* dealing with the structure of the earth and the nature of minerals, mixing what modern scientists would label mineralogy and geology. Both terms were mostly used by scholars such as Peithner, *Erste Gründe der Bergwerkswissenschaften* (1769).

[11] Even in the seventeenth century, most *Augenscheinkarten* were 'not yet based on exact geodetic surveys', for it was simply not their main purpose. See Horst, 'Kartographie und Grundstückseigentum in der Frühen Neuzeit' (2014), p. 373; Horst, 'Alpine Grenzen auf frühneuzeitlichen Manuskriptkarten' (2018), pp. 49–70.

courts. The Captain-general and his peers tried to balance the short-term greed of investors with sustainable exploitation of underground riches. As a matter of fact, the very concept of *Nachhaltigkeit* (sustainability) would be coined by Schönberg's successor.[12]

Abraham von Schönberg was born during the Thirty Years War, and was still a baby as the third siege of Freiberg occurred. He studied at the universities of Jena and Wittenberg before embarking on a European *grand tour*. He subsequently learned mining sciences in the Ore Mountains and was immediately appointed at the mining council in 1663, as many members of his family before him.[13] Schönberg would be an active mining official for the next 48 years, from 1676 onwards at the head of the administration, relentlessly pushing for reforms. He was also a major intellectual figure who promoted the culture of mining sciences, exchanging with G.W. Leibniz or Ehrenfried Walther von Tschirnhaus (1651–1708). Schönberg saw himself as the mayor of an underground city, in which the major axes were the galleries draining ground water. All deep mining pits had to be connected to these arteries in order to avoid drowning, leading to Schönberg's remark that 'drainage galleries are the heart and the key of mountains'.[14] A major problem was that the Captain-general only had a sketchy vision of what was happening in such subterraneous towns. Contrary to Brown's description of mines as a perfectly arranged network, operations had developed over centuries without any overarching plan, as mining companies followed the sinuous veins. Even a diligent observer could not keep in mind a panorama of the hundreds of kilometres of shafts and galleries, operating on several levels, let alone ascertain their relative positions. In order to convince the Elector of the seriousness of the situation, Schönberg needed compelling arguments, and this quest ultimately led to his cartographic project.

The original duty of subterranean surveyors was to provide measuring skills on site, both in the mining pits and above ground, an activity that lay at the intersection of handicraft and legal work. The optimal case was the consensus, encountered when all stakeholders – investors, local officials, and the higher administration – were brought on site and witnessed a survey. Even in Schönberg's times, a mining master had to 'be present in the mountains every day', and to 'diligently visit himself the mine workings and drainage galleries'.[15] These visitations (*Befahrungen*) were meant to ensure trust in the

---

[12] The concept of *Nachhaltigkeit* was formally introduced by Hans Carl von Carlowitz (1645–1714), who took up Schönberg's position after his death, in his *Sylvicultura œconomica* (1713).
[13] A biography of Abraham von Schönberg can be found in Schmidt, 'Die Familie von Schönberg und das Sächsische Oberbergamt' (2004), pp. 58–61.
[14] Schönberg, *Ausführliche Berg-Information* (1693), p. 190: 'Weil die Erb-Stöllen das Hertz und Schlüssel der Gebürge sind'.
[15] Ibid., p. 22: 'Auff denen Gebürgen sich täglich befinden … Die Gebäude und Stöllen selbst fleißig befahren.'

administration. Decisions were still taken based on visual inspections – a direct descendant of the medieval *Augenschein* – because senses were thought to provide an unfiltered access to reality. Visual inspections were common legal procedures in most of Europe and by no way specific to mining. They were, for instance, used in draining operations, in England or in the Netherlands.[16] When direct testimonies were not possible, subterranean surveyors strove for an immediate replication of surveys. Lengths and directions, recorded in the mines using wax rings, could be quickly reproduced at the surface. In other cases, a surveyor was compelled to work not only in the presence of two jurors (*Geschworene*), but also up to 'four witnesses' who could later testify of what they had seen.[17] In the region of Tirol, measures were re-enacted 'with compass, level, and cord', on frozen lakes during winter, or on the 'long meadow' south of Innsbruck in summer.[18] The underground world was temporarily replicated at scale, on a plane surface, a conventional procedure used in many areas of practical geometry. During the silver rush, technical activities hardly mentioned maps, which only became standardized tools in the second half of the seventeenth century.[19]

A major hurdle faced by Abraham von Schönberg when he took charge of the mining administration was thus the lack of records. Decades of wars, economic crises, and floods had led to the abandonment and crumbling of complete sections, so that many pits could not even be inspected any more. Prior to his appointment, one was left to visit 'as far as possible' the abandoned workings 'with the help of several old and seasoned miners, who knew the particulars of these galleries'.[20] In that context, mining officials began to seek media to embody or transmit their observations over generations. Galleries were increasingly interconnected on several levels to drain water: representing the relative height of pits and drainage galleries was becoming crucial. Some abandoned pits might surely be made profitable if only one knew exactly where to invest and what to do. One of Schönberg's very first proposals was thus to make it mandatory to organize visitations before shutting down a pit. The mining master had to 'describe in a specific book, with all circumstances, why it has been abandoned', indicating 'the dimension of the vein, the hardness of stone, what [kind of metals] the ore contains, how deep the pit is, how

---

[16] See Streefkerk, Werner and Wieringa, *Perfect gemeten* (1994).

[17] See Span, *Sechshundert Bergk-Urthel* (1636), §44 p. 18 and §110 p. 38; Löscher, *Das erzgebirgische Bergrecht des 15. und 16. Jahrhunderts* (2003), p. 93: '§ von zeugen neben die lochtein zu seczen.'

[18] Kirnbauer, 'Die Entwicklung des Grubenrißwesens in Österreich' (1962), pp. 74–75; Krumm, 'Visualisierung als Problem' (2012), p. 243: 'Schin, wag und maß.'

[19] See Kirnbauer, 'Die Entwicklung des Grubenrißwesens in Österreich' (1962), pp. 90–94.

[20] SächsStA–F, 40168 Grubenakten Marienberg, Nr. 512, letter from 19 March 1663: 'mit zuziehung ezlicher alten Berckwergs erfahrnen Leuthen, denen solche Stöllen bewandtnüs bekandt … fortzukommen möglichen befahren'.

many working spots are, where and in which hour [of the compass] they are, and how far they have been brought'.[21] The issue of immediate visualization was gradually being replaced by that of information storage. Administrators now bitterly complained about what they saw as negligence of their forefathers. Settling disputes by directly reproducing measures on frozen lakes might have felt slick and convenient at the time, but 'these surveys had disappeared yearly with the ice'.[22] Technical drawing on paper emerged relatively late, as one solution – along with painting, models, and reports – to this cluster of problems.[23]

### 'For No One Can See through Stone...'

A need to supplement visits had existed since the mining boom of the late fifteenth century. Even so, replacing direct observation by an artefact, be it a map or anything else, was a formidable challenge. The first solution were the mining reports, or *Gruben-Berichte*. Already in the sixteenth century, surveyors could be asked to send their reports – written 'by their own hands' – to the administrations 'so that such [reports] can be stored there for future intelligence'.[24] These tedious documents, painstakingly written by unlearned officials struggling for thoroughness, sought to turn both the structure of galleries and their dimensions into words. Mining laws compelled surveyors to 'give in writing what they have measured and how deep one should dig'.[25] Reports and textual descriptions, however, were hardly adapted to deal with quantitative and spatial issues. As mines went deeper, surveyors began to supplement them with so-called *Abriße* (sketches), a vague category used at the time both by miners and by painters. An *Abriß* was mainly a synonym for the terms *iconographia* or *delineatio*, but could also refer to a drawing, a project, or even a proper map.[26]

---

[21] Schönberg, *Ausführliche Berg-Information* (1693), p. 37, §75: 'in ein sonderlich Buch / mit allen Umständen / verzeichnen / warum die Aufllassung geschehen / wie mächtig die Gänge / wie feste das Gestein / was die Anbrüche gehalten / wie tieff das Gebäude / auch was für Oerter / und wohin / in was vor Stunde / und wie weit selbige getrieben'.

[22] Kirnbauer, 'Die Entwicklung des Grubenrißwesens in Österreich' (1962), p. 75: 'dergleichen Schinnen jährlichen mit dem eyss vergangen'.

[23] Pitz, *Landeskulturtechnik, Markscheide- und Vermessungswesen* (1967), p. 7.

[24] SächsStA–F, 40001 Oberbergamt Freiberg, Nr. 3477, f. 27r, § 6: 'unter seinen eigenen Handschrift übergeben, damit solche alda zu künfftiger nachrichtung hinterlegt werden können'.

[25] *Bergk-Ordenung / des Durchlauchtigsten / Hochgebornen Fürsten und Herrn / Christianen / Hertzogen zu Sachssen* (1589), p. § 17: 'was sie ziehen / schrifftlich vorzeichent geben / wie tieff man zu sincken'.

[26] Hübner, *Curieuses Natur-Kunst-Gewerk und Handlungs-Lexicon* (1712), p. 6. Uta Lindgren uses the neologism 'map-related drawings' (*kartenverwandten Zeichnungen*) to refer to these sketches in Lindgren, 'Maß, Zahl und Gewicht im alpinen Montanwesen um 1500' (2008), p. 348.

First attempts at representing mines went in two different, mostly unrelated, directions. Recording the limits of concessions could be done by reproducing the marking stones on paper (see Figure 2.2 above). Such sketches were merely used to record the relative position of concessions and to ensure that no malicious miner would move the stones. Second, *Abriße* were also used to record the relative height of galleries for the purpose of draining ground water. In both cases, the representations were usually not made to scale.[27] Drawings were thus an offspring of the mining reports: the information was simply handled and stored in a more convenient way. Instead of listing the length of galleries or the depths of mining pits in a paragraph, one would align them on schematic straight lines (see Figure 5.2). Alternatively, surveyors condensed bits of written information – depth, length, but also the hardness of stones or the richness of ore – in small bubbles, or pieces of paper, glued to the main document. The layout of the sketches' elements only roughly reproduced the geographical positions of pits, galleries, and veins. Such cartographic instruments are fundamentally different from modern maps. They could not be measured or worked on as technical tools, but simply offered an overview of a pit, including some quantified information. Their generally rudimentary aspect notwithstanding, they could be efficient for certain tasks. In fact, their use is still attested in the early eighteenth century, when one author referred to them as 'the basic way' of recording information, 'which can be done by those who are not surveyors'. By then, the *Abriße* were mostly used by foremen who did not master elaborate technical drawing. It was of 'poor use' to advanced engineers, but still 'gave a better idea of the workings' than pictureless reports.[28]

The renewal of cartography in sixteenth-century Europe did not immediately affect the activity of underground surveying. Mathematical map-making was an occupation that mainly suited professional printers or scholars, respected by rulers for being teachers or university professors.[29] Such important persons were only summoned to mining districts to settle important disputes or to please the highest authorities. The ensuing maps would then be drawn by professional painters, often using the bird's-eye view appreciated at the time. Such objects might sometimes serve as 'substitute to the usual protocols', according to art historian Ziegenbalg, but 'were not highly precise' and above all represented landscapes, not the mines themselves.[30] More often than not, they were show-pieces for rulers, not working tools used in the daily running of mines.

---

[27] See Nehm, 'Über die Anfänge der Rißlichen Darstellungen' (1951), pp. 18–23, 124–137.
[28] Rösler, *Speculum metallurgiae politissimum* (1700), p. 88: 'Dieses ist nur eine gemeine Weise, daß einer dergleichen / der auch kein Markscheider ist / verrichten kann … schlechten Nutzen zu haben vermeinet … so geben sie doch bessere Einbildung der Gebäude.'
[29] On the general history of cartography in the Holy Roman Empire, see Stams, 'Die Anfänge der neuzeitlichen Kartographie in Mitteleuropa' (1990), pp. 37–105.
[30] Ziegenbald, 'An Interdisciplinary Cooperation' (1993), pp. 313, 323.

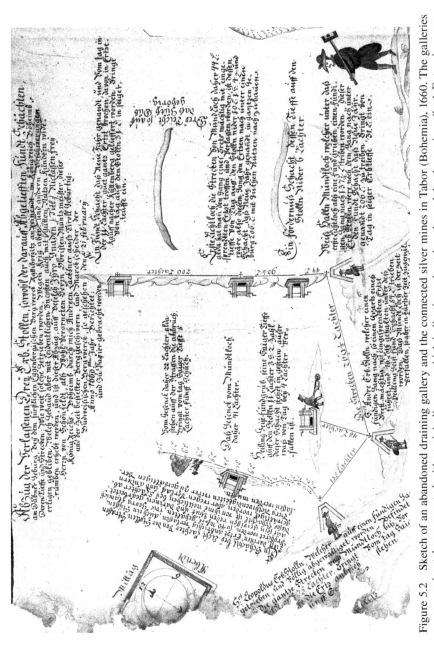

Figure 5.2 Sketch of an abandoned draining gallery and the connected silver mines in Tabor (Bohemia), 1660. The galleries are approximately represented with straight lines, on which indications about lengths and depths are inscribed. Extensive commentaries on the state of mine workings have been written on the document by the Tirolean surveyor, Andras Lackber. Courtesy Státní oblastní archiv v Třeboni (Báňská správa Hory Ratibořské, sign. F-1, inv. nr. 369).

When Duke Julius of Brunswick asked the painter David von Hemmerdey to realize a four-part painting of his mining district in 1572, the command specified that he should do:

all this with his most beautiful, lovely colors, and then, after completion, to decorate each piece with gold and silver paint and to hang them in concordance in four different frames, in order to obtain a good, deep architectural perspective, so that it would be a princely adornment in the ducal chamber and apartments.[31]

Another promising technique to visualize the underground was the use of models. Once again, subterranean surveyors were out of their area of expertise and had to collaborate with other practitioners. In the Harz region, surveyor Buchholtz was asked to produce a 'more realistic' overview of a complex mine, using a '*representatio corpore in solido*, that is, a model of the workings'. To this aim, he had to contact 'a carpenter' and produced a wood piece for which he received a significant amount of money.[32]

Sketches, paintings, and models: none of these solutions proved satisfactory from a technical point of view until a mathematically valid method was designed. In fact, testimonies indicate that representing the underground was considered a daunting task. In the early seventeenth century, mining master Peter Gerhardt organized the visitation of a clogged drainage gallery in Eisleben in order to assess the extent of the problem, hoping to 'realize a small survey and a drawing [*Abriß*]'. Although it is not clear if Gerhardt meant a sketch or a proper map, he soon acknowledged that he was not up to the task: 'this requires someone [who knows] not only the mines, but especially the *arithmetica*, *geometria*, the painting and map drawing, neither of which I am sufficiently taught and skilled in'.[33]

As late as 1708, when their use was already firmly established, surveyor Beyer felt the need to clarify the basic concepts of plans and sections in his personal teaching manuscript. His vivid explanation relied heavily on images and comparisons. We might infer that the concept of a technical map was still unfamiliar to at least some of his students:

---

[31] Excerpt from an exchange in March 1572, quoted in Pitz, *Landeskulturtechnik, Markscheide- und Vermessungswesen* (1967), p. 74: 'mit ihren schönsten, lieblichsten Farben alsdann nach Erheischung eines jeden Stücks mit einen gemalen Gold und Silber zieren und verhöhen auf vier unterschiedlichen Tafeln zuhauf concordierend in artigen, tiefen architectischen Rahmen perspective zu verfertigen, damit es eine fürstliche Zierde im herzoglichen Gemach und Zimmer werde.'

[32] NLA WO, 33 Alt, Nr. 414, *Extract Protocolli*, 28 March 1681: 'eigentlicher zu haben ... wo möglich *per representationem corpore in solido*, alß ein *Model*'. Two years later, J.C. Buchholtz received fifty thalers, equivalent to the yearly wage of a common miner, for the model (see the *Extract Protocolli* of 10 December 1683).

[33] Landesarchiv Sachsen-Anhalt, F8, Bb Nr. 16, report of a visit made on 12 March 1614: 'to kleine mensur Unde Abriß bringen ... dieses ein solche Person erfordert, welche nicht allein des üblichen Bergwerges, besonders auch der Arithmetica, Geometria, Mahlkunst und abreißnens, denne Ich keines gnügsam gelernet und erfahren'.

A vertical section (*Seyger-Riß*) is so to speak the cutting of a physical figure, or as if the earth was cut from top to bottom really perpendicularly and stood as a wall, upright in the air, so that one had the inner configuration – its width, height, and breadth – right in view and could measure [on it] with a compass and a scale.[34]

Beyer's description features an important element in the culture of a mining region. A popular saying claimed that 'no one can see through stones', and yet it was exactly what mining maps now offered to do.[35] Presenting a complex artefact as a direct access to the reality of the underground was obviously an oversimplification, both from a mathematical and from an epistemological point of view.[36] However, Beyer had good reason to emphasize that studying a mining map was similar to looking – through the earth – directly at the mine working: the analogy made it easier to argue that a map could accurately replace, or at least supplement, direct observations.

### The *Freiberga subterranea*

Visual depictions of the underground underwent dramatic transformations in the second half of the seventeenth century, as subterranean geometry came to rely on systematic data collection. In the preceding chapter, I have described how Balthasar Rösler introduced a series of innovations that quickly spread within a mostly handwritten tradition. His fundamental idea was to record measurements in a systematic way and to offer a homogeneous resolution of the multifarious problems arising in mining. In the domain of mining maps, the new method swiftly led to a unified approach, in which a plan and the corresponding section were combined (see Figure 5.3 and 5.4).

In this new framework, maps could represent the underground and record visitations. More importantly, they also served as a planning tool for future works, both politically and technically. By measuring accurately a distance on the map with a compass, then checking against its scale, a mining master could retrieve older data or ascertain the distance between two galleries he sought to connect. Previous sketches had been scattered with texts, leading to confusion: 'Because everything is drawn into each other, it seems to be very unclear',

---

[34] TU BAF – UB XVII 12, proposition 37, f. 124r: 'Ein Seyger-Riß ist gleichsam ein Durchschnitt einer Cörperlichen Figur od auch als wenn das Erdreich recht perpendiculariter von Oben an bis unten auß den Grund durchschnitten wäre und gleich einer Mauer außgericht in der Höhe stünde, damit wenn die innenwandige Gestalt nach der Breite Hohe und Dicke recht vor Augen habe und mit einen Cirkel aussen Maaßstab meßen könne.'

[35] Petri, *Der Teutschen Weissheit* (1605), vol. 2, p. 280: 'Es kan niemand durch den Stein sehen.'

[36] As a matter of fact, Beyer's explanation was also mathematically inaccurate. Drawing a *Seyger-Riss*, or section, was not as simple as cutting a slice of a cake. It corresponds in fact to 'a projection of the [mine] workings on a plane running parallel to the main longitudinal direction of the mine', according to Brough, *A Treatise on Mine-Surveying* (1913), p. 285.

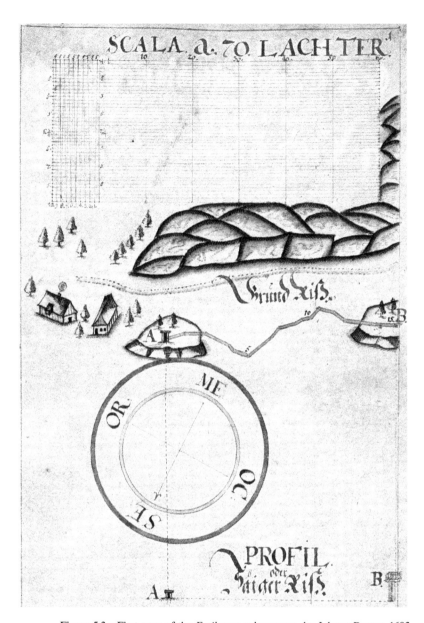

Figure 5.3    First page of the *Freiberga subterranea*, by Johann Berger, 1693. The reader is presented with a combined representation. The plan, or *Grund-Riß*, is seen 'from above', contrary to what the bird's-eye view drawing seems to imply. It is articulated with a vertical section, the *Profil* or *Saiger-Riß*. The pathway of galleries can be followed from A to B on both maps. The scale (*SCALA a. 70 LACHTER*) is respected throughout the whole document, to unify the measurements and facilitate the collection of information. Courtesy Sächsisches Staatsarchiv, Bergarchiv Freiberg, 40040, Nr. G5006, f. 1. Photo: Thomas Morel.

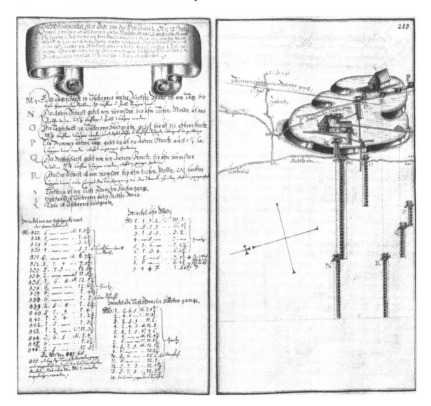

Figure 5.4   *Freiberga subterranea*. Portions of the mining pits *Zscherper* and *Consolation of Israel* are represented in a combined plan and vertical section (right). On the left side are the three data sets used to draw the map, together with commentaries about the specifics of the mine workings. Courtesy Sächsisches Staatsarchiv, Bergarchiv Freiberg, 40040, Nr. G5003, f. 213. Photo: Thomas Morel.

complained a contemporary. In his textbook, surveyor Voigtel thus recommended to draw combined plans and sections 'whereupon nothing is written, and to rather [place] the written report in another place, not far from the map'.[37] Map-making demanded a strict discipline from surveyors, while the drawing part itself required new technical skills.

When he was appointed in the mining administration, Abraham von Schönberg immediately welcomed this new technology, although the reasons

---

[37] Voigtel, *Geometria subterranea* (1686), p. 120: 'weiln alles unter einander hinein gezogen sicherweiset / es sehr undeutlich zu seyn scheinet ... worauff nichts geschrieben, vielmehr aber der schrifftliche Bericht / an einem sonderlichen Orte / von dem Risse ein wenig entfernt.'

behind his move are not entirely clear. It might have come from his university education, where similar techniques were taught in courses on military architecture. Maybe he directly interacted with Balthasar Rösler, who was at the time a veneered mining master in Altenberg. For the ambitious young Schönberg, embracing maps might also have come out of sheer necessity, as an affordable attempt to revive the sluggish mining economy.[38] Four years after his appointment to the mining council, he was the driving force behind an important decree. From now on, the Saxon Elector required all active mine workings to have a proper plan:

It is known to you how many mines in Our country occasionally grind to a halt, and for many of them there are no reports as to why they were once abandoned. That is why, for the benefit of Our descendants, we have found it both necessary and useful that reliable maps, of the inner building and veins of the mines still operating in Our country, be deposited in our mining offices.[39]

Balthasar Rösler and his son Christian were asked several times to draw maps of the most important mine workings and related veins. They seem to have indeed begun this exhausting work, but both unfortunately died within a few years. By that point, Schönberg (now Captain-general) was already pleased to have gathered 'quite a few maps altogether'.[40] A most important task, however, was still wanting: there was no large map of the district, presenting the relative position of all individual pits. Schönberg's overarching ambition had always been to map the whole Freiberg region along the long drainage galleries (*Wasserstollen*). All mine workings had to be related to these sinuous galleries, from which water-wheels and pumps kept the deepest pits dried and exploitable. In 1784, a severe drought ravaged the Ore Mountains. Within a few weeks, all reserves had been depleted and water ponds were empty. Ironically, water shortages resulted in massive flooding in the mines. Without water from the surface, it was not possible to power the machines that drained ground water. The dry summer also prompted bad harvests that were followed

---

[38] It has rarely been explicitly noted, but a major advantage of practical mathematics compared to other early modern planning methods, or reform tools, are their relative affordability (trial and error methods or the hiring of foreign experts were, for instance, very expensive).

[39] Lünig, *Fortgesetzter Codex Augusteus* (1772), vol. 3, p. 1343, edict (*Reskript*) from 20 December 1667: 'Die Verfertigung richtiger Abrisse, von jedem gangbaren Gruben- Gebäude, betreffend … Euch ist wissende, wie viele derer Berg-Gebäude in Unserm Lande hin- und wieder erliegen, da bey vielen keine Nachrichtung zu befinden, warum dieselben vor dieser Zeit aufläßig worden … so nöthig als nützlich erachten, daß über die annoch treibende Bergwerke in Unsern Landen, gewisse Abrisse der innerlichen Gebäude und Stroßen gefertiget, und bey denen Berg-Aemtern beygeleget würden.'

[40] SächsStA–F, 40001 Oberbergamt Freiberg, Nr. 3631, f. 1v: 'nicht weniger Abrisse über all'. In the same document (dated 1672), Schönberg seems to ask the two Röslers for a global map of the district focusing on the drainage galleries. Balthasar, however, died in the following year, and his son shortly after him.

by a rigorous winter, endangering the fragile recovery of the region. Abraham von Schönberg repeatedly asked the Elector to renovate the mining infrastructure but failed to convince the Dresden bureaucrats.[41]

A few years after posing for a portrait with a mining map and a pair of compasses, Abraham von Schönberg then commissioned Johann Berger to produce an overview of all mine workings, pits, veins, and machines in his central district of the Ore Mountains. Berger, who is only remembered, when at all, for his magnificent *Freiberga subterranea*, was of modest extraction. His father was a simple miner who sat in the local court (*Gerichts-Schöppen*) of his village, close to Freiberg. Berger went to secondary school but was not wealthy enough to pursue his studies. He became a miner at the age of sixteen, but his weak constitution could not support it. At the age of twenty-four, he thus entered the low mining administration, acting as clerk in a neighbouring mine and spent ten years in various subordinate administrative positions.[42]

The timing is telling: in the early 1680s, Johann Berger quickly rose to significance precisely as his patron Schönberg renewed his efforts in map-making. In 1684, Berger trained in subterranean geometry and assaying using the usual apprenticeship system. He also learned 'field surveying, with the fortification and the optics *ex fundamento*' with Gottfried Christian Braun, a mathematics student.[43] He finally took lessons in 'drawing and painting' with the Freiberg artist Andreas Warnitz. Now in possession of all the technical skills needed for a large-scale mapping enterprise, Berger was soon appointed foreman, and became in the same year juror (*Berg-Geschworener*). In 1686, he was finally named inspector of mines (*Einfahrer*) and had to visit mines 'often, both in the morning and in the afternoon, also at night, during shifts and outside of it'.[44] He would assist the local mining master throughout his visitations, his visual testimony lending gravitas to the decisions, very much in the spirit of medieval laws described above. When visiting a mine, inspectors such as Berger gathered two kinds of information. Some were practicalities: was there enough tallow for the miner's lamps? Enough leather for the pumps? Was the expensive blasting powder used sparingly? More fundamental was ascertaining the gallery's direction, checking the structure of the masonry, and similar tasks.[45] This second type of data remained largely

---

[41] Jobst and Schellhas, *Abraham von Schönberg, Leben und Werk* (1994), p. 107.

[42] Berger was not a *Bergschreiber* or *Gegenschreiber* – relatively important positions – but a simple *Zechen-Schreiber*. Most of his biography is known through a secondary source, Grübler, *Ehre Der Freybergischen Todten-Grüffte* (1731), vol. 2, pp. 236–237.

[43] Ibid., pp. 236–237: 'bey Gottfried Christian Braunen, *Matthem. Studioso*, die Erd und Feld-Meß-Kunst, nebst der *Fortification* und *Optica ex fundamento* erlernte'.

[44] The profession of mine inspector is described in these terms in Schönberg, *Ausführliche Berg-Information* (1693), pp. 50–51: 'zum öfftern so wohl früh / als nachmittag / auch des Nachtes / ausser und in der Schicht'.

[45] Ibid., *Ausführliche Berg-Information*, p. 51: 'Was die Kunst-Steiger / nach Anzahl der Sätze / an den Künsten vor Leder / Unschlitt / und anders bedürffen … Ob und wie mit dem Pulfer umbgegangen.'

unchanged from one visit to the next: as an inspector, Berger was thus well placed to grasp the value of maps for the long-term administration of mining districts.

## A Technical and Political Work

In addition to his inspection work, Berger patiently surveyed all major galleries and 149 related mining pits from 1687 to 1693, eventually collecting tens of thousands of angles.[46] This mammoth job required incredibly hard work, to which he sacrificed his 'health and spare time', in his own words. Two years after completing his masterpiece, Johann Berger would die at the age of 46. His youth spent working in the mines and the seven years of surveys needed for the *Freiberga subterranea* – both of which he saw as 'blood-souring work' (*blutsauere arbeit*) – had exhausted him.[47] The 410–sheet manuscript of this cartographic work was dedicated to his patron and, in a paragraph of the long subtitle, Berger specified the ambitions of his work: 'One can clearly and briefly see 1) how current mine workings are organized and how far they have been driven, where 2) new workings could hopefully be directed, according to miners' guesses, as well as 3) where old abandoned ones can be revived.'[48]

The work itself is ordered as a large book, composed of a handful of maps, each of which runs continuously over dozens of pages. It is the unadorned work of a faithful miner, without introduction, methodological statement, nor precise instructions about the intended use. The only refinement was a manuscript frontispiece (Figure 5.5) that closely mirrored Nicolaus Voigtel's *Geometria subterranea*, published a few years earlier. It conveyed the fact that Berger's cartographic work was grounded on the new subterranean geometry, while alluding to a consistent movement directed by Schönberg, as will be detailed in the next section.

The structure of the *Freiberga subterranea* reflects the mindset of a technician. Since all relevant mining activities had to be connected to the drainage galleries, the reader follows them and is successively presented with all connected buildings, machines, and veins. One instantly discerns how the various pits connect to these galleries, as the author followed Voigtel's advice to separate the maps from the report. Each folio follows a similar setting (see Figure 5.4): on the right-hand page, a section of the continuing map is presented, while the left-hand

---

[46] SächsStA–F, 40040 Fiskalische Risse zum Erzbergbau, G 5003, title page 'mehr denn 29. biß 30 000 Winckeln'.

[47] SächsStA–F, 40040 Fiskalische Risse zum Erzbergbau, G 5003, title page: 'ungemein Fleiß'; c verso: 'Gesundheit und freygelaßene Stunden'.

[48] SächsStA–F, 40040 Fiskalische Risse zum Erzbergbau, G 5003, title page: '1. Wie gegenwertige Berggebäude beschaffen, und wie weit selbige getrieben. Wo 2. Bergmännischen Vermuthen nach höffliche Neue-Gebäude auszurichten, auch 3. alte Auffläßige wieder rege zumachen sind kurz und deutlich sehen kann.'

Figure 5.5    The frontispiece of Berger's *Freiberga subterranea* is a hand-written reproduction of N. Voigtel's *Geometria subterranea*, published seven years earlier (compare this figure with Figure I.1). Courtesy Sächsisches Staatsarchiv, Bergarchiv Freiberg, 40040, G5003, frontispiece and title page. Photo: Thomas Morel.

page generally contains related data sets with captions and commentaries. One observes the sinuous structure of galleries, while the vertical ladders represent the shafts, down to the dotted line indicating the deepest drainage gallery.

Berger's cartographic work illustrates the transition towards a mathematized cartography of the underground. In the sixteenth century, mining customs referred to surveying procedures and reports in loosely defined terms: 'A surveyor shall describe clearly and precisely to the mining master the surveys of mines (*Marscheidtzüge*).'[49] The vernacular word *Markscheidezug*, which once

---

[49] Löscher, *Das erzgebirgische Bergrecht des 15. und 16. Jahrhunderts* (2003), pp. 188–189, § 183: 'Marscheidtzüge müß die marscheider all gründlich sambt den örtungen clar und deutlich dem bergkmeister beschrieben geben.'

referred to the pathway of a mine, was now given a complementary, purely mathematical meaning. It now stood for the data set that accurately represented the gallery (see Figure 5.4). The evolution eluded most contemporaries, as well as modern historians, for several reasons, not least because in everyday life words were often used with their ancient, non-technical meanings. More fundamentally, surveyors still took pains to draw maps in a figurative way, occasionally depicting water-wheels and miners at work, smelting huts, or local castles. A glimpse at the preserved drafts of the *Freiberga subterranea*, however, immediately reveals a technical piece of work (see Figure 5.6). Each data point is numbered and accurately located on paper, all figurative elements being added later on.

The *Freiberga subterranea* was a milestone that served as a model for later mining maps. Unlike earlier sketches, modern cartographic works were tools on which one could work with a pair of compasses, for such maps reproduced the three-dimensional structure of the underground. Understanding the combined plan and sections was a complex operation, and trying to plan further operations from this visualization was even more ambitious. Given the outsize importance of subterranean geometry, these objects had to be popularized and made understandable to the many. In the last decades of the seventeenth century, maps thus suddenly began to be mentioned in mining sermons and literature. This didactic aim explains why Beyer described vertical sections as 'a wall, upright in the air'. We can assume that this is the reason why Berger took painting and drawing lessons with master Warnitz, and still used numerous figurative elements. Large maps like the *Freiberga subterranea* had to be user-friendly, sometimes even at the expense of thoroughness. For instance, all points were mathematically drawn, but Berger only wrote down less than a tenth of the angles as data sets. Similarly, he had included the raw data in full detail in some of the earliest drafts but ultimately changed his mind. The final version simply presents lengths and hours of the compass, things that most miners could understand.[50] Compared with some other maps Berger had drawn for the administration, this work shows a real effort to produce a technical tool that would also be illuminating to non-specialists.[51]

From the content and design of the *Freiberga subterranea*, a few assumptions can be made about its intended audience. One needed a good knowledge of the mines and be familiar with cartography in order to understand it, but it was not

---

[50] SächsStA–F, 40040 Fiskalische Risse zum Erzbergbau, G 5006, f. 16–22, presents, for instance, a series of 131 angles with complete data sets, equivalent to several days of work. In the final version (G 5003, f. 19–24), this raw data has been replaced by much lighter information.

[51] The painstaking work to simplify the representation can be traced from the early drafts (G 5005 and G 5006) to the final version, which is significantly pared down.

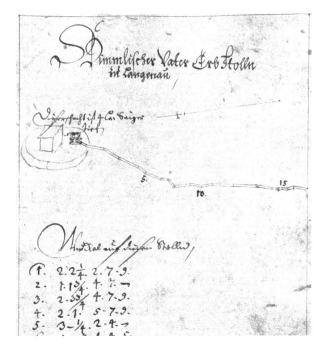

Figure 5.6    Draft plan of the *Holy Father* gallery in the *Freiberga subterranea*. The numbered 'angles of this gallery' (*Winckel auf diesen Stolln*) are listed in the lower part. Surveyor Berger used this data to pinpoint with a dry-point compass the exact path of the gallery, labelling for clarity the data points 5, 10, and 15. The figurative lines indicating the gallery, the hut, and the winch were added afterwards for illustration purposes. Courtesy Sächsisches Staatsarchiv, Bergarchiv Freiberg, 40040, G5005 f. 109. Photo: Thomas Morel.

necessary to be a surveyor. In practice, maps could be used by a vast range of people, from local members of the mining offices to the highest administrators in Dresden, not forgetting the often irresolute mining investors. A detailed district map was thus precisely the kind of tool that Captain-general von Schönberg needed to advance his reform agenda. To revive the local economy, he had recommended two main measures, both being costly operations: drowned mines should be connected to the drainage galleries, and promising sites should be systematically explored. Conveniently, both recommendations are expressed throughout Johann Berger's *Freiberga subterranea*. Commentaries frequently instruct the reader about the distance left to dig 'before one achieves the connection', in which case prospective works were indicated on the map as dotted lines.[52] Clusters of blank pages, where promising spots were located, show that his cartography was meant to be continued by the next generation.[53] Combining his experience of mineralogy with precise surveys, Berger provided guidance about the specific measures to implement. Unlike previous works, the *Freiberga subterranea* was not limited to existing mine workings and offered proposals for new investments. At one place, for instance, where an exploratory shaft had encountered ore veins in the depths, Berger computed their inclination and used his map to ascertain zones where one should step up prospecting: 'in this area, various rich inclined veins have taken their direction'.[54] Using his own map as an overview, he was in the position to ponder whether connections made sense, or where to search for ore. 'There are veins the thickness of up to a fist, with pyrites, blend, and some galena in it', Berger wrote about an abandoned pit that might be revived, 'but the extraction to the surface is very difficult, for the shafts are inclined'. In this specific instance, his conclusion was severe: 'For my part, I doubt that one can recoup the costs.'[55]

Far from being a transparent description of the underground, a close reading reveals that this map is a multifaceted artefact, in which technical finesse was blended with political motivations. In one instance, Berger's description begins as a surveying report: 'If this gallery was driven 100 *Lachter* [ca. 200 m] further into the mountain, one would certainly find silver ore veins.' Berger's tone then changes: 'I doubt, however, that this will happen under the current court of appeal's secretary Tobias Lichtenegger.'[56] Tobias Lichtenegger (1630–1693) was a high-ranking official in the Dresden bureaucracy and an adversary of Abraham von Schönberg. Schönberg, as we have seen, fought for the autonomy of the mining administration and direct access to the *Kammer* – the inner circle of the Elector ruling the country – while Lichtenegger and the court of appeal claimed to have authority on mining disputes.

In Schönberg's push for reforms, political and technical goals were thus intertwined. Political support from the Elector was necessary to finance costly

---

[52] SächsStA–F, 40040 Fiskalische Risse zum Erzbergbau, G 5003, f. 70: 'ehe Mann durchschlägig werden wird'.

[53] In the Harz Mountains, a general map also featured blank pages. The local administrator similarly explained that it was 'arranged in such a way, with space left, to be continued from year to year'. NL HA Dep. 150 K Acc 2018/700 Nr. 3, report from 6 July 1663.

[54] SächsStA–F, 40040 Fiskalische Risse zum Erzbergbau, G 5003, f. 134: 'umb diese gegend unterschieden Edle stehende und flache Gänge ihre streichen haben'.

[55] SächsStA–F, 40040 Fiskalische Risse zum Erzbergbau, G 5003, f. 82: 'stehen zwar ein halbauch ein fäustel mächtig Gänge, darein bricht Kies, blende und wenig Glantz … dieweil nun die förderung bis am tag, wegen der flachen schächte, sehr schwehr fället … Alß zweifele ich meines orts, daß mann die Kosten darinnen hauen kann.'

[56] SächsStA–F, 40040 Fiskalische Risse zum Erzbergbau, G 5003, f. 369: 'Wenn dieser Stolln solte annoch 100. lachter in das Gebürge getrieben werden, so würde mann wohl Silberhaltige gänge antreffen, ich zweiffele aber daß es bey den itzigen Herrn *Appellation* Gerichts-*Secretario*, Tobiae Lichtnegern, geschehen wird.'

investments in 'the heart and the key of mountains', the draining galleries. Technical overview was no less crucial to the Captain-general, since he risked irremediably damaging his reputation if he wasted the Elector's money. Berger's *Freiberga subterranea* offered a wide range of geometrical and visual arguments to make his patron's case. It was a detailed, yet purposely accessible work to achieve both goals at once. The map reads like a plea to finance major mining projects, while offering the technical road map to carry it out. The huge scale presented on the first page of the work promised an accuracy of the order of one inch in the mines, to convince the reader that it was indeed a perfect overview of the mines (Figure 5.3).[57] Moreover, this general map also served as a guide to find one's way into the wealth of cartographic works already gathered by Abraham von Schönberg during his three decades at the head of the mining administration.[58] Conceived as a full volume – its pages are bounds – Berger's masterpiece was quite literally a thesaurus, a gate to present to the Dresden bureaucracy the reform plans of his patron Abraham von Schönberg.

**Maps and Books as Reform Tools**

Just as Berger performed his first surveys in 1687, Abraham von Schönberg was sending a long *pro memoria* to the Elector Johann Georg III about the poor state of Saxon mines, alerting him to the risk of 'extreme impotence and ruin'.[59] Seven years later, as the mapping project was successfully drawing to a close, Schönberg's agenda was still stuck in the quicksand of the Dresden politics. Existing veins were almost exhausted, Schönberg alerted once again, and without considerable investments and prospecting, their output would steeply fall. The Elector was sympathetic to his efforts, but more preoccupied by his rivalries within the Holy Roman Empire. Mines were seen as a cornucopia of metals and the estates were reluctant to invest in order to ensure sustained revenues in the future. A key provision to revive the industry was the publication of an updated mining law, but it kept stumbling on the question of jurisdictions. If a miner or a smelter was involved in 'matters of insults and other *delicta*', should he be judged by a mining court, as Schönberg defended, or by a general court, as for instance his opponent Lichtenegger claimed?[60] In a politically volatile context, how could he best ensure political support

---

[57] In practice, neither the scale nor the maps were precise enough for this level of accuracy.
[58] Johann Berger himself simultaneously drew large-scale, more detailed versions of crucial areas.
[59] Jobst and Schellhas, *Abraham von Schönberg, Leben und Werk* (1994), p. 85, letter from 26 July 1687: 'äusserste Unvermögen und Ruin'.
[60] See Schönberg, *Ausführliche Berg-Information* (1693), p. 210, where he defended his authority on 'diejenigen *Injurien*-Sachen / und andere *Delicta*' as a matter of '*propter connexitatem*'. The general conflict about jurisdiction is detailed in Bernhardi, *Drey Fragen über die Berggerichtsbarkeit im Königreich Sachsen* (1808), see pp. 220–222 on the matter of insults.

for broader mining reforms? Underground geometry seemed to offer a satisfying answer. Schönberg strove to build a new social framework, in which the technology of mining maps came to symbolize both the expertise and the autonomy of his mining administration.

Frustrated by the political status quo, Schönberg gradually engaged the offensive on another, technical front. His elusive mining law was meant to ensure trust in the legal system and to 'spur a better desire to build' (*eine bessere Baulust*) in the investors.[61] Unable to achieve it with legal tools, he decided to promote knowledge as a basis for a new rational development of the mines based on customs. Bypassing his political adversaries, Schönberg simply published his reform projects as a book – for which no vote or approval was needed – hoping that it would become established. His *Complete Information on Mining* contained detailed information about surveying, smelting, and all mining offices, and was soberly dedicated to the *patriae* (homeland). It was published in 1693, the very year Berger completed his *Freiberga subterranea*. The frontispiece, to which Schönberg added a descriptive sonnet to make the message even clearer, reveals that these works were two sides of the same coin (see Figure 5.7).

The 53-year-old Schönberg represented himself and his fellow officials on the left side, wearing ritual hatchets and miners' lamps. Nature was represented by a woman sitting on the right, identified by her hat as a 'metal-pregnant fortress' to be conquered – the lion at her feet illustrating both her strength and danger.[62] Miners bowed before her, humbly begging for the keys granting access to hidden metallic riches. To unlock the secrets of nature, they counted on the familiar surveying instruments represented at their feet: the suspended compass, the semicircle, and the pair of compasses. Such instruments were to be mastered through the 'quick-flying science', represented in the middle with winged helmet, holding a triangle and a quadrant.[63]

Alternatively cajoling and threatening, the *Complete Information* was clearly meant as a promotion of political reforms, openly luring investors by stating emphatically: 'Reader, here you have the right information'. At the same time, mining officials were warned that 'no one [could] exculpate himself with ignorance anymore', now that it had been published.[64] Schönberg also updated the list of duties ascribed to mining officials, and mining plans were now given a central function. His instructions on

---

[61] Schönberg, *Ausführliche Berg-Information* (1693), introduction: 'Derer Gewercken wegen / ihnen zum Theil ein bessere Baulust zu erwecken'.
[62] Ibid., Explanation of the copper plate by a double and alternating sonnet: 'Die Metallen-Schwangre Veste.'
[63] Ibid., 'Die Flugschelle Wissenschaft ... Bergwercks Wissenschafft die Gröste / Und die beste Kunst in Feld.'
[64] Ibid., *Ausführliche Berg-Information*: 'Leser / hier hastu davon Rechte Information ... sich keiner mit der Unwissenheit zu entschuldigen'.

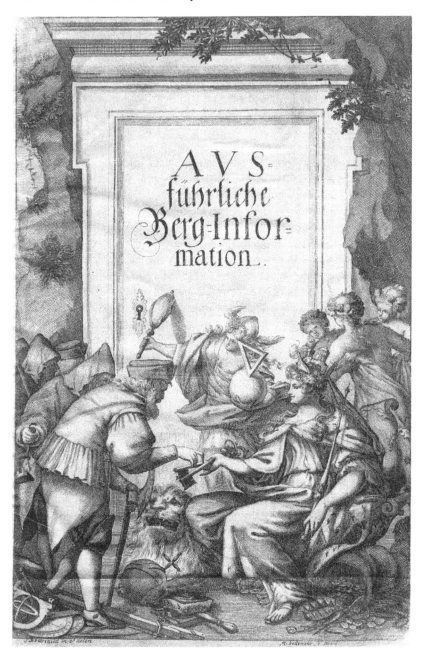

Figure 5.7   Frontispiece of Abraham von Schönberg's *Ausführliche Berg-Information* (Complete Information on Mining, Leipzig, Fleischer, 1693). Courtesy ETH-Bibliothek Zürich, Rar 1387.

map-making were precise; he was indeed competent on the subject, having himself drawn mining maps.[65] In his mind, the technical skills required to produce a map were not to be separated from a more general supervision of extractive operations:

After having measured the dimensions of a mine, [a surveyor should] produce each time an accurate map ... on which all traversing veins – and their hours – should be diligently included, in duplicate, one sent free of charge to the central mining office, the other one sent to shareholders against commission, and at the same time report in writing where and how deep one should dig, prospect, how far to deflect, to direct shafts on another, to work upwards or to connect.[66]

The *Complete Information* was an editorial success. Even more interesting for us is the network of publications that developed around it. A few years earlier, Schönberg had supported the publication of Voigtel's *Geometria subterranea*, the work being dedicated to him.[67] The astute choices of frontispieces and dedications made clear that these three important works – his legal compendium, Voigtel's mathematical textbook, and Berger's general map – all had the same political goal. Taken together, they offered a road map for improving the industry of the Ore Mountains, presenting the extent of the technical and economic problems while promoting practical geometry as their solution.

Schönberg also sponsored in 1700 the publication of manuscripts gathered by the late surveyor Balthasar Rösler, and sorted by his grandson Golberg, under the title *Speculum metallurgiae politissimum* (Brightly Polished Mirror of Mining). It was an encyclopaedic work on prospecting, extraction, assaying, and smelting methods, in which underground surveying was mentioned only in passing in a three-page section.[68] Its title, frontispiece, and dedication now sounded like a warning to the Elector and its council: as long as 'the underground affairs have been blessed, the ones above ground have prided themselves and benefited from it'.[69] It was now time to return the favour and support

---

[65] Jobst and Schellhas, *Abraham von Schönberg, Leben und Werk* (1994), fig. 31 (at the end of the volume).

[66] Schönberg, *Ausführliche Berg-Information* (1693), §4, p. 111: 'Nach verrichteten Zügen iedesmahl einen richtigen Abriß / darauff alle übersetzende Gänge und ihre Stunden fleißig mit eingebunden seyn sollen ... / einen ohne Entgeld in das Ober-Berg-Ambt / den andern denen Gewercken um die Gebühr ausstellen / und darbey schrifftlich melden / wo / und wie tieff man sincken / ansitzen / wie weit man auslängen / Schächte auff einander oder nach richten / über sich brechen / oder durchschlägig machen soll.'

[67] Voigtel, *Geometria subterranea* (1686), dedication: 'To the well-born Herr Abraham von Schönberg ... my highly respected Master and highly esteemed Patron.'

[68] In editing Rösler's book, Golberg presented subterranean geometry in the first section of the fourth book, pp. 86–88.

[69] Rösler, *Speculum metallurgiae politissimum* (1700), Zuschrifft: 'damahls das *superius*, wie das *inferius* gewesen / behaupten könnte / folglich auch / da die unterirdischen *Affai*ren glückseelig gewesen / die Oberirdischen sich dergleichen zu rühmen gehabt / und sich wohl befunden'.

the struggling Ore Mountains. Many passages of the *Mirror of Mining* are so close to Schönberg's thoughts and writings that one wonders if the surveyor influenced the statesman or vice versa. Despite being written by a subterranean surveyor, the textbook did not tackle the issue of maps from a technical point of view, but as a matter of policies. 'Many thousand *florins* can be saved', one reads in the *Mirror*, by drawing maps and consigning minute reports of the mines. Subterranean geometry was especially useful, he continued, 'for those who invest in mines, but cannot visit them in person'; in such cases, 'one draws mine workings, underground and at the surface, on a special map for them'. Even mining experts working underground 'cannot be completely sure without the map of the *Markscheider*'.[70]

This strategy of publications shows a powerful figure minutely pursuing his agenda over several decades. Although Abraham von Schönberg is generally not mentioned among the major figures of the *Kameralismus* – the German science of administration – his course of actions was a model for policy making at the time.[71] He encouraged all sympathetic figures to publish and ensured that political, technical, and operational aspects were tackled coherently.[72] The authors he patronized all emphasized the technical efficiency of maps and sometimes openly lured investors, as for instance Paul Jacob Marperger, who chose to entitle his book *Das neu-eröffnete Berg-Werck* (The Reopened Mine, 1707).[73] These publications supported Schönberg's claim that the power and finance of the mining administration should be enhanced. The ability of the mining offices increasingly rested on combined maps, presented as the only reliable – albeit relatively new – technology. In spite of political obstacles, the *Oberberghauptmann* found within his administration original resources to advance his agenda.

Legal and technical considerations on the necessity of visual inspections in the mines mirrors the scientific debates of the seventeenth century about the use of new instruments – the telescope and the microscope. Could one really rely on 'artificially produced nature'?[74] The direct testimonies of the senses

---

[70] Ibid., p. 88: 'können bey solchen Zeiten viel 1000. fl. ersparet werden ... welche Bergwerck bauen / aber selbst in die Grube nicht fahren können / daß man ihnen Gruben- und Tage-Gebäude uff einen Special-Abriß bringe ... ohne des Markscheiders Abriß keine rechte Gewißheit haben.'

[71] On the German *Kameralismus*, a synthesis article is Sokoll, 'Kameralismus' in the *Enzyklopädie der Neuzeit*, vol. 6 (2007), pp. 290–299.

[72] Schönberg's publication campaign had indeed begun with the publication of the *Institutiones metallicae* by Georg Caspar Kirchmaier (1635–1700). See Kirchmaier, *Institutiones metallicae* (1687).

[73] Marperger, *Das neu-eröffnete Berg-Werck* (1704). The connection between Marperger and Schönberg is established in Tenfelde, Bartels and Slotta, *Geschichte des deutschen Bergbaus* (2012), vol. 1, p. 472.

[74] For an introduction to the imposing literature on this central aspect of seventeenth-century science, see Cohen, *The Scientific Revolution: A Historiographical Inquiry* (1994), pp. 183–198, '3.4. The New Science and the Creation of "Artificially Produced Nature"'.

were mediated by mining maps just as they were mediated by the new instruments of physics. On the one hand, a map allowed one to look through stone and allowed to be seen what had been hitherto hidden from human eyes. On the other hand, recording dimensions and putting them on paper was a highly selective procedure that highlighted some relations and eclipsed others. As with mechanical instruments, several decades would pass until the maps were fully standardized and their original imperfections overcome. Moreover, our analysis of the *Freiberga subterranea* shows that the cartography of underground riches was not meant to be neutral. Geometrical techniques were mobilized to advocate for investments in deep-mining infrastructure. As a technical activity, cartography could not be separated from its political interpretation; this was especially true when maps were used as a planning tool, as was increasingly the case. Politically and culturally, cartographic works had to be legitimated before they could play a significant role as a new, technical instrument. This process was not specific to the mines and explains why, in Western Europe, 'atlases and maps were not the familiar instruments of administration that they were to become' before the late seventeenth century.[75] In Saxony, multiple efforts in this direction were coordinated and spearheaded by Captain-general von Schönberg.

A striking counterexample from France, whose mines regulations developed outside the cultural influence of the Holy Roman Empire, will illustrate this point.[76] Although early modern France suffered no shortage of brilliant mathematicians, and mining maps indeed existed, they were not generally accepted as appropriate tools. As late as 1761, almost three generations after the *Freiberga subterranea*, a legal dispute over the mine of Pont-Péan (Brittany) revolved entirely around maps. Jean-Pierre Duhamel, an engineer sent by the king, was in the first instance flatly denied the right to draw a map by the local investor. Duhamel's ability to draw a precise map was not contested – incidentally, he would author the first treatise in French on subterranean geometry. His opponent disputed, however, that the 'dry study of geometry' could in any way help manage mine workings.[77] 'Would it be necessary to be a *géomètre* to know if there is ore in one place?', he asked ironically, soberly adding: 'It is well-known that a map only draws a mine in its passive state', and could not be useful for managing operations.[78] This argument was not the isolated babble

[75] Wandel, 'Maps for a Prince' (2019), p. 175.
[76] For an overview of mining culture in France, see Demeulenaere-Douyère and Sturdy, *L'enquête du Régent, 1716–1718* (2008).
[77] Bibliothèque départementale d'Ille-et-Vilaine, C 1485, f. 108: 'l'etude seche de la geometrie'.
[78] Bibliothèque départementale d'Ille-et-Vilaine, C 1485, f. 117: 'seroit-il donc necessaire d'être géomètre pour scavoir s'il existe du minéral dans un lieu?'; f. 117v: 'on scait bien qu'un plan trace passivement l'etat d'une mine'.

of a frustrated, harebrained investor. If Duhamel's map was ultimately drawn, both parties agreed that 'it could not be legally used in any way' to resolve the dispute.[79] This example underscores how the efficient use of geometry in the German-speaking world, far from being natural or self-explanatory, was indeed a social construction.

## Administrating Data

The new technology of combined mining maps deeply transformed the visualization of the underground. In the Holy Roman Empire, Central Europe, and Scandinavia, plans and sections quickly gained acceptance as valid instruments to monitor the mines.[80] Scarcely used in the early seventeenth century, they soon came to embody the discipline: 'a subterranean geometry without engravings is a dead thing' sentitiously wrote a scholar a few decades later, apparently unaware that virtually all those maps were hand-drawn, not printed using copper plates.[81] Once artistic depictions presented to rulers, maps now firmly obeyed a logic of administration, in which standardization was of utmost importance. To be of durable use, each surveyor's cartographic technique had to be sufficiently systematized to be used by his colleagues and still be understood decades or even centuries later. In the Harz region, the Goslar mining office once complained that data collection and map drawing were 'still so personally colored that a subterranean surveyor could not find his way into the *observata* of another one'.[82]

Moreover, the growing number of plans soon made it mandatory to organize their storage, to ensure that different representations of a pit could be simultaneously worked on and fruitfully compared. In Saxony, no unified cartography cabinet existed when Schönberg was first appointed in 1663. There was no shortage of documents, beginning with the dozens of maps and sketches made by Balthasar Rösler since the beginning of his career.[83] Most of these resources, however, were uneven in their quality. Mining reports of the sixteenth century were mixed with

---

[79] Bibliothèque départementale d'Ille-et-Vilaine, C 1485, f. 150: 'sans qu'il puisse être opposé en façon quelconque'.

[80] On the use of maps in the territories of the Austrian crown, see Kirnbauer, 'Die Entwicklung des Grubenrißwesens in Österreich' (1962), pp. 60–129. On visualization and the Swedish mining administration, a recent resource is Fors and Orrje, 'Describing the World and Shaping the Self' (2019), pp. 107–128. The use of maps in *Nueva España* seems to have been less immediately accepted, see Nuria, 'Underground Knowledge' (2016).

[81] Lehmann, *Kurze Einleitung in einige Theile der Bergwerks-Wissenschaft* (1751), introduction: 'eine Marckscheide-Kunst ohne Kupfer ist eine todte Sache'.

[82] Quoted in Pitz, *Landeskulturtechnik, Markscheide- und Vermessungswesen* (1967), p. 196: 'noch so sehr persönlich gefärbt, daß sich ein Markscheider *in des andern observata nicht zu finden wüßte*'.

[83] Jobst and Schellhas, *Abraham von Schönberg, Leben und Werk* (1994), p. 129.

sketches, while only a fraction of the pits was mapped using modern methods. The five decades of his career saw a swift normalization of mining maps. The financial aspect of map-making is essential, for it ultimately conditioned the acceptance of the new system by practitioners. Schönberg's 1667 decree stipulated that maps should be sent free of charge to the administration, but surveyors were tough negotiators. They usually received a modest fixed salary and were mostly paid-by-the-task, for instance to set marking stones or ascertain a vein's direction. By collecting maps, the administrations aimed precisely at reducing the need for frequent surveys, potentially drying up their already meagre source of income. In most mining states, a compromise was gradually reached about a new mode of payment '*pro studie et labore*'.[84] Instead of being paid for specific actions, surveyors managed to obtain a compensation for the 'arduous work' of collecting data. In practice, each surveyor produced an invoice containing his 'surveying fees' (*Markscheidergebühren*) as they sent the 'free' maps to the administrations. It included the number of angles that had been recorded, the lengths and duration of the operation, but also the price of paper and inks. Today, these documents help us understand the exact courses of surveys, as for example in Clausthal, where surveyor Rausch listed his costs for 'a horse and a carriage' or his 'assistants'.[85]

Maps had once been considered the property of the engineer who had drawn them, and used to be stored in a surveyor's own house. In Saxony, Hans August Niemborg described the chaos of his 'mapping activities and maps to draw, accumulating by the day', while he had 'already a whole cabinet full of these'. Years later, he requested an apartment from the Elector, alerting that the some 800 maps he had accumulated were 'about to be completely spoiled by humidity and rot'.[86] Right after being appointed Captain-general, Schönberg bought a house in Freiberg for the mining office, offering a proper storage for maps.[87] It was used, among other things, to store Johann Berger's *Freiberga subterranea*, of which we know that it was still used half a century later.[88]

Controlling where maps were stored was crucial, since mining offices were increasingly dependent on the data they contained. When Conrad Christian Elster, mining surveyor in the Harz Mountains, considered that he was definitely

[84] NLA HA Dep. 150 K. Acc 2018 / 700, Nr. 3.
[85] NLA BaCl, Acc. 8 Nr. 300, f. 2r-v: 'Pferd und Karriol … Gehülfe'. See also Schönberg, *Ausführliche Berg-Information* (1693), p. 113: The exact price of a survey was to be determined by the mining master and the jurors, 'since there is more effort in some place than in others'.
[86] SächsStA–D, 10026 Geheimes Kabinett, Nr. Loc. 01254/06, f. 6r: 'täglich sich häuffender Lander-Verrichtungen undt zuverfertigen habenden Land-*Mappen*, derer ohne dies bereits ein ganzer Schrank voll vorhanden'; f. 41r: 'durch Feuchtigkeit und Mohr kunfftig hingänzlich verderben'.
[87] Jobst and Schellhas, *Abraham von Schönberg, Leben und Werk* (1994), p. 57.
[88] A copy was made around 1750 (signature SächsStA–F, 40040, G 5004), which means that it was still in use at that point.

not paid enough, he used a mission to Silesia as a way to abscond, and entered the rival Prussian administration. Elster had foreseen that his family, still in the Harz, would be taken as hostages in response to his desertion. He had thus carefully devised a plot and hidden all the mining maps he owned, his catalogues, and even his field books. Without the maps that allowed them to see through stone, the mining administrations were quite literally blind and had no choice but to fold. Elster managed to get his family sent to Prussia and even extorted 300 thalers from his former employer before revealing the location of his treasure.[89] The anecdote explains why the first move of mining administrations, when a subterranean surveyor or a mining master passed away, was to 'seal the mining maps and other belongings found in the house', if possible 'immediately after the death' to prevent any theft or loss.[90]

Once centralized, maps became a vital component of the planning process, which now took place in the mining offices. Abraham von Schönberg increasingly relied on them to settle legal disputes. This can be illustrated, for instance, with a case that unravelled in Schneeberg in 1686. It revolved around complex issues about the precedence of the plaintiff and the nature of two ore veins exploited by rival companies. The concession limits had once been set 'by visual observation', but renewed visits organized by the local mining office stayed inconclusive. Schönberg took the case in his own hands and sent Martin Hornig, subterranean surveyor in Freiberg, to discuss with the parties and 'draw a map about the contested mine workings'. The Captain-general then reviewed the various files and received from surveyor Hornig 'a verbal explanation of his map', which allowed him to quickly reach a decision.[91] Stored in the mining offices, maps were used by the higher officials as forecasting tools, enhancing their ability as projectors. Perusing a mining map was not only quicker than a visitation; it gave the projector another perspective and helped him 'to better imagine his district', as the *Mirror of Mining* argued.[92] Since combined plan and sections were now accurate, a good part of the planning work could be carried out on the map, by protracting angles and distances.

---

[89] Pitz, *Landeskulturtechnik, Markscheide- und Vermessungswesen* (1967), pp. 208–209.

[90] NLA HA, BaCl Hann. 84a, Nr. 6698, Markscheiderregistraturen und -inventarien, Zellerfelder Revier 1711–1809, *Extract Protocolli* dated 8 February 1720, §18: 'die im Sterbhause befindliche abriße und was dazu gehörte so versiegelen ... sogleich ... nach Absterben'. See for another example NLA HA, BaCl Hann. 84a, Nr. 6698, 8 February 1720. The surveyor Johann Just Schreiber is dying of a 'nasty contagious disease'. The administration wants to seal (*versiegeln*) his house to ensure that none of his maps and field books will disappear, but balances the contagious risks of visiting him.

[91] SächsStA–F, 40015 Bergamt Schneeberg, Nr 1448, f. 76r: 'einen riß über die streitigen gebäude fertigen zu laßen'; f. 66r: 'mündlichen Erklährung seiner, *as Acta* gebrachten Grund-Riße über des leztens'.

[92] Rösler, *Speculum metallurgiae politissimum* (1700), p. 87: 'damit er sich seine Refieren desto besser einbilden kan'.

Beyond the emblematic figure of Abraham von Schönberg, maps and data came to be systematically used to monitor the use of natural resources, most prominently mining and forestry, of the Holy Roman Empire.[93] In the Duchy of Brunswick, a student of Balthasar Rösler went on to produce general maps of the local districts in the aftermath of the Thirty Years War.[94] These monumental maps were several metres long and were kept rolled up in the central mining office. In the words of the local *Berghauptmann*, maps of a complete district 'one can demonstrate to anyone outside the pits, as if ocularly [*ad oculum*], how mine workings present themselves even in the deepest depths'.[95] There, too, surveyors were soon required to send copies of their map to the administration, and to keep '*protocollum* and surveys book'.[96] Similar decrees about 'the record keeping of all mining maps and [survey] books' were promulgated by the Habsburg monarchy in 1709.[97] Everywhere, visits reports were now supplemented, and sometimes superseded, by maps.

In the final years of his administration, Abraham von Schönberg was able to enforce a last reform of underground geometry, almost unnoticed but highly significant. Its trace can be found in an austere manuscript buried in the mining archive of Freiberg. Just as Berger was finishing his monumental cartography, his patron appointed a new surveyor, Paul Christoph Zeidler (1660–1719). In the book's dedicatory letter, Zeidler thanked the mining administration for financing his apprenticeship – by then a common practice – and took the customary oath to be a faithful servant of the mining state. The significant novelty is found in the last sentence of his introduction: 'And for everything to be accurate and useful for others, it should be diligently *protocolli*ret and recorded in the present book.'[98] The rather monotonous piece of work scrupulously recorded over hundreds of pages all the data tables collected by

---

[93] For instance, a general map of the Harz forests was established in 1680 by two surveyors, Henningo Groscurt and Johann Zacharia Ernst. See Bei der Wieden and Böckmann, *Atlas vom Kommunionharz in historischen Abrissen von 1680 und aktuellen Forstkarten* (2010).

[94] This student was Daniel Flach (ca. 1625–1694). His general map of the Clausthal district, drawn in 1661, is 9.5 metres long and 1 metre high.

[95] NLA HA, Dep. 150 K. Acc 2018 / 700, Nr. 3, introduction: 'einem Jedweden außerhalb der gruben, gleichsam *ad oculum* remonstrirt werden kann, wie die Gruben auch im tieffesten sich praesentiren'.

[96] The repeated orders of the administration and the opposition of surveyors can be read here: NLA HA, BaCl Hann. 84a, Nr. 6684.

[97] Decree of 1709 'Zur sicheren und verläßlichen Evidenzhaltung sind von dem gesammten Bergbaue Mappen und Bücher auf nachstehende Art zu errichten'. See Schmidt, *Chronologisch-systematische Sammlung der Berggesetze der österreichischen Monarchie* (1835), vol. 5, document from 30 June 1709. On cartography in the Hapsburg monarchy, see Kasiarova, *Bergbau- und Hüttenwesenvergangenheit der Slovakei in kartographischen Quellen* (2010).

[98] SächsStA–F, 40047 Winkelbücher, Nr. 1429, f. 1, 1693: 'Damit aber alles in guter Richtigkeit undt andern brauchen möge so soll es in gegenwärttiges Buch fleißig *prodocolli*ret und eingetragen werden.'

Zeidler during his career. Similar field books came to be labelled angle books or observation books (*Winkelbücher, Observationsbücher*). Subterranean surveyors had previously sent only invoices and finished maps, but the underlying data, from which new maps could be produced, used to be strictly theirs. At the turn of the eighteenth century, the mathematical data recorded by surveyors came to be considered as belonging to the administration. The annual *Observationsbücher* were now stored in the new building of the mining office in Freiberg, rather than in the possession of the individual surveyor. Within a generation, Abraham von Schönberg had not only tightened his grip over standardized maps, but also obtained the underlying raw data from which they had been drawn.

*

Making the underground visible or palpable was a crucial step in the process of making it fully manageable. The long textual descriptions of mining reports had to be abbreviated, made coherent and manageable, as the underground was hardly reducible to words. Visualization had been traditionally performed by reproducing surveys on frozen lakes, painting mining landscapes, or constructing models, but these methods hardly allowed for precise, quantifiable considerations. In the second half of the seventeenth century, the development of mining maps finally solved the conundrum and quickly became a pre-eminent skill of subterranean surveyors. The crucial advantage of this new cartographic language was to offer both a graphical depiction and an accurate mathematical tool.

As centralized administration developed, maps helped enhance their control over mineral resources, both in space and in time. Maps thus became a memorial of the mines, considerably more precise than the laboriously drafted visit protocols. The development of combined representations was a technical achievement for subterranean surveyors, and one that was largely independent from the contemporary evolution of cartography. Their growing accuracy notwithstanding, the success of mining maps was in no way a natural development from preceding conditions. The rise of mathematically produced maps as specialized reform tools, far from being restricted to mining, can be observed in several other regions of Europe. It happened along similar timelines for the Dutch *polders*, the draining of the English Fens, the digging of the *Canal du Midi* in the South of France, and the metal mines of Central Europe.[99]

In Saxony, Abraham von Schönberg was instrumental in the social acceptance of mining maps. The *Freiberga subterranea*, patiently produced by Johann Berger, illustrates how the new underground geometry, enhancing the administrative power of mining offices, promoted political and technical

[99] See Johnston, *Making Mathematical Practice* (1994), p. 246; Streefkerk, Werner, and Wieringa, *Perfect gemeten* (1994), p. 10; Willmoth, *Sir Jonas Moore* (1993); Mukerji, *Impossible Engineering* (2009).

Figure 5.8    Captain-general of the mines Georg Friedrich von Schönberg (1586–1650), above, and Caspar von Schönberg (1621–1676), below. The original portraits are on the left, while the copies later ordered by Abraham von Schönberg, holding respectively a suspended compass and a semicircle, are on the right. Courtesy Schönbergsche Stiftung. Photo: Ingo Ladleif von Schönberg.

reforms of the mining states. In early modern societies, however, artefacts were not self-explanatory. While customary methods of visual inspections had long been an integral part of the vernacular culture of mathematics, learning how to work with a combined map was a long process. Abraham von Schönberg thus put his weight behind the new mining geometry, hoping to reap political gains from this emerging technology. He repeatedly chose to be portrayed as a *Markscheider*, using his pair of compasses to perform measurements on a mining map.[100]

As an astute administrator, von Schönberg very consciously promoted geometry; at the same time, he was careful not to emphasize the novelty of the technique. The Captain-general wisely acted as if it had always been part of the surveying tradition to the point of redrawing history. As his portrait was being drawn, Abraham also asked the painter to remake the portraits of several of his predecessors that hung in the central mining office (see Figure 5.8 and 5.1). Caspar von Schönberg (1621–1676), previously holding fine gloves, was now depicted with a semicircle in hand, just as a modern surveyor would be. Caspar's father Georg Friedrich von Schönberg (1586–1650), instead of a precious gem, suddenly held a suspended compass – an instrument that did not even exist in his own lifetime![101] Depicting modern instruments in the hand of ancient administrators was meant to ease the acceptance by emphasizing a fictional continuity, downplaying the radical novelty of his reforms. By making the maps and instruments visible on the portraits of its forefathers, the Captain-general astutely made the new subterranean geometry inconspicuous. Just as the mathematization of underground nature was accelerating, Abraham von Schönberg pretended that it had always been the case.

---

[100]  Besides the 1681 portrait analysed above (Figure 5.1), there is a 1711 etching realized by Martin Bernigeroth conserved at the University of Leipzig (*Bildnis des Abrahamvs a Schönberg*).

[101]  On the first series of portraits, see Fritzsch, 'Die Schönbergporträts der Bergakademie Freiberg' (1962), pp. 311–317. On the second series, see Schmidt, 'Die Familie von Schönberg und das Sächsische Oberbergamt' (2004), pp. 32–66.

# 6 How to Teach It?

## Finding the Right Direction

During my seventeen years of teaching duty in Clausthal, first as *conrector* and then as *rector*, I have taught mathematics to the youth; among them, I even had some capable minds. However, I cannot remember of a single one who would have gone from this school to mining or construction, and would thus have brought usefulness from the theory of mathematical sciences.

Calvör Henning, teacher and Bergverständiger, 1763*

This chapter describes the challenges met by the mining culture of mathematics during the eighteenth century, its evolution, and how it finally came to be taught around a new kind of institution, the mining academies. The *Bergakademie* Freiberg was founded in 1765 to train officials, engineers, and scientists; several others soon followed around the world, most notably in Schemnitz and Clausthal. The popularity of academies, I argue, partly relied on a painful but in the end successful integration of subterranean geometry, and more generally of mathematics and natural sciences, into a coherent curriculum. These institutions pioneered 'a fortunate synthesis of school and workshop culture' that would be widely imitated by later institutions.[1]

Despite a seemingly obvious success, their role has been questioned on various grounds. In some cases – the *Bergakademie* Berlin for instance – a prestigious name turned out to be little more than a 'precarious and unstable' series of courses.[2] The prestigious Parisian *École des mines*, a powerful institution, famously did not end the French reliance on German-speaking experts; even the *Bergakademie* Freiberg has lately been presented as a purely political endeavour detached from technicalities.[3]

---

* Henning, *Acta Historico-Chronologico-Mechanica* (1763), p.7: 'Ich habe ins 17. Jahr in meinem Clausthalischen Schulamte, erstlich als *Conrector*, hernach als *Rector*, die Jugend in der *Mathesi* mit unterrichtet, und darunter gar fähige Köpfe gehabt ... Ich weiß mich aber nicht auf einen einzigen zu besinnen, der sich von solchen aus der Schule zum Berg- oder Zimmerwerk begeben, und dabey mit solchen mathematischen Wissenschaften, vermittelst der Theorie, Nutzen geschafft hätte'.
1 Brianta, 'Education and Training in the Mining Industry, 1750–1860' (2000), p. 283.
2 Klein, 'Ein Bergrat, zwei Minister und sechs Lehrende' (2010), p. 463: 'prekären und unstabilen'.
3 On the short-lived *École des mines* (1783–1788), see Birembaut, 'L'enseignement de la minéralogie et des techniques minières' (1964), pp. 365–418, esp. pp. 397–403 on the small number of

How to properly teach mathematics for practitioners? The question, we will see, plagued mining administrations for most of the century. Brick-and-mortar institutions now appear the natural venue and are thus wrongly assumed to be the only viable system. In the eighteenth century, however, states' needs, local resources, and political considerations were multifarious. In mining regions, there was a widespread – and largely justified – defiance towards universities, and more generally 'all the institutes that taste too much like school'. The very idea to use higher science to improve the extraction of ores was not fully consensual, and was seen by some as 'ridiculous', if not a 'concocted chimera'.[4]

The institutionalization of subterranean geometry, ore assaying, or machine building presented the mining states with new challenges as each subdiscipline grew in complexity. In the following sections, I describe how a consensus was gradually reached. Everyone agreeing that teaching these crafts in the mines was not enough, the mining administrations had to get more actively involved. Besides, subterranean surveyors, machinists and engineers now needed to be trained in a coherent and standardized manner. Highly disputed, however, was the most adequate way to combine specific on-site instruction with general theoretical elements. 'Codification', Jochen Büttner has noted, 'necessarily involves selection, abstraction and, more often than not, the emendation of practical knowledge with knowledge from other domains.'[5]

This chapter tries to offer a balanced account of the creation and early decades of the Freiberg mining academy by focusing on the teaching of mathematics, which has traditionally been neglected in historical analyses. It dissects how a compromise on the best way to train engineers was sought in the form of collaborations between mining officials and academic professors. In fact, new institutions did not instantly revolutionize the teaching of mathematics and natural sciences. The status of higher mathematics, for instance, long remained a sticking point. The elaborate system of manuscripts described in the previous chapters continued for several decades, symbolizing the enduring authority of practising *Markscheider* on both students and professors. But these institutions were not empty shells serving unavoidable political purposes. Graduates from the *Bergakademie* Freiberg were indeed able to combine practical experience and theoretical innovation, something that – more often than not – eluded the university-trained mathematicians or cameralists of the time.

In a first section, a description of how mining sciences were taught in the early eighteenth century underlines the inner coherence of the companionship system. Faced with undeniable shortcomings, the mining administration tried

students, the lack of a library, and the absence of coordination. The thesis of the *Bergakademie* Freiberg being a largely empty shell is developed in Wakefield, *The Disordered Police State* (2009), in the introduction and ch. 2.
[4] Zimmermann, *Ober-Sächsische Berg-Academie* (1746), p. 46: 'alle Anstalte, die zu sehr nach der Schule schmecken ... lächerlich ... ausgeheckte Grillen'.
[5] Büttner, 'Shooting with Ink' (2017), p. 118.

to adapt the existing frame instead of overthrowing it. I then turn to a decisive, yet virtually unknown figure, mining master Johann Andreas Scheidhauer (1718–1784). An autodidact, he acquired a remarkable knowledge of the most recent mathematical theories, meticulously recorded in dozens of personal notebooks. Scheidhauer's curiosity allowed him to successfully combine higher mathematics with the daily issues of mining. Working behind the scenes, he promoted a new teaching of subterranean geometry just as the *Bergakademie* opened his doors. In the early years of the academy, apprenticeship was then preserved and slowly integrated in a more public academic training. Mining officials, many of them with a deep knowledge of mathematics, tried to accommodate a broad teaching of elementary arithmetic and geometry with the rapid development of a specialized engineering knowledge. Rather than engaging in fruitless disputes between theory and practice, a new generation was trying to find the best way to teach useful mathematics.

### The Mining *Universitas* and Its Limits

When Johann Berger died in 1695, shortly after completing the monumental cartography of the *Freiberga subterranea*, his pupil August Beyer held a funeral speech. Beyer lauded his master, who for decades had been a revered teacher, in the following manner:

> Here lies the university
> Of all mining and smelting sciences
> It is here that councils of other towns
> And whole countries get instruction,
> To direct mining operations
> If they are to work properly
> And from here the paragon is expressed,
> Of mining and other arts.[6]

While Beyer indeed used the word *Universität*, no such thing existed in Freiberg at the time. When a request was indeed addressed to the Saxon King some decades later, asking to have 'an *Augustus Universität* founded' in the city, it was quickly dismissed because the need for such an institution was hardly felt.[7] A network of mining experts and officials already lived in the capital of the Ore Mountains and acted as a *universitas montanorum*, a 'voluntary association of individuals with collective will and means of exercising

---

[6] Herrmann, 'Bergbau und Kultur' (1953), pp. 26–27: 'Hier ist wohl die Universität / Aller Bergwerks- und Schmelzkünste / Hier erholn sich Rates andre Städte / Ganze Länder, wie die Dienste / Und das Bergwerk müsse sein / Recht bestellt, soll's treffen ein / Und von dar verschickt der Ausbund / Des Bergwerks und andrer Künste.' The full poem is transcribed in Herrmann, 'Ein bergmännisches Gedicht aus der Barockzeit Freiberg' (1971), p. 210.

[7] Jobst and Schellhas, *Abraham von Schönberg, Leben und Werk* (1994), p. 51, attributes this demand to August Beyer, without evidence.

a common goal'.[8] The training of subterranean surveyors and assayers, as we have shown, was regulated from at least the end of the sixteenth century. Berger himself had taught these arts privately within the system of contracts and manuscript mentioned in Chapter 4. For instance, a course written down by a student bears the inscription '*Anno 1694–1695. Johann Gottlieb Voigt. LL Studiosus*'. This manuscript, Voigt assured, had been proofread by Berger who had eventually delivered the usual certificate.[9] Administrative sources and certificates routinely used the term of *privatim*, as one did in most universities of the time. There too, the bulk of teaching – besides a handful of public elementary courses – happened in small groups, paying private lessons, at the end of which students received an attestation.

In 1702, the training system existing in Freiberg was formalized by Captain-general Abraham von Schönberg, as part of his general reform of the mining administration described in the previous chapter. A permanent grant system was put in place to 'introduce some young people to the learning of mining sciences, the arts of smelting and underground surveying, and the like'.[10] This evolution, significant as it is, should not be overstated. Training still took place in surveyors' houses and was directly put into practice in neighbouring mines. Nor was the system fully adapted to local needs. A few years after its introduction, there were already complaints about overfilling. In 1713, David Gottlob Lehmann complained that a hard-won grant had given him 'a sure hope' of quickly securing a job in a mining office. This expectation, however, was not realized. Alluding to 'his previous practice of law' to underline his versatility, he wondered why 'it had been more than six years' that he had not 'received any position'.[11]

The situation was similar in most mining centres, and more generally in all European countries trying to institutionalize their technical training. In the Austro-Hungarian Empire, the yearly sum of 1500 Gulden was provided not only for the training of mining students, but also for 'mathematical instruments, and a few books from distinguished authors … as well as numerous mining compasses, two surveying instruments [*Schünzeuge*], and an assaying scale with its weights'.[12] Still, future mining officials, known as *Expenktanten*, could only be appointed

---

[8] Graulau, *The Underground Wealth of Nations* (2019), pp. 160–161. The evolution of the concept of *universitas* in early modern German is addressed in Walker, *German Home Towns* (1971), ch. 2. More specifically on mining cities as *universitas montanorum*, see Bornhardt, *Geschichte des Rammelsberger Bergbaues von seiner Aufnahme bis zur Neuzeit* (1931), pp. 62–63.

[9] Voigt's certificate is reproduced in Herrmann, 'Ein bergmännisches Gedicht aus der Barockzeit Freiberg' (1971), p. 209.

[10] See Baumgärtel, 'Von Bergbüchlein zur Bergakademie' (1965), pp. 114–121; Sennewald, 'Die Stipendiatenausbildung von 1702 bis zur Gründung der Bergakademie Freiberg 1765/66' (2002), pp. 407–429.

[11] SächsStA–F, 40001 Oberbergamt Freiberg, Nr. 2310, f. 35r: 'bereits geraumer Zeit … desto sicherer Hoffnung … Von meiner bisherigen *Praxi Juris* … bereits über 6. Jahre … ohne daß ich einige *Function* erhalten'.

[12] This was summarized in a 1735 *Instruction*, Schmidt, *Chronologisch-systematische Sammlung der Berggesetze der österreichischen Monarchie* (1835), vol. 5, pp. 474–475. Grants were

when positions were freed.[13] The issue was similar in France, where the civil engineering school of *Ponts et Chaussées* (which also educated students for mining careers) ran into trouble for training too many students. Such students, literally labelled *surnuméraires* (supernumerary), had to wait for a vacancy, sometimes for decades, in order to get a position.[14] The modest size of mining administrations in the eighteenth century was barely compatible with a sustained training system, let alone a full-fledged institution.

A more fundamental flaw in the mining education was the absence of a structured initial training. A contract signed in Eisleben in 1742, for instance, went beyond the proper mine surveying. On top of the usual sum, the local surveyor was promised a *douceur* of 70 thalers for presenting the following content: '1) decimal arithmetic and the five operations of arithmetic, 2) the simple rule of three (*directa* and *inversa*), the compound rule of three 3) the measurement of lines 4) the measurement of surfaces, or *planimetria*, 5) the measurement of volumes, 6) the *trigonometria*.'[15] Apprenticeship was thus diverted from its initial hands-on approach and ultimately amounted to an elementary arithmetic and geometry course. The idea of compensating for the lack of basic training was a sensible one, as such training was sorely lacking, but failed to address the real issue. Surveyors, who had themselves learned by doing, were hardly well-placed to teach elementary mathematics. Around the middle of the century, voices asking for reforms were getting louder: 'The old times are gone', wrote a higher Saxon official, 'one needs work, one needs investments, and one also needs reason if one wants to save both.'[16] In the Duchy of Brunswick, the mining administration was increasingly reluctant to deliver teaching grants. 'It would come down to sheer hazard' to invest time and money into a young man not previously prepared, since so 'few young people can be taught with *accuratesse*.'[17] Remarkably, these calls for more theory and reasoning were not ushered by scholars but came from within the mining administration.

---

established in the important mining towns of the Austro-Hungarian Empire, such as Saint Joachimsthal, Schmölnitz, Orvice, and Idria.

[13] This system was formalized by an instruction published in 1676. On the early seventeenth century, see Mihalovits, 'Die Gründung der ersten Lehranstalt für technische Bergbeamte in Ungarn' (1938), p. 10. About the 1676 instruction, see Schmidt, *Chronologisch-systematische Sammlung der Berggesetze der österreichischen Monarchie* (1835), vol. 5, p. 239.

[14] See Birembaut, 'L'enseignement de la minéralogie et des techniques minières' (1964), pp. 365–418.

[15] SächsStA–F, 40001 Oberbergamt Freiberg, Nr. 3477, f. 10r: 'ihn dahero in folgenden Speciebus 1) der Decimal Rechnung nach allen 5 Speciebus [9v] der Arithmeticae 2) der Regula de tri Simplae directa und inversa ingleichen der Regula de tri composita und regula focutatis 3) der Eythymetria 4) der Epydometria oder Planimatria 5) der Stereometria 6) der Trigonometria.' The name of this apprentice was Johann Carl Richter.

[16] Zimmermann, *Ober-Sächsische Berg-Academie* (1746), p. 40: 'Diese Zeiten sind vorbey, die reichen Anbrüche unterm Rasen sind weggehauen, aber herum ist nichts mehr zu finden, man braucht also Arbeit, man braucht Kosten, man braucht auch Verstand, wenn man beides ersparen will.'

[17] Wolfenbuettel, NLA WO, 2 Alt, Nr. 19675, f. 6r: 'indem es auf reinen *hazard* ankommen würde' (letter from 1 June 1750); f. 5v: 'weniger jungen Leuten mit *accuratesse* beygebracht werden'.

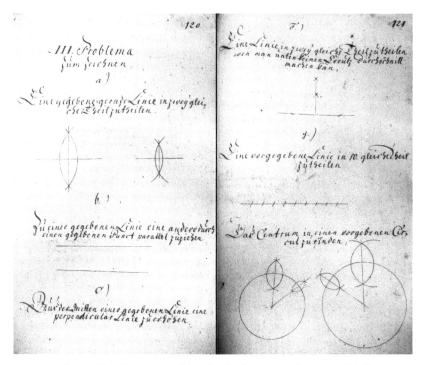

Figure 6.1  'Drawing problems' solved by an applicant for a Freiberg mining grant. The applicant is asked to draw parallel and perpendicular lines, to divide a given segment into equal parts and to find the centre of a given circle, among other things. These standardized exercises were part of the examination of 'those who want to learn the arts of assaying and subterranean geometry'. Courtesy TU Bergakademie Freiberg/Universitätsarchiv, UAF – OBA 6, f. 120r – 121r. Photo: Thomas Morel.

A couple of years later, the King of Saxony reacted by publishing a new mining regulation, introducing a preliminary test for 'those who want to learn the arts of assaying and subterranean geometry'.[18] The examination had three parts: writing, computing, and drawing (see Figure 6.1). In concrete terms, the administration wanted to ensure that all candidates entering an apprenticeship mastered the four operations, knew the rule of three, and could aptly use their ruler and compass. The regulation contained a second major provision: If subterranean surveyors had *de facto* become professors, they had to be designated as university professors were. A selection system was introduced, strikingly close to the hiring process in place at the neighbouring *Universität* of Leipzig.

---

[18] UAF, OBA 6, f. 120r–121r: 'Specimina vor Diejenigen, welche die Probier- und Marckscheide Kunst erlernen wollen.'

From then on, the mining administration selected three qualified candidates, made an ordered list and determined the favourite, who would be officially appointed by the ruler. Although mostly formal, the system proved at least effective in deflecting flagrant nepotism.[19] If the administration was unsure, as happened in 1747 for the position of surveyor in Johanngeorgenstadt, the local mining master would invite the three candidates, bring them into a gallery, and 'instruct them to perform a survey, and draw the corresponding map', thus testing their skill with real-life 'duties' (*Aufgaben*).[20]

In addition to the administrations' efforts, particular initiatives flourished, as many learned officials actively fostered innovations and the circulation of knowledge. Johann Friedrich Henckel (1678–1744), a well-known chemist and physician, became assessor in Freiberg in the 1730s. He owned a chemistry *laboratorium*, a collection of minerals, and regularly trained young people. One of his students, Johann Gottlieb Kern (d. 1745), wrote a *Bericht vom Bergbau* (Report on Mining), a widely circulated manuscript textbook that was eventually printed after the creation of the academy.[21] Each mining district had an official surveyor who regularly trained apprentices. Against this background, it is clear that 'a comprehensive training in mining was already available' in the mid-century, and similar, if less developed systems, existed in all major mining regions.[22] Rather than being a sudden development, the creation of a mining academy in Freiberg appears as a rationalization and gradual improvement of existing initiatives.[23]

How could a curious young man find his way in this seemingly fragmented training landscape? An anonymous writer, ready to help, published in 1747 the *ABC of Mining Sciences, Framed in Questions and Answer to Instruct the Youth*.[24] The thin booklet provided definitions for the basic notions of metallurgy, mineralogy, and underground surveying, together with a reading list. Once the reader had acquired these rudiments by himself, it indicated how

---

[19] Candidates would present their certificates, but also '*specimine*' of their works, or letters of reference. This nomination system mirrors the selection of professors at the neighbouring University of Leipzig, as can be read in Morel, *Mathématiques et politiques scientifiques en Saxe 1765–1851* (2013), pp. 36–41, 66–68, 91–98.

[20] SächsStA–D, 10036 Finanzarchiv, Nr. 32833 Rep. LII, letter from the higher mining administration to the king, 6 May 1747: 'ein Zug durch den Bergmeister Christian Salomon ZEIDLER aufgeben und ieder solche vor sich zu verrichten, so wohl darüber ein Riß zu fertigen, anweisen laßen'.

[21] Kern and Von Oppel, *Bericht von Bergbau* (1772).

[22] In Schemnitz, for example, a *Bergschule* (mining school) was created in the first half of the century, see Konečný, 'Die montanistische Ausbildung in der Habsburgermonarchie, 1763–1848' (2013), pp. 95–101. Other historians have underlined that those early institutions often revolved around a single teacher, in this case Samuel Mikoviny, and withered away after his death. See Kamenicky, *Baňícke školstvo na Slovensku do založenia Baníckej akadémie v Banskej Štiavnici* (2006).

[23] Sennewald, 'Die Stipendiatenausbildung von 1702 bis zur Gründung der Bergakademie Freiberg 1765/66' (2002), p. 410: 'eine umfassende montanistische Ausbildung ... bereits möglich war'.

[24] See Baumgärtel, 'Von Bergbüchlein zur Bergakademie' (1965), p. 83.

to begin with the active part of training. The *ABC of Mining Sciences* recommended 'to look around in mining pits, in the smelting huts … to visit the assaying chambers and the subterranean surveyor's *Museum*, but also to frequent and converse diligently with humble, clever, and skilled mining and smelting people, from both high and low extraction'.[25]

I think that such a system – which might retrospectively appear esoteric compared to more centralized institutions – was obvious for contemporaries. The *ABC of Mining Sciences* was a quasi-official introduction to the rich *universitas*, the community of experts who lived and worked in the Ore Mountains of Saxony. The idea of a brick-and-mortar institution, rightly considered as very expensive, was still a distant dream, while the indisputable shortcomings of the system were addressed piecemeal rather than given a thorough overhaul. Most of its future features, including the collections of minerals and instruments, the libraries, and laboratories, already existed. They simply belonged to private persons who saw themselves as *Bergverständiger*, that is, 'knowledgeable of mining'.[26] A sense of common good was indeed present, as most of them were mining officials. In this framework, the role of the state was a regulatory one, ensuring that enough resources were available for selected individuals at any given time.

The defiance of the *Bergverständiger* against learned knowledge was real. When pastor Christian Ehrenfried Seyffert (1683–1729), who had been teaching religion and Latin for years, offered to open a 'small mining academy' in the 1720s, he was promptly rebuffed.[27] Mindful of possible ambiguities, the administration suggested to use the label *Erudita Societatis Metallicae* (learned mining society) instead. To warrant the name of a mining academy, his prospective students should 'spend much more time' in surveying mines and assaying ore, and do this under the supervision of an 'officially appointed surveyor'. More specifically, one deplored that the putative teacher was not 'himself skilled in mining sciences', meaning that his scholarship was not directly usable in the mines, and the project ultimately aborted.[28] In the first half of the century, the mining administration thus essentially saw itself as a regulator. Defending the traditional apprenticeship amounted to claiming that running a mine or carrying

---

[25] Anonymous, *Das ABC der Bergwercks-Wissenschaften* (1747), p. 5: 'in Bergwercken, an den Hütten, in Brennhaus, Sayger- Vitriol- Schwefel- und andern Hütten, sich umsiehet, die Probirstube, und das *Museum* der Herrn Markscheider besuchet, und, mit gescheiden, klugen und erfehrnen Berg- und Hüttenleuten, hohen und niedrigen Standes, fleissig umgehet, und *conversir*et.'

[26] Adolf Beyer, for instance, had a natural history cabinet and was considered as 'ein grundgeschickter und gelehrter Bergwercksverständiger' (*Briefe über eine Reise nach Sachsen*, by Georg Andreas Will, 1785, p. 48).

[27] See SächsStA–F, 40001 Oberbergamt Freiberg, Nr. 1362, 1726. Here f. 3r: 'Eine kleine bergwercks-*Academie* anzulegen'.

[28] SächsStA–F, 40001 Oberbergamt Freiberg, Nr. 1362, 1726, f. 24v: 'selbst in der bergwergs wißenschaft geübt seyn müße'; f. 25r: 'weit mehr zeit daran spendiren müßen … ordentlich bestelten Marckscheidern'.

out a solemn survey could not, as a last resort, be learned in a book. Mining officials and useful scholars were not hostile to scholarly knowledge *per se*, but defended the specificity of their own training system. Surveyor August Beyer, in his teaching manuscript, underlined the unique skills his profession:

And I am quite ready to admit that a scholar would have presented many things better, especially from the natural and mathematical sciences. But is it not equally true that we would have waited in vain for a scholar [to write] a practical book of this kind?[29]

The inadequacy of this system regarding the initial training, however, was getting hard to overlook. In Clausthal, the major mining city of the Harz region, there was only one secondary school. Its *Rector*, Calvör Henning, was a university-trained theologian. He also happened to have a personal inclination for mining sciences, publishing for instance an acclaimed *Description of the Mining Machines in the Upper Harz*.[30] Looking back at his long career in 1763, Henning complained of a woefully incoherent system. 'Those who intend to become miners', he deplored, 'generally do not stay long enough in the existing schools', where they only learned the rudiments of Latin and religion. 'The only ones who learn *mathematics* are those who want to study, finish school, and then go to university.' In universities, however, they would only learn theories, not 'the experience of *praxis*'. Henning was convinced, however, that one should 'go from the practice to the theory' while learning crafts.[31] These statements are all the more remarkable given that their author was himself a scholar, teaching in a Latin school. The venerable rector could hardly be suspected of being biased when asking for a new mining school:

In mining cities, young people should be instructed in *Mathesi* and *Mechanic* in schools. This purpose would be achieved best if a special mathematical school was set up, in which the most capable and awake minds – among those who want to become miners and carpenters – were trained during their youth. Every week, during the few hours they can break from the work they have already begun, they would be introduced to the elements of *geometry, trigonometry, static,* and *mechanic,* as well as *aerostatic, hydrostatic,* and *hydraulic*.[32]

---

[29] Beyer, *Gründlicher Unterricht von Berg-Bau* (1749), introduction: 'Und ich will ganz gerne eingestehen; daß ein Gelehrterbesonders aus der Natur-Lehre und Mathematischen Wissenschafften manches vielleichtbesser fürgetragen haben könnte. Alleine ist es denn nicht auch wahr? Daß wirein Practisches Buch in dieser Art vergeblich von einen Gelehrten erwartet hätten.'

[30] Henning, *Acta Historico-Chronologico-Mechanica* (1763).

[31] Ibid., p. 7: 'Diejenigen, die sich auf das Bergwerk … wollen legen, sich gemeiniglich in den gewöhnlichen Schulen so lange nicht aufhalten … nur diejenigen, die studieren wollen, und zu dem Ende die Schule fortsetzen, die *Mathesin* lernen, und damit wol auf der Akademie fortfahren … es ihnen an der Erfahrung in *Praxi* fehlt … aus der *Praxi* zur Theorie kommen.'

[32] Ibid., p. 7: 'in den Schulen auf den Bergstädten die Jugend in der *Mathesi* und *Mechanic* solte angeführt werden … [dieser] Zweck würde noch eher zu erhalten stehen, wenn insbesondere eine mathematische Schule aufgerichtet würde, darin die fähigsten und aufgeweckten Köpfe von denen, die Berg und Zimmerleute werden wollen, in der Jugend einige Studen in der Woche, die sie von ihrer schon angetretene Arbeit abbrechen können, in der Gründen der *Geometrie, Trigonometrie, Static* und *Mechanic,* auch der *Aerostatic, Hydrostatic* und *Hydraulic* … unterrichtet würden.'

In the absence of such an institution, many gifted young miners had to learn the elements of mathematical theories by themselves, hoping to later use this knowledge for the improvement of mining.

### Lost in Transition: Johann Andreas Scheidhauer

Mining master Johann Andreas Scheidhauer (1718–1784) was neither a subterranean surveyor nor a professor of mathematics. He nevertheless played a major role in the evolution of mathematics teaching and in the debates surrounding the foundation of the *Bergakademie* Freiberg. His contribution was hitherto unknown, and he is also absent from modern festschrifts and written accounts, for reasons that will be outlined in the following pages. Scheidhauer is remembered as a loyal and efficient mining master (*Bergmeister*), but also happened to be a gifted mathematician who tirelessly tried to bridge the gap between works of eighteenth-century theoreticians and the practical issues encountered in metallic mines. As such, he is a transitional figure who promoted new analytical methods in subterranean geometry, but also in machine theory and hydraulics.

Scheidhauer was born in 1718 in Johanngeorgenstadt, a small city in the Ore Mountains of Saxony.[33] Son of a mining clerk and grandson of a mining master, he was destined to enter the local administration. In the eighteenth century, it was becoming increasingly frequent for such persons to spend a few semesters in a university, at the time the only institution where mining legislation was taught. Scheidhauer studied law at the University of Leipzig, where he also received a couple of certificates for courses in physics and mechanics. In 1743, he was appointed *Adjunkt* to his father, an unpaid position merely meant to secure his succession, soon receiving an attestation of his skills '*in causis metallicis*, as well as *civilibus* and *criminalibus*'.[34] He succeeded his father a decade later, at which point he had already learned assaying and subterranean geometry, based on the manuscripts that we know circulated in this district.[35] Although he had been in contact with various German scholars in the early 1750s – being for instance informed about water wheels experiments performed at the Berlin Academy of Science – his passion for mathematics is only attested from 1757 onward.

That year, Scheidhauer began a series of notebooks in which he collated excerpts from numerous textbooks and articles, allowing us to reconstruct

---

[33] On the life and career of Scheidhauer, see Klinger and Morel, 'Was ist praktisch am mathematischen Wissen?' (2018), pp. 271–276.

[34] UAF, OBA 181, f. 234r, attestation dated 4 September 1745: 'sowohl *in causis metallicis*, als auch *civilibus* und *criminalibus*'.

[35] At the end of the seventeenth century, a subterranean surveyor of Johanngeorgenstadt, Johann Scherez, wrote his own version of the standard *geometria subterranea*. At least three copies have been preserved, now in TU BAF, WA, XVII 15, 333 and 636. Scheidhauer kept with him a manuscript, in which surveys made in 1738 to 1749 were preserved (see TU BAF – UB, Nachlass Scheidhauer, 300m, unpaginated).

precisely his pathway through theoretical mathematics.[36] Scheidhauer applied the typical administrative skills of collecting, ordering, and collating to the study of abstract sciences. He naturally began with the arithmetic, borrowing from cheap merchant handbooks. His elementary geometry was excerpted from the *Mathematical Garden of Pleasure and Utility*, a popular introduction from Johann Leonhard Rost, while his surveying manual was the *Mathematical School for War and Peace* of an (otherwise) unknown major Gruber.[37] The abundant vernacular literature available at that time meant that it was relatively easy for outsiders like him to teach themselves the rudiments of these disciplines. Notably, these early readings were no standard introductions used in universities, such as Wolff's *Elements* of mathematics, although it is hard to say if Scheidhauer simply relied on the works he had to hand, or indeed found these unscholarly books more adapted to his purpose.

The mining clerk did not stop here. We soon witness him working his way through calculus using Maria Agnesi's *Instituzioni analitiche* (1748), the standard of the time, or learning advanced topics such as spherical trigonometry by reading Leonard Euler. Scheidhauer did not neglect the most recent theories of hydraulic and hydrodynamic, perusing the publications of Karsten and Bernouilli, works that he would later use in the mines for his own experiments on water-wheels.[38] This painstaking work of excerpting in notebooks – for which Scheidhauer even devised his own register – is exceptional by its extent and organization. It also offers a rare glimpse into the way gifted practitioners could, outside of the usual university curriculum, become acquainted with higher mathematics and scholarly knowledge.

A solid mathematical culture allowed Scheidhauer to tackle the issues of subterranean geometry in a more systematic way than most of his predecessors. He developed his central ideas in a textbook – not intended for publication – entitled *Contributions to Subterranean Geometry* (*Beyträge zur Markscheidekunst*).[39] 'The main purpose of these *Contributions*', Scheidhauer wrote in the introduction, 'is actually to find by computations the determination of all the information that can be required from a subterranean

---

[36] The collected papers of Scheidhauer (his *Nachlass*) are preserved in the *Altbestand* of the Bergakademie Freiberg (TU BAF – UB, Nachlass Scheidhauer).

[37] See TU BAF – UB, Nachlass Scheidhauer, box 330e, the notebook *Geodesiae* (1757) is borrowed from Johann Sebastian Gruber, *Neue und gründliche mathematische Friedens- und Kriegs-Schule* (Nürnberg, 1697), caput IX, pp. 75–93. His notebook *Von der Geometrie* (1757) reproduces Johann Leonhard Rost, *Mathematischer Lust- und Nutz-Garten* (Nuremberg, 1745 [1724]), pp. §174–§251, including the drawings.

[38] See TU BAF – UB, Nachlass Scheidhauer, 300d, vol. 1; the *Instituzioni analitiche* were copied from a Latin exemplar, just as Euler's *Principia trigonometriae sphericae* had been. The most recent works by Karsten, Bernoulli, Winck, and Euler were collected in boxes e, f and g, older works by Newton in box d.

[39] Scheidhauer's works on subterranean geometry are collected in boxes m1, m2 and n–o.

surveyor'.[40] His ambitious goal was thus to achieve the transformation of the discipline. Since the introduction of data tables, the typical *Markscheider* was not bound to the mining pits any more. His duties had shifted from the mere manipulation of cords and compasses to the drawing of scaled mining maps, on which measures were made and plans were developed. Scheidhauer criticized these last steps because of the 'low reliability and several mistakes' that came from using physical instruments. He proposed to circumvent maps altogether and replace them by analytic methods – in his words 'an accurate and reliable *calculum*' – applied directly to the recorded data. He envisioned an administration in which mining maps would be used simply for planning and administrative purposes, and ceased to be mathematical tools for surveyors. Furthermore, analytical methods had 'a wider range of application': generality was obviously a value to which he attached great importance.[41]

Scheidhauer's proposal was sensible, but it hardly solved the problems faced by subterranean geometry at the time. The lack of initial training meant that some surveyors already had difficulties with the existing methods, as evidenced by the complaints that mining companies lodged in several districts.[42] Such individuals could not easily be convinced of the urgent need for even more ambitious, abstract, and analytical tools. Attempts to address the problem by centralizing the examination of all subterranean surveyors in Freiberg, capital of the Ore Mountains, had aborted as the other districts staunchly defended their autonomy.[43] Before Scheidhauer's ambitious plan could even be considered, several fundamental issues had to be addressed to ensure the homogeneous quality of surveys. In the Harz, the vice-mining master Röder complained, for instance, about the poor condition of trainees: one of them had indeed received a teaching grant, 'but he lack[ed] good *Instrumenten* and mathematical books'. The apprentice had 'inherited a *compass* from his father, but it cannot be used reliably'.[44] Another promising young boy could not afford to stop working to

[40] TU BAF – UB, Nachlass Scheidhauer, 300m, *Beiträge zur Markscheidekunst*, introduction: 'Die Hauptabsicht dieser Beyträge, gehet eigentlich dafür, die Bestimmung aller Angaben, die von einem Markscheider gefordert werden können, durch Rechnung zu finden.'
[41] Universitäts- und Landesbibliothek Tirol in Innsbruck, Cod. 1187, f. 3r–3v: 'vieler Unzuverläßigkeit und Fehlern abzunehmen ... einen richtigen und zuverlässigen *Calculum* ... einen größeren Umfang sich erstrecken'.
[42] In 1753, for instance, the mining administration of Altenberg complained about their local surveyor, Gottfried Wilhelm Grellmann (SächsStA–F, 40006 Bergamt Altenberg, Nr. 1480).
[43] From 1738 onwards, prospective assayers in Saxony had to take an examination in Freiberg. Attempts to extend the measure to subterranean surveyors failed when one of them, G.G. Seibt, underlined that 'subterranean geometry could be bound neither to a person nor to a place'. Several letters discuss the issue in UAF, OBA 5, 146r–150v.
[44] NLA WO, 33 Alt, Nr. 414, *Pro Memoria* by Johann Christoph Röder, 11 February 1786, unpaginated: 'er fehlet denselben aber an guten *Instrumenten* und *matematischen* [sic] Büchern ... zwar von seinen Vater einen Compas erwerbet / allein er kann davon keinen zuverläßigen Gebrauch machen'.

learn the craft. The teaching fees, even if paid by the administration, were only a fraction of the total cost. When the young Ilse left his mining village to 'stay for one year with the subterranean surveyor Rausch in Clausthal', in order to 'perfect himself in this art', he could not afford to pay 'for burden, accommodation, and teaching' and asked the administration for help.[45] Röder finally suggested a more advantageous 'assistance' scheme for young people than simply covering the training fees. The administration could 'purchase the instruments of underground surveying ... once and for all, at the ruler's costs' and lend them to trainees until they could afford to buy new tools with their own incomes. It would then be enough to give 'every trainee a grant of 10 to 12 thalers to buy a set of drawing instruments and mathematical books'.[46]

Around the middle of the century, a consensus was being reached: the existing grants were not enough, and mining administrations had to play a more active and central role in the training of technicians. To ensure the efficiency and tighten their grip on mining districts, rulers came to assume not only the teaching fees but also related costs. Depending on the region, the administration could pay for books and instruments, or finance trips abroad to observe the latest innovations. When an important official died, the mining office purchased his collection of manuscripts, or minerals, for the common benefit of the local district. Taken together with the new examinations, such measures extended the nascent teaching system, but noticeably fell short of a brick-and-mortar institution. Still, I think that these mostly uncoordinated movements in various states largely laid the ground for the creation of the first mining academies.

The precipitating event was the Seven Years War (1756–1763). This modern war triggered a wave of institutionalization all over Europe, mostly in technical domains such as engineering and medicine. In the mining states, the aim was to improve the productivity of extraction, in order both to increase fiscal revenues in the short term and to ensure more regular receipts.[47] Saxony implemented a *Restaurationskommission* in 1762 to propose radical reforms and restart the economy. State's commissioner of mines Friedrich Anton von Heynitz (1725–1802) initially suggested a train of measures in line with preceding reforms. The grant system would be expanded once again, this time to finance three-year apprenticeships and travels for selected students. The

---

[45] NLA WO, 112 Alt, Nr. 1777, 2r: 'ein Jahr lang bey dem Markscheider Rausch zum Clausthal aufhalte ... um sich in dieser Kunst völlig zu *perfectioniren* ... für Last, Quartier und Lehrgeld'.
[46] NLA WO, 33 Alt, Nr. 414, *Pro Memoria* by Johann Christoph Röder, 11 February 1786, unpaginated: 'Beyhülfe für solche Leute, zu Anschaffung der Markscheider Instrumente ... ein für alle mahl zum unterricht in Markscheiden ... jeden Lehrling 10 bis 12 thlr beyhülfe zu anschaffung eines Reißzeuches und Matematischen Büger gegeben wirde.'
[47] On the aftermath of the Seven Years War and its influence on mining policy, see Baumgärtel, *Bergbau und Absolutismus* (1963), pp. 68–95; Wakefield, *The Disordered Police State* (2009), pp. 26–48; Klein, *Technoscience in History: Prussia, 1750–1850* (2020), pp. 10–22.

proposal evolved and finally came to include the foundation of an academy in Freiberg, capital of the Ore Mountains.[48] The academy was to be open to both Saxons and paying *Ausländer* (foreigners). It initially employed two teachers: Christlieb Ehregott Gellert (1713–1795) had already been privately teaching chemistry and metallurgy, while a newcomer, Friedrich Wilhelm Charpentier (1738–1805), was appointed professor of pure and applied mathematics. It is very telling that one of the two chairs was dedicated to mathematics, whose influence on mining might not be as obvious as, for example, mineralogy.[49] Mathematics would indeed play a crucial, yet ambiguous role from the very first years of the new institution. This will be presented in detail in the next section, but let us first go back to Johann Andreas Scheidhauer.

In 1765, the very year the *Bergakademie* was founded, Scheidhauer was appointed mining master in Freiberg.[50] The coincidence has never been noticed, precisely because his significance as a mathematician was unknown. Why was Scheidhauer, whose theoretical skills were well known, not simply rewarded with the position of professor of mathematics instead? Not only had he taught himself the most recent results of his time, but two decades of experiments on mechanics, hydraulics, and more generally applied mathematics, made him a perfect candidate to blend theory and practice. Charpentier, on the other hand, was alien to the Ore Mountains, having studied law and mathematics at the University of Leipzig. He was so unfamiliar with mining sciences that he had to sign up as a student, even as he was appointed official professor of mathematics!

Charpentier was chosen instead of more seasoned experts, I think, because those were already playing a part in the renewal of mining sciences. As a mining master, Scheidhauer was a key person of the *universitas*, the local community of mining experts. Moreover, appointing him *Bergmeister* meant that he now had a position of authority much higher than any academic chair could possibly offer. From there, he contributed to several reforms that, considered in conjunction with the foundation of the academy, reveal the coherence of the reform movement. Two years after the opening of the academy, it was decided that all surveying instruments would be made and sold in Freiberg. The same year, it was forbidden to train subterranean surveyors anywhere but at the newly founded mining academy. These two decisions are both political and technical, underlining the will to standardize measurement practices: from now on, all the *Markscheider* would undergo the same training and use similar

---

[48] The foundation of the *Bergakademie* is analysed in Baumgärtel, 'Die 200-Jahrfeier der Bergakademie. Probleme und Aufgaben' (1962), pp. 77–86; Weber, 'Bergbau und Bergakademie' (1985), pp. 79–89.

[49] Weber, for instance, has wondered why there was no chair of mechanics, see Weber, *Innovationen im Frühindustriellen deutschen Bergbau und Hüttenwesen* (1976), pp. 158–159.

[50] SächsStA–F, 40012 Bergamt Johanngeorgenstadt, Nr. 949 describes his possessions and all the documents he left in Johanngeorgenstadt, where he had been mining clerk (*Bergschreiber*).

instruments.[51] The following year, an attempt was launched to unify the length units used in the various mining districts. In order to determine what would be the official *Lachter*, Scheidhauer used both his authority as mining master and his skill as mathematician. In July 1772, the new standard was promulgated and exemplars were sent to all local mining offices.[52] Such reforms could only be proposed or enforced from a position of power.

## Mathematics at the Mining Academy

Seventeen years after the creation of the Freiberg mining academy, Charpentier was still the professor of pure and applied mathematics. In 1782, he wrote the preface of a textbook on subterranean geometry that had been composed by his most gifted student, Johann Friedrich Lempe. Charpentier used the opportunity to offer a straightforward and positivist account of its early history:

And so subterranean geometry stayed, for those who had to perform it, in the usual craft usage [*handwerksmäßigen Behandlung*], until this most valuable institution – the mining academy that was built here in 1765 – gave to everyone who had the capacity and desire of thinking, the opportunity not only [to master] the principles of subterranean geometry but also its complete scope, through the learning of mathematics and other auxiliary sciences.[53]

In order to really appreciate the influence of an institution, we can obviously not be satisfied with the self-congratulations of its professors. Charpentier's account ignores both the slow maturation of subterranean geometry described in the previous chapters and the difficulties inherent to the practical use of theories. Conversely, a recent reappraisal from Andre Wakefield presents the academy as an empty shell, in which the '*practical sciences* were, in their way, just as idealistic and romantic as Novalis's blue flower or Young Werther's yellow vest', summing up lapidary with the words: 'maybe we have taken science too seriously'.[54] These two positions are equally problematic: positivism and social-constructivism both tend to be 'put forward in none but either/or

---

[51] SächsStA–D, 10036 Finanzarchiv, Loc. 32833 Rep. LII, f. 100, decree from 26 November 1768.
[52] It should be noted, however, that these new measures do not seem to have prevailed in daily use but remained a 'scientific standard'. The challenging road towards a unified standard unit can be followed in SächsStA–F, 40001 Oberbergamt Freiberg, Nr. 2805 (period 1772–1830).
[53] Lempe, *Gründliche Anleitung zur Markscheidekunst* (1782), p. 10, introduction by Charpentier: 'Es blieb also die Markscheidekunst immer noch bey denen, die sie ausüben sollten, in der gewöhnlichen handwerksmäßigen Behandlung, bis durch die Preißwürdigsten Anstalten, die im Jahr 1765 hier errichteten Bergwerksakademie die Gelegenheit allgemein wurde, wodurch sich ein jeder, der Fähigkeiten und Lust zum Denken hatte, durch Erlernung mathematischer und anderer Hülfswissenschaften, nicht nur von den Gründen der Markscheidekunst, sonder auch von ihrem ganzen Umfange … selbst überzeugen konnte.'
[54] Wakefield, *The Disordered Police State* (2009), pp. 25, 34–35.

terms', to quote Floris Cohen.[55] Wakefield's interpretation and Charpentier's account, despite being separated by more than two centuries, rest on a similar misconception. A monolithic institution is presented in splendid isolation, and the narrative of its foundation is framed in such a way as to support one's view: either that it instantly transfigured previously backward practices, or conversely as a political instrument without any scientific ambition or concrete influence.[56] As a matter of fact, the present book has already shown how the teaching of mining sciences had been highly monitored from at least the sixteenth century. The mining sciences, including their mathematical parts, never existed *in abstracto*: they had to be both politically and concretely efficient to warrant their rulers' support.

It is more fruitful, I argue, to consider technical academies as a long-term evolution of venerable administrations. In the late eighteenth century, the *Bergakademien* were relatively modest enterprises that only gradually developed into the modern institutions of today. In Freiberg, there were only two, later three, chairs in addition to the mining officials who had traditionally passed the knowledge on to the next generations. Very much as in the preceding century, the major mining centres were communities of knowledgeable experts, the *Bergverständiger*. In the early years of the academy, there was no clear separation between the teaching institution and the mining administration in a broader sense. The lessons took place 'in the house of mine captain von Oppel', who had offered not only 'an auditorium', but also 'a library and a room [for the] mineral collection'.[57] Professor Gellert had already been giving lessons for decades and was simply given a new title, acting both as a professor and as chief administrator of foundries and forges. Scheidhauer followed a similar pattern, teaching subterranean geometry – albeit unofficially – from his position of mining master.[58] The chief innovation was to bring together in one place maps, books, mineral collections, and various experts. It facilitated communication and collaborations between members of the *universitas* and the learning youth, while granting the mining administration a greater control on the teaching content and offering a coherent curriculum.

The evolution of the grant system has not received the attention it deserves, having too often been seen as an archaism, or a stopgap measure pending a 'true'

---

[55] Cohen, *The Scientific Revolution: A Historiographical Inquiry* (1994), p. 231. Ursula Klein has underlined a similar 'historiographical problem' about the understanding of mining sciences. For further references, see Klein, *Nützliches Wissen* (2016), pp. 160–161.

[56] For an attempt to balance the political and scientific ambitions in the early years of the *Bergakademie*, see Weber, 'Bergbau und Bergakademie' (1985), pp. 86–88.

[57] UAF, OBA 236, f. 99r: 'Und daß der in des Herrn Oberberghauptmanns von Oppel Hause darzu genommen Plaz bereits dargestelt vorgerichtet sey, daß nicht allein *Auditorium* Stuffen-Saal, Bücher- und Modell-Kammer, in völliger Ordnung.'

[58] On the career of Christlieb Ehregott Gellert as mining official, see Habashi, 'Christlieb Ehregott Gellert and his Metallurgic Chymistry' (1999), p. 32.

or 'modern' institution, whatever this should mean.[59] In the mid-eighteenth century, however, the grand system had become a keystone of the mining administration. Most higher officials initially trained as technicians – assayer, subterranean surveyor or both – before slowly climbing the ranks of the administration, as for instance Scheidhauer had done. Apprenticeship and grants could not be replaced at once by lectures and textbooks. Instead of a sudden overhaul, the mining academy built on the existing system, which was closely articulated with a new theoretical teaching. Subterranean surveyors and assayers became officially teachers, and were satisfied to receive a salary instead of irregular grants: 'a yearly *fixum* has its advantages compared to uncertain and often cancelled revenues'. Now paid by the academy, they had to train three selected students every year 'in the necessary mining knowledge' and 'present them with the mining maps belonging to the institute'. The 'instructing visits' in the mining pits took place every Monday and Tuesday, making up for a sizeable portion of the training.[60] This free training in subterranean geometry was of course restricted to the Saxons, excluding the *Ausländer*. Foreigners could enrol at the mining academy, but could only access the public teaching on mathematics and chemistry, not the more coveted hands-on training in surveying and assaying.[61] All things considered, the key improvement in the early years of the academy concerned the initial training. New courses in 'pure mathematics', physical sciences, and 'metallurgical chemistry' were now publicly given by professors to ensure that students made the most of their concomitant practical instruction.[62]

It now becomes clear why Charpentier had been hired as professor in spite of his lack of practical experience. His scholarly background was exactly what the mining administration was looking for, and his lessons were definitely not meant to replace the existing on-site training. The 'thorough learning of mathematics and physics' in a classroom addressed, a report to the school council indicated, the needs of aspiring miners 'who, though they have good dispositions and proper diligence, are otherwise not prepared by education'.[63] Such lessons, moreover, could also help the installed mining officials. In the first

---

[59] It is all the more surprising given that Sennewald has shown the importance of this grant system in Sennewald, 'Die Stipendiatenausbildung von 1702 bis zur Gründung der Bergakademie Freiberg 1765' (2002), pp. 407–429.

[60] UAF, OBA 236, f 54r: 'Ein jährliches *Fixum* hat seine Vorzüge vor ungewissen und öffters ganz wegfallenden Zugängen … in der einem Bergnöthigen Kenntnis von Gruben-Bau'; f. 80v-81r: 'zum *Instituto* gehörigen Markscheider Rißen'.

[61] Mining officials did not even imagine publicly teaching subterranean geometry (UAF, OBA 236, f 80v–81r, § *Die Anweisung zum Marckscheiden*).

[62] See the printed *Avertissement* that was circulated at the time (a copy can be found in UAF, OBA 236, f. 121).

[63] UAF, OBA 246, f. 264v: 'gründliche Erlernung der Mathematik und Physik … die zwar gute natürliche Anlage haben und gehörigen Fleiß anwenden, sonst aber duch Erziehung nicht vorbereitet sind'.

academic year, vice-surveyor Christian Gottfried Kießling frequented the class of pure mathematics.[64] Kießling had learned his craft in the mines thanks to a teaching grant fifteen years earlier. Witnessing a seasoned expert – 'forty years old, of average stature, and wearing a *perruque*' – go back to school shows that the need for a structured theoretical training was real.[65] This underrated fact – teaching elementary mathematics to mining officials already in place – formed a major influence of technical institutions in the short term. Next to Kießling, several foremen and jurors (*Schichtmeister* and *Geschworene*) also learned the rudiments of 'arithmetic, algebra, and trigonometry' with Professor Charpentier. These public lectures for students and mining officials 'eager to learn' (*lehrbegierig*) took place on Wednesday and Saturday morning. During the first year, Charpentier used Christian Wolff's *Elements of Mathematical Sciences*, the standard university textbook at the time.[66] However, he soon modified his notes to accommodate the remarks of students, including more interactions with 'the practical applications', and 'the influence of mathematical knowledge on the theory of machines and subterranean geometry'.[67]

This two-tiered teaching system was an original, yet efficient, innovation that improved the existing system without endangering what was already working. On the one hand, the *Bergakademie* cemented the status of Freiberg as a teaching centre, while on the other hand the specific organization of underground surveying was preserved. Students visiting the institution in the following decades confirm this dual organization. From the travel story of Dominicus Beck, written almost a generation after the foundation of the academy, one learns that Professor Werner lectured mineralogy from his textbook *On the External Characters of Fossils*. Professor Gellert did the same using his *Elements of Metallurgical Chemistry* and Professor Charpentier, as we have seen, adapted printed textbooks. 'Subterranean geometry, though', added Beck, 'is lectured from a manuscript which is said to indicate special methods. Even though I have not come to see it yet, I hope to obtain it soon'.[68] Even the system of certificates was prolonged, attesting of a parallel system of diplomas

---

[64] Kießling is listed in UAF, OBA 236, f. 55v–56r.

[65] Kießling's name appears in the list of *Stipendiaten*, likely in the year 1750; see UAF, OBA 6, f. 245v. He had received 100 thalers to learn both surveying and assaying. Kießling also made good use of another innovation of the academy: he was financed for an observation trip to the neighbouring Duchy of Anhalt-Cöthen, see UAF, OBA 182, f. 30r. I borrowed his physical description from his *laissez-passer*.

[66] Morel, *Mathématiques et politiques scientifiques en Saxe 1765–1851* (2013), pp. 157–163.

[67] UAF, OBA 237, f. 194r: 'die practische Anwendung … der Einfluß den die mathematische käntnis [*sic*] in das Maschinen Wesen und die Marckscheider Kunst habe'.

[68] Beck, *Briefe eines Reisenden von \*\*\* an seinen guten Freund zu \*\*\** (1781), p. 232: 'Die Markscheidekunst aber wird aus einem Manuskripte gelesen, welches besondere Methoden angeben soll. Zwar habe ich selbes noch nicht zu Gesicht bekommen; ich hoffe aber es mit nächstem zu erhalten.'

at the academy: a new institutional environment notwithstanding, the scribal teaching system had survived.[69]

The foundation of mining academies thus appears as a compromise between competing needs and ambitions in the German states of the late eighteenth century. The exact organization of classes and topics evolved quickly in the early years, suggesting that a lot was actually adapted and negotiated.[70] Besides the obvious need to train reliable technicians and officials, two competing tendencies seem to have influenced the teaching of mathematics. On the one hand, the mining council primarily wished to offer a more solid foundation to future foremen, assayers, and surveyors. In that respect, a modest institution delivering few theoretical courses was a convenient supplement to the existing system. This generalized initial training did wonders to improve the training of technicians and quickly expanded. Following the mining academy, several preparatory mining schools opened in the Ore Mountains during the 1770s, forming a complete system adapted to the exact needs of mining.[71] On the other hand, mining master Scheidhauer and a few individuals pushed for a much more ambitious approach. In their idea, subterranean surveyors, machinists, and more generally all mining officials had to be trained as engineers with a robust mathematical background. There was of course not clear delimitation, and the two goals indeed advanced in parallel, but obvious points of disagreement existed. Should the theoretical lessons, for instance, precede the practical training? Should the curriculum include more than just the usual applications to mining and include calculus, or could the student's time be employed in a more useful way?

## The Challenges of Teaching Higher Mathematics

A remarkable development in mathematics teaching happened at the end of the eighteenth century. Under the leadership of Johann Friedrich Lempe (1757–1801), courses became theoretically more ambitious, while paradoxically focusing more closely on the improvement of mining. Coming from a modest family, Lempe had begun his career as a simple miner (*Bergmann*), but was deemed brilliant enough to receive a grant from the mining academy in 1773. Archives indicate that he studied higher mathematics privately with Charpentier, which incidentally is another sign that what was actually taught did not correspond to the printed programmes. His professor soon attested that 'he [could] understand and read with benefit the works of Euler, Kästner,

---

[69] See, for example, UAF, OBA 244, p. 76r–77r, for a certificate made in 1783, almost two decades after the foundation of the academy.

[70] On the content of the mathematics teaching and its evolution up to the mid-nineteenth century, see Morel, *Mathématiques et politiques scientifiques en Saxe 1765–1851* (2013), pp. 141–252.

[71] Kaden, *Das sächsische Bergschulwesen* (2012); Morel, ‚Usefulness and Practicability of Mathematics‘ (2016), pp. 49–62.

or Karsten' – the leading mathematicians of their time. Thanks to this support, Lempe obtained a second grant and went to the University of Leipzig to deepen his knowledge of higher mathematics.[72] In the meantime, he also perfected himself in subterranean geometry, not with the local surveyor but with mining master Johann Andreas Scheidhauer. Once again, there is no trace of this activity in the official curricula, nor indications in the narrow frame of the academy's archive: the mining master acted here as a member of the larger *universitas*.[73] Scheidhauer's crucial influence might very well have remained unrecognized, had his pupil not turned out to be a very prolific writer. Lempe published several books and dozens of articles on mining mathematics. In these publications, he repeatedly praised Scheidhauer for his efforts in mathematizing mining sciences:

> Mining master Scheidhauer is to be credited with many improvements of subterranean geometry … although these are not yet generally known in print … Anyway, mining master Scheidhauer has offered in a manuscript written for himself more complete solutions for most of the problems that one encounters in subterranean geometry, using formulas, but mostly without proofs.[74]

Lempe followed in his master footsteps in presenting bold, ambitions methods that could improve mining practices. Finding the right direction to achieve this, however, was a difficult endeavour. The two men – together with a handful of their contemporaries – were firmly convinced that mining sciences should be fully mathematized. This was especially true for the most important one, subterranean geometry, but also increasingly for practical hydraulics and machine theory. At the same time, they feared empty scholarly discourses and were focused on concrete achievements. This attitude, characterized above as a logic of administration, explains why Scheidhauer wrote thousands of pages for himself and his fellow miners but did not print anything. Neither did the mining master fully develop an overarching reflection on his goals or offer a theoretical system of mining mathematics. For decades, Scheidhauer stuck to his credo, tirelessly collecting data and trying to come up with analytic

---

[72] UAF, OBA 236, f. 72 about his grant request, and f. 105: 'so weit gekommen, daß er die Schriften eines Eulers, Kästners, Karsten u.a. mit Nutzen lieset, und verstehet'. On Lempe's biography, see Morel, *Mathématiques et politiques scientifiques en Saxe 1765–1851* (2013), pp. 164–186. Yearly reports about his Leipzig activities are listed in f. 201–203 and f. 233.

[73] In fact, the only mention of Scheidhauer I could find in the archive of the academy confirms that he taught on a regular basis, but expresses only a reluctant support (UAF, OBA 242, f. 14r): 'We continue to approve the constraints that have been found necessary by mining master Scheidhauer, concerning the students that have been entrusted to him, and have to let him perform extrajudicial settlements and other similar duties, that he can chose one or the other mining student, as frequently as he pleases, as long as these [students] are not too distracted from their main academic works and duties.'

[74] Lempe, *Gründliche Anleitung zur Markscheidekunst* (1782), p. 178. Elsewhere, he directly thanked Scheidhauer for 'the communication, that he had the kindness to offer me' (p. 2).

formulas matching it. His student Lempe, who in 1784 replaced Charpentier as professor of mathematics, was in the position to publicly influence the coming generations. The existence of an official institution, a public space carrying its own values about mining sciences, acted as a catalyst. Over the years, he introduced at the academy a theoretical teaching of subterranean surveying and a course on the theory of machines, both based on his own works. Even his public lectures on elementary arithmetic and geometry shifted considerably, as Lempe designed specific textbooks such as the *Miner's Arithmetic* (1787).

Vowing to teach useful knowledge was a classical *topos* in the eighteenth century, but Lempe was serious when he claimed, in his yearly reports to the academic council, to focus on the 'applications to mining'.[75] He sent his students into the mines with real life problems, asking them to model the situations and offer their solutions, on the basis of which they were evaluated. The results present us with a mixed bag of all issues that can somehow be modelled using numbers and geometrical figures. These exercises range from the optimization of the shape of mining buckets (*Kübel*) used to carry the ores to experiments on the strength and deformation of timber.[76] Lempe's approach was deeply pragmatic and oriented towards problem-solving. The best solution could sometimes rely on elementary geometry: in 1786, student Richter was asked to ascertain the volume of the dike of a water pond. Other issues might be more ambitious and require a full modelling of the situation. When Lempe decided to tackle the thorny issue of the optimal speed of mining tubs (the so-called *Hünde*), he sent a handful of students into neighbouring mines and factories. They gathered real-life data about the weight of tubs, the length and duration of rides. Depending on their abilities, they produced anything from simply computing the speed of a *Hund* to using calculus in order to ascertain its optimal speed, and consequently prescribed the right number of trips a miner should do during his eight-hour shift.[77]

The attitude towards calculus illustrates the professor's general approach: most practitioners abhorred it, while scholars often saw its use as a *sine qua non* for true science. Lempe offered a very pragmatic middle way. On the one hand, he claimed in a learned journal to 'show how ridiculous is the prejudice, that the science of higher mathematics could not be useful to the subterranean

---

[75] Lempe, *Bergmännisches Rechenbuch* (1787). See UAF, OBA 246, f. 233r: his yearly reports tout the ambition to teach mathematics in a useful way: 'so viel mir möglich war, Anwendung auf den Bergbau zu machen gesucht'.

[76] This first assignment was a favourite of Lempe. He had several students work on it, and their work likely influenced the new legislation on the shape of mining bucket issued in 1788 (see the official leaflet in SächsStA–F, 40001 Oberbergamt Freiberg, Nr. 3578 *Regulativ der Maaße*).

[77] UAF, OBA 246, yearly report on the academic year 1786. That year at least three students worked on the problem. Mothes went to the 'Heavenly Lord' (*Himmelsfürst*) mine, Löbel to the 'New Luck and the Three Oak Trees' (*Neuglück and 3 Eichen*), Scheidhauer – a younger relative of the late mining master – to another pit of the 'Heavenly Lord' (243v–248v; 249r–252v; 253v–255v).

surveyor'.[78] When relevant or interesting, he used it for matters seemingly so trivial as computing the volume of an elliptical-shaped mining bucket. If a simpler way could be found, however, Lempe usually favoured elementary methods that appealed more directly to his fellow technicians.

Lempe's promotion of higher mathematics among practitioners was ultimately a mixed success. The ambitious idea of turning subterranean geometry and machine theory into fully analytical sciences was a bit steep for many contemporaries. It was not easy to convince practitioners to 'stop measuring' and 'begin to compute', as Oppel provocatively advocated half a century before.[79] Lempe was supported by mining master Scheidhauer and by Charpentier, who in the meantime had become Captain-general, in his push to improve the mathematical curriculum in successive steps. His attempts to ascertain the shape of buckets or the optimal speed of mining tubs were not theoretical quibbles, but were firmly grounded in specific debates happening in the Ore Mountains. These calculations prolonged the reform program driven by F.A. von Heynitz and supported the standardization agenda of the mining administration.[80]

Lempe's publications in subterranean geometry also won him the respect of fellow officials, but only the best students were in fact able to use these results in the mines. His collaborations with mining officials were mostly successful, but the evolution of practices could only be incremental. Calculus was taught and used by a handful of selected students, while the average foremen or the older *Markscheider* still relied on time-tested methods and August Beyer's manuscript, now available in print.[81] This outcome was less a failure, however, than the natural inertia of practices. A full mathematization of subterranean geometry only succeeded in the mid-nineteenth century when Julius Weisbach (1806–1871), engineer and professor in Freiberg, introduced the modern mine theodolite. Still, Charpentier, Scheidhauer, and Lempe had proven that higher mathematics could sometimes be truly useful.

In the short term, the *Bergakademie*'s most tangible success was a large diffusion of elementary mathematics. Thanks to the introduction of public

---

[78] Lempe, 'Auflösung einer Aufgabe aus der Markscheidekunst' (1783), p. 188: 'indessen zeigt doch diese Abhandlung klar und deutlich, wie lächerlich das Vorurtheil ist, daß dem Markscheider keine Lehren der höhern Mathematik brauchbar wären'. The piece on the mining bucket appeared in Lempe's own *Magazin der Bergbaukunde*, vol. 8, 1791, pp. 110–116.

[79] Von Oppel, *Anhang der Anleitung zur Markscheidekunst* (1752), p. §930: 'Nun aber höret ein Meßkünstler auf zu messen, wenn er alle Werkzeuge weggelegt. Und wenn er zu messen aufhöret, so höret er entweder auch auf einen Meßkünstler abzugeben, oder er fängt an zu rechnen und suchet aus wirklich gemessenen oder willkürlich angenommenen Größen andere unbekannte zu finden.'

[80] Von Heinitz had tried to standardize the shape of mining buckets and introduce modern mining *Hünde* to replace older vehicles, but encountered strong resistance. See Baumgärtel, *Bergbau und Absolutismus* (1963), pp. 53–55.

[81] Morel, 'Five Lives of a *Geometria subterranea* (1708–1785)' (2018), pp. 207–258.

courses and of a coherent curriculum, initial training quickly improved with far-reaching consequences. Johann Friedrich Freiesleben, the subterranean surveyor who now taught the Freiberg students, summed up the contribution of theoretical courses with the words: 'if students are entrusted to the training in subterranean geometry only when they have a thorough understanding of its content, I will simply have to deliver the practical methods to them'.[82] The initial training was not useful *per se*, but because it had been tailored to its audience by the professors of the academy. The early history of mining academies still offers a lot of room for interpretation: local initiatives flourished and were rapidly evolving, thanks to dynamic cross-cultural exchanges ranging from Spain to Russia.[83] As the example of mathematics illustrate, however, the true issue was not about politics, but about scientific policies. How best to bring together a solid tradition of practices and a culture of theoretical investigation? Johann Carl Freiesleben, student of the academy and a friend of Alexander von Humboldt, summed it up perfectly in his travel diary. 'Mining sciences', he wrote in 1795, were on the verge of reaching a 'scientific treatment'. This had become possible first and foremost 'since the miners themselves appeared as authors'.[84]

By establishing their own combination of theory and practice, mining academies entered into direct competition with universities. Once a place reserved to erudition and scholarship, the German *Universitäten* had taken a turn, symbolized by the foundation of the University of Göttingen in 1737. They vowed to be more directly useful and train the rapidly growing number of state's servants, while continuing to serve as scientific references.[85] Universities and mining academies, in the late eighteenth century, competed for the highest authority in several domains. Leipzig and Wittenberg first resisted the opening of a chair dedicated to mining law in Freiberg, and their monopoly on legal studies was only broken in 1786, when the discipline was taught at the *Bergakademie* for the first time. The rivalry was even fiercer in practical mathematics: which institution would be the reference when it came to use mathematics in a technical setting? This general question took an unexpectedly tangible turn in the 1770s, when a new water-column engine was introduced in the mining district of Marienberg, Saxony.

---

[82] UAF, OBA 244, f. 247r, report by Johann Friedrich Freiesleben: 'wenn Scholaren zur Erlernung des Markscheidens nur anvertraut werden, die den Inhalt derselben gründlich verstehen, ich mit selbigen weitere keine andere als practische Markscheider-Unterweisung und Bestätigungen, vorzunehmen haben wurde'.

[83] See §3.3, *The diffusion of the German and Austro-Hungarian Model* in Brianta, 'Education and Training in the Mining Industry, 1750–1860' (2000), pp. 279–283.

[84] Freiesleben, *Bemerkungen über den Harz* (1795), pp. iii–iv: 'Bergbaukunst ... wissenschaftliche Behandlung ... seit dem Bergleute selbst als Schriftsteller auftraten.'

[85] Hilde de Ridder-Symoens and Rüegg, *A History of the University in Europe* (1996), vol. 2, pp. 142, 469.

## The Engineers and the Professor

During the winter of 1776, the young Friedrich W.H. von Trebra was assessing a technical problem. He had been appointed mining master in the corrupt and heavily indebted district of Altenberg a few years before. Trebra arrived with an ambitious plan to relaunch the extraction of galleries that were once promising, but had been abandoned and flooded since the Thirty Years War (1618– 1648).[86] Promising results had led him to invest in a new machine to pump ground water out of the *Father Abraham* mining pit. It was powered by an innovative water-column engine designed by his machinist Johann Friedrich Mende. The engine was meant to be cheaper, smaller, and more efficient than the venerable water-wheels, but for the moment it 'worked poorly and faltered very often'.[87] The mining master had already disbursed on that project most of the advance received from the central administration. He needed money, and rightly feared his superiors would be angered if they learned from a third party that he had spent too much on a machine whose steering design was still under discussion. Local investors, running out of patience, were now asking to abandon the risky experiment and to build a traditional water-wheel instead. Although he had hoped to quickly solve the problem before sending his report, Trebra felt compelled to reach out to the mining council in Freiberg.[88] He openly shared his thoughts, expecting either a backing or a rebuff, but something surprising happened instead: the council forwarded his message to the cameralists of the central administration in Dresden, the *Kammerkollegium*.

Bureaucrats from the capital subsequently decided to approach Johann Ernst Zeiher (1725–1784), professor of higher mathematics at the University of Wittenberg. Zeiher was instructed to 'go to Marienberg … observe this machine closely, and indicate how to improve it and bring it into useful condition, and to give to machinist Mende the necessary instructions'.[89] In doing so, the council acted according to the logic of university-trained administrators, convinced that a practical issue could be solved using theory alone. If some of them had ever read about mining mathematics, it was probably the Latin *Geometria subterranea* of Johann Friedrich Weidler (1691–1755), a predecessor of Zeiher in Wittenberg. In this learned treatise, Weidler chastised the mining officials as 'inexperienced in the mathematical sciences', blaming them for being 'unclear,

---

[86] See his visitation reports and early plans in SächsStA–F, 40013 Bergamt Marienberg, Nr. 1438, f. 20–21, 125–142.

[87] Gätzschmann, 'Zur Geschichte der Wassersäulenmaschinen in Sachsen' (1873), p. 3: 'ziemlich mangelhaft, stockte sehr häufig, machte sich stets wiederholende Nachhülfen, Reparaturen und Abänderungen nothwendig'. On the introduction of this new technology, see for instance Baumgärtel, *Bergbau und Absolutismus* (1963), pp. 57–58.

[88] See SächsStA–F, 40168 Grubenakten des Bergreviers Marienberg, Nr. 909, f. 31–89.

[89] SächsStA–F, 40168 Grubenakten des Bergreviers Marienberg, Nr. 909, f. 94r: 'nach Marienberg begeben … sothane Machine gehörig in Augenschein nehmen, auch wie solche zu verbeßern und in brauchbaren stand zu setzen, dem Kunstmeister Menden die erforderliche Anweisung geben'.

and not thorough enough'.[90] The polymath Johann Gottfried Jugel (1707–1786), a cameralist in the service of the Prussian state, similarly expressed a bleak view of engineers, compared to 'coopers', 'wainwrights,' and 'illiterate millwrights', some of whom 'had never seen or heard of *geometria* and *mechanica* in their lifetime'.[91] Having learned that mathematics was an 'auxiliary' or 'helpful' science developed by scholars, the bureaucrats of the capital thought that no one could be more qualified than a prestigious professor.

The encounter between the mine engineers and the university professor, however, turned out to be a farce. Professor Zeiher lectured his interlocutors with grand philosophical principles about unity and the inner power of machines, but was reluctant to even enter the mine to consider the problem. Machinist Mende – who in the meantime had improved his steering mechanism – was not in the least impressed by the theoretical skills of his visitor. He felt insulted to be lectured on his own area of expertise. If not for the respect due to his superior, he would have 'dismantled all of his freshly invented steering mechanism, and then let it to the *professor* to mathematically invent it again'.[92] Mende, who was a specialist of *Maschinenlehre* (machine theory) did not see mathematics as a discipline to be studied at university only, but as a set of skills acquired through years of experience. Mining master Trebra described the mathematician Zeiher as unable to articulate any useful proposition. The professor was depicted as a profiteer only interested in 'his young girlfriend' – who to his surprise could not enter the mines – as well as in the concert and dancing evening that followed.[93] His analysis of the machine was limited to vague claims that the improved mechanism indeed worked better than the original one – something hardly surprising, and for anyone to see. Neither the mine engineers nor the central administration received the promised report of the professor's visit for which, Trebra sardonically noted, he had been handsomely paid.[94]

In the late eighteenth century, countless examples illustrate the persistence of a gap between scholars and their bookish knowledge on the one side, and engineers or technicians on the other. Trebra and Mende did not mince

---

[90] Weidler and Fuchsthaler, *Anleitung zur unterirdischen Meß- oder Markscheidekunst* (1765), p. 9: 'in den mathematischen wissenschaften unerfahren … undeutlicher und nicht gründlich genug'.

[91] Jugel, *Gründlicher und deutlicher Begriff von dem gantzen Berg-Bau-Schmeltz-Wesen und Marckscheiden* (1744), p. 103: 'ungelehrte Müller … die doch ihr Lebtag weder von der *Geometrie* noch *Mechanica* etwas gehöret noch gesehen haben'; p. 104: 'Hutmacher … Böttger … Wagner'.

[92] Von Trebra, *Bergmeister-Leben und Wirken in Marienberg 1767–1779* (1818), pp. 231–232: '[er] würde da alles von seiner neuesten Erfindung der Steuerung wieder zertrümmert, und dann dem Professor überlassen haben, es mathematisch wieder zu erfinden'.

[93] Ibid., *Bergmeister-Leben und Wirken in Marienberg 1767–1779*, p. 231: 'Ob er seine Begleiterin (eine junge Freundin) könne mitnehmen … in die Grube hinein, das wird nicht gehen.'

[94] SächsStA–F, 40168 Grubenakten des Bergreviers Marienberg, Nr. 909, f. 98r-v; pp. 115–116.

their words against Professor Zeiher. Conversely, academics were prompt to critic practitioners and to label them as nearly illiterate. On another occasion, machinist Mende had been described by a visiting French *savant* in a most unflattering way: 'nature had made him a mechanic [*mécanicien*]; it is true that study had brought little to his knowledge, but a long practice and many constructions had enabled him to apply his natural gifts'.[95]

This story, however, is much more interesting than a simplistic opposition between a frivolous scientist and two down-to-earth practitioners. Professor Zeiher thought that he was perfectly competent to deal with practical matters, or at least pretended to be interested in it. A few years later, he would namely give a speech in Latin about the utility of mathematics, explaining why 'an erudite knowledge of mathematics [was] not sufficient'.[96] On the other hand, neither Trebra nor Mende were mere craftsmen. Von Trebra had spent his *triennum* at the University of Jena, studying 'law, mathematics, and cameral sciences with the best professors'; he would eventually become associate of the Göttingen Academy of Science. Mende's origins were more modest. Son of a miller, he had nevertheless 'been about to study mathematics at the University of Leipzig', but the file-cutting machine model he had submitted in order to get a grant was refused.[97]

The two mining officials had met at the Freiberg academy, where they belonged to the very first promotion. In the following year (1767) the founder of the academy, Commissioner Heynitz, brought his two gifted students on a trip to the Harz mines. The 'primary goal' of the trip was to observe newly introduced water-column engines, which 'might profitably be used in the Saxon mountains'.[98] When Mende proved able to build a 'much simplified', yet similar machine back home, he was on the spot appointed *Maschinenmeister* in Marienberg.[99] His attempts to scale up the machine he had observed in the Harz, although successful, proved extremely difficult and triggered the farcical visit of Professor Zeiher. The two Freiberg alumni thus possessed a solid sense of practical mathematics and went on to cultivate it throughout their careers. Steering a water-column engine was indeed arduous, and Mende offered successive improvements to fix the related issues, each new method having its own shortcomings.

The culture that developed in mining academies was thus quite different from the theories taught in contemporary universities. In creating the mining

---

[95] Daubuisson, *Des mines de Freiberg en Saxe* (1802), vol. 2, p. 51.
[96] Zeiher, *De studio mathematico eruditis non satis commendando* (1784).
[97] Von Trebra, *Der Bericht von F.W.H. von Trebra über den sächsischen Bergbau zwischen 1766 und 1815* (1998), p. 6.
[98] Ibid., p. 7: 'Zum Hauptgegenstand für diese [Reise], setzte der Herr von Heinitz die … Wassersäulenmaschine, welche wie man glauben konnte, vielleicht auch in den sächsischen Gebirgen, wegen mehrerer Leichtigkeit im Aufstellen, mit Vortheil anzuwenden sein möchte'.
[99] Wappler, 'Oberberghauptmann von Trebra und die ersten drei sächsischen Kunstmeister Mende, Baldauf und Brendel' (1905), p. 130: 'sehr *simplificiret*'.

academy, Heynitz's *Restaurationskommission* had envisioned it as a way to improve methods and machines.[100] As the two examples of this chapter – subterranean geometry and the water-column engine – show, this was not a mere rhetoric, but a well-crafted plan. After Mende, a new generation of engineers was even more successful.[101] The example of the *Bergakademie* inspired a string of new institutions, from the academy of forestry in Tharandt (1816) to the technical, soon polytechnic school of Dresden (1828). They trained the engineers that propelled Saxony at the forefront of the Industrial Revolution, mixing hands-on experience with textbooks written by local mathematicians.[102]

A key feature of these new institutions was the active collaboration of teachers, students, and engineers around concrete issues. The young von Trebra knew Scheidhauer well, both being mining masters, and greatly admired his elder: this was the kind of mathematician with whom he could envisage working. A few years after Professor Zeiher's unfortunate visit, Trebra belonged to the commission of mining officials who recommended that Scheidhauer – 'a good mathematician and seasoned miner, as is well known' – head to Marienberg to work with machinist Mende. 'On the basis of all the existing data and previous experiences', the two men should 'now produce additional calculations, on the basis of which a scientific report be issued' in order to assess the generalization of water-column engines to the rest of the Ore Mountains.[103] In other words, there was no general distrust of theory, but a serious prejudice against the existing universities. Scheidhauer being too sick to travel, he was replaced by the younger Lempe, who was still a student at the time. The machinist and the mathematician engaged in successful collaborations and published several articles about engineering methods, machine theory, and underground surveying.[104] In their mission to advance the mathematization of mining, the professors of the *Bergakademie* could thus count on a network of ancient students, many of which went on to become engineers.

---

[100] For a contrasting account of Trebra's actions as mining master in Marienberg, in which science is presented 'not [as] a useful tool for technical experts', but as 'the ideology of a desperate fisc', see Wakefield, *The Disordered Police State* (2009), pp. 46–48.

[101] The biographies of Karl Gottfried Baldauf (1751–1811) and Christian Friedrich Brendel (1776–1861) are given in Wappler, 'Oberberghauptmann von Trebra und die ersten drei sächsischen Kunstmeister Mende, Baldauf und Brendel' (1905).

[102] On the *Forstakademie* Tharandt and the *Technische Bildungsanstalt* in Dresden, see Morel, *Mathématiques et politiques scientifiques en Saxe 1765–1851* (2013), pp. 253–336.

[103] SächsStA–F, 40001 Oberbergamt Freiberg, Nr. 2479, f. 1r: 'ein bekannter guter *Mathematicus* und *practi*scher erfahrener Bergmann'; 1v: 'aus allen sich sodann ergebenden *datis* und der bisherigen Erfahrung zusammen genommen nun ferner Berechnung vor sich anzustellen, auch dem zu folge als dann ein Wissenschaftliches Gutachten abzugeben'.

[104] See for example in Lempe's *Magazin der Bergbaukunde*, vol. 4, 1787, 'IV. Beschreibung einer von dem Churf. Sächs. Kunstmeister Herrn Menden bey dem Chursächsischen Bergbaue eingeführte Art, irreguläre Schächte und Strecken, oder überhaupt jeden irregulären Raum in der Grube durch Lehreinstriche körperlich aufzunehmen'; vol. 13, 1799, 'VIII. Einige Aufsätze veranlaßt durch die Mende'schen Hubsätze.'

This example shows once again how trying to distinguish between a 'utilitarian' and a 'political' goal of early mining academies is anachronistic. In the late eighteenth century, mathematics was a common language to mining officials, and the finest minds among them were both technical experts and skilled cameralists.[105] Building on the solid tradition that had preceded the academy and accompanied its early decades, a new kind of useful knowledge, largely based on practical mathematics, developed and quickly outpaced the university tradition. This approach was summed up by Friedrich Gottlieb Busse (1756–1835), who succeeded Lempe in the chair of mathematics. Significantly, one of his first actions was to publish a book on the water-column engine, in which he reflected on the 'association between theory and practice'. According to him, the success of the approach developed in Freiberg came from the conscious combination of 'theoretical considerations' with the 'scientific use of the general experiences'.[106]

*

Teaching a mathematics truly useful for the running of ore mines was a daunting task that underwent important transformations during the eighteenth century. In this chapter, I have tried to present a nuanced view of this process, providing a legitimate reappraisal of the Freiberg mining academy. In the main mining regions, numerous engineers, officials, and 'hybrid experts' already formed a *universitas*, a community of mining science.[107] The age-old apprenticeship system was an ever-adapting process, but full-fledged institutions overseen by the state's administration only materialized in 1765, in the aftermath of the Seven Years War. Mining academies represented a highly visible step in a gradual process, but did not in any case trigger the 'radical extinction of the old mystery-mongering', as positivist accounts have argued.[108] Nor could these new institutions be self-sufficient: academies long maintained a bond with the local community, preserving for instance the companionship

---

[105] Wakefield, *The Disordered Police State* (2009), pp. 46–47, erroneously draws an opposite conclusion from the appointment of Trebra in Marienberg (maybe because his analysis only relies on a few pages in Trebra's autobiography, and does not include any archive material to offer a more balanced account).

[106] Von Busse, *Betrachtung der Winterschmidt- und Höll'schen Wassersäulenmaschine* (1804), pp. xi, 20: 'Verbindung zwischen Theorie und Praxis … theoretische Betrachtungen … wissenschaftliche Benutzung allgemeiner Erfahrungen.' In ch. 5 (pp. 31–52), Busse presents observations made by Lempe and Baldauf on several Mende engines.

[107] About the concept of hybrid experts, see Klein, 'Hybrid Experts' (2017), pp. 299–302. Klein also uses the related concept of 'scientific-technological experts' in *Technoscience in History: Prussia, 1750–1850* (2020).

[108] Mihalovits, 'Die Gründung der ersten Lehranstalt für technische Bergbeamte in Ungarn' (1938), p. 10: 'Die Fachbücher haben zwar die Macht der alten Geheimtuerei geschwächt, aber die radikale ausrottung der letzteren können wir noch mehr der systematischen staatlichen Fachausbildung verdanken.'

system of surveyors. Their success, however, was not readily transferable, which is why they only succeeded where vigorous teaching traditions had been present.[109]

Eighteenth-century engineering faced a formidable challenge in trying to reconcile two diverging goals: standardizing the teaching of elementary concepts while trying to legitimate a broad use of advanced mathematics. Scheidhauer and Lempe, both mining officials with a solid mathematical culture, endeavoured to replace geometrical methods and piecemeal solutions with analytical tools.[110] They tried to navigate their way between the Charybdis of efficient, yet limited methods, and the Scylla of soothing, yet impracticable scholarly discourses. Abraham Gottlob Werner, then director of the *Bergakademie*, summed up this dilemma as follows: 'higher mathematics', he noted, 'is admittedly only rarely necessary, but in some cases then really so'.[111]

Nowhere was the tension between the exigence of a solid initial training and the temptation of a complete mathematization of nature more palpable than in the case of Johann Friedrich Scheidhauer. Self-taught, the mining master was well placed to know what it meant to learn elementary mathematics by oneself. Yet the seemingly infinite possibilities of science prompted him to apply higher mathematics to most mining problems of his time. Scheidhauer taught selected students of the mining academy, but stayed first and foremost a mining master. How could such an important figure, whose preserved *Nachlass* offers a unique view on eighteenth-century technology, stay unnoticed from historians? I think that it boils down to the false dichotomy I previously alluded to. For positivistic historians seeing the new mining academies as a revolution, Scheidhauer is an outlier, whose activity predates and develops beyond the academic system. Conversely, his unquestionable achievements and influence also refute the narrative of mining academies being mere political arenas.

Although the mining academies were dedicated to the advancement of ore extraction and mining sciences, lots of attention and a great deal of resources were devoted to mathematics. The prism of practical mathematics thus provides original and interesting insights about the development of modern technical institutions, and the essential tension between engineers and professors. In the short term, the *Bergakademien* ensured a solid initial training, tackling

---

[109] A collection of essays dealing with the institutionalization of mine training is Schleiff and Konečný, *Staat, Bergbau und Bergakademie* (2013), more specifically the second (pp. 95–192) and fourth (pp. 251–352) parts. On the Habsburg, in which the state's control was even more important (and to some extent hindering), see Konečný, 'Kameralisten, Bildungsreformer und aufstrebende Bergbeamte' (2015), pp. 66–89.

[110] For an example of the contribution of Freiberg mathematicians to applied mathematics and mine engineering, see Stoyan and Morel, 'Julius Weisbach's Pioneering Contribution to Orthogonal Linear Regression (1840)' (2018), pp. 75–84.

[111] UAF, OBA 10, f. 132v–133: 'höhere mathematik, welche zwar nur in seltenen, aber doch wirklich in einichen Fällen nötig ist'.

the previous lack of coherent curriculum head-on. In the longer term, as the teaching of fundamentals was taken over by lower mining schools, the efforts of Scheidhauer, Lempe, and their successors paved the way for a thorough science-based engineering. The success of technical schools showed the ability of practitioners to promote a rational engineering, based on mathematics and yet departing markedly from the discourses of university professors. The culture of mathematics peculiar to mining regions crystallized around the new institution, where mining master von Trebra, machinist Mende, and a new generation of officials were trained. It would soon merge into the modern engineering of the nineteenth century, to the establishment of which they had patiently contributed.

# 7 'One of Geometry's Nicest Applications'
## Crafting the Deep-George Tunnel (1771–1799)

When Jean-André Deluc arrived in the city of Clausthal on 24 October 1776, he was immediately struck by its uncommon landscape: 'one sees, here and there, large molehills interrupting the vegetation. These are mine entrances and their halls, or rather heaps of rubble.' Deluc had travelled all the way from London, where he was reader to Queen Charlotte (1744–1814).[1] He had brought a portable barometer, eager to test a formula he had devised to compute the height of places from barometric observations. After decades of mountainous experiments, mostly among the high summits of his native Switzerland, he now wished to go down metal mines. In a letter to the Royal Society of London, of which he was a fellow, Deluc explained why he had chosen the mining pits of the Harz: 'These I knew were extremely deep; and it made me very desirous to try in them my rules for measuring heights by the barometer.'[2]

The Captain-general of the mines, Baron Friedrich Wilhelm von Reden (1752–1815), had been warned of the arrival on an important visitor. He thus took Deluc on his tour of neighbouring mines, assisted by the subterranean surveyor Samuel Gottlieb Rausch (d. 1778). In his *Letters* to the Queen, Deluc described in detail the well-ordered exploitation of mines, but noted that he was brought 'not to the bottom of the mine, but to the level at which water cannot any more be pumped out'. It was by then clear that 'a deeper *Draining Tunnel*' would be needed before operations could resume and go deeper (see Figure 7.1).[3]

---

[1] Deluc, *Lettres physiques et morales sur l'histoire de la terre et de l'homme* (1779), vol. 3, p. 189: 'On aperçoit çà et là comme de grandes taupinières qui interrompent cette verdure. Ce sont les entrées des mines, et leurs halles ou monceaux de décombres.' On Jean-André Deluc's appointment as Queen Charlotte's reader, see Hübner, *Jean André Deluc 1727–1817* (2010), pp. 88–97.

[2] Deluc had been promoted to a Fellow of the Royal Society (FRS) in 1773, upon his arrival in London. His first letter concerning 'Barometrical Observations on the Depth of the Mines in the Hartz' was read on 20 March 1777. Although Deluc wrote in French, the letter was published in the *Philosophical Transactions* together with an English translation (which will be used here, unless otherwise stated).

[3] Deluc, *Lettres physiques et morales sur l'histoire de la terre et de l'homme* (1779), vol. 4, pp. 559–560: 'non tout au fond de la Mine, mais au niveau des eaux que l'on ne pompe plus … une *Galerie d'écoulement* plus profonde'.

213

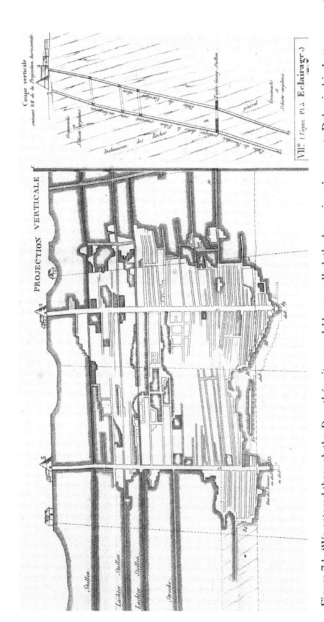

Figure 7.1 'We entered through the *Dorothée* pit around 11 a.m., all clothed as miners', recounts Deluc in his *Letters upon the History of the Earth*. The Dorothea (3, left) and Caroline (2, right) mining pits were among the deepest and more productive of the district. Draining tunnels are visible on the left side, the lowest horizontal line depicting the Deep-George tunnel. From Héron de Villefosse, *De la Richesse Minérale*, Paris, 1819, vol. 2, plate 8 (details). Courtesy ETH-Bibliothek Zürich, Rar 9699.

This was Deluc's first encounter with the Deep-George tunnel, named after George III, King of Great Britain and Duke of Hanover. Although the tunnel was still in its planning phase at the time, the Captain-general was kind enough to show him where its mouth (entrance) might be located. There, he could compare his barometrical observations with the 'geometrical measures' of surveyor Rausch. Deluc was so impressed by the project that he came back two years later. Construction work for the new draining tunnel had then begun under the supervision of two surveyors. The natural scientist wondered once again at the 'trust that miners have in their art', and how well this 10 km-long project had been planned:

All of this was born on paper, and will be carried out in the deep obscurity of the earth's entrails, where each of our miners will only have in front of him the stone that he digs by the light of his lamp ... This is thus certainly one of *geometry*'s nicest applications; and when the miner is proud of his art, this shall not surprise me. This gallery will be named George, and it will be a beautiful landmark, worthy of the ruler who has ordered this work for the good of the inhabitants of the Harz.[4]

Deluc's scientific interests were here paired in a timely way with a compliment to Queen Charlotte's husband, who had recently approved the financing of this project. From his two articles sent to the Royal Society about the Harz mines, one grasps that the single thing that impressed him most was subterranean geometry. He marvelled at length about the 'exactness' of the surveyors' 'geometric measures' and 'the astonishing works which are undertaken in consequence of them'.[5]

Deluc was a prominent natural scientist of his time. Born in Geneva, he had settled in London and was foreign associate to the *Académie des Sciences* in Paris. He designed his own instruments and undertook extensive travels to understand the general laws of the earth and its atmosphere.[6] Why was Jean-André Deluc so impressed by subterranean geometry? His interest illustrates a more general attitude of eighteenth-century scientists about nature, exploring mountains, mines, and inhospitable regions. Deluc's testimony is special inasmuch as the author precisely described the encounter of two worlds. In the Harz, he met mining officials who had spent their whole careers in the local administration. Their differences notwithstanding, the engineers and

---

[4] Ibid., vol. 4, pp. 624–625: 'Tout cela prend naissance sur le papier, & va s'exécuter dans la nuit profonde des entrailles de la Terre, où chacun de nos Mineurs n'aura jamais devant lui, que le Rocher qu'il creufe à la lueur de fa lampe. Cependant ils se rencontreront enfin deux à deux ... Voilà donc certainement une des plus belles applications de la *Géométrie*; & quand le mineur eft glorieux de fon Art, je ne faurois m'en étonner. Cette galerie fe nommera GEORGE; et ce fera un beau Monument, digne du SOUVERAIN qui a ordonné cet ouvrage, pour le bien des Habitans du *Hartz*.'

[5] Deluc, 'A Second Paper Concerning Some Barometrical Measure in the Mines of the Hartz' (1779), p. 495.

[6] Two recent works offer a comprehensive view of Deluc's life and numerous activities: Hübner, *Jean André Deluc 1727–1817* (2010) and Heilbron and Sigrist, *Jean-Andre Deluc: Historian of Earth and Man* (2011).

the natural scientist shared common interests, as they all sought to ascertain heights, standardize measurement methods, and tackle issues of depth. Moreover, they constantly tried to improve their instruments and developed methods to check their observations and calculations.

The construction site of the Deep-George tunnel opens a window on the relationships between scientists and mining officials at the onset of the Industrial Revolution. In the mining states of the Holy Roman Empire, engineering was reaching its full potential. Geometry had become a powerful planning tool, used to systematically design and assess the best solutions. This marked change from the seventeenth century was not restricted to subterranean geometry: from chemistry to forestry, a new 'quantifying spirit' was at work.[7] But in relation to mining, it was on a formidable scale as more and more plans of new draining galleries were initiated. Construction sites displayed the new role assigned to subterranean surveyors. Their maps, measurements, and projections were now precise enough to avoid any major delay or modification. Geometry's role was not merely to adapt to circumstances any more, as it now defined the very frame in which political decisions would be considered, taken, and executed.

This chapter recounts the digging of the Deep-George tunnel, from the first sketches in 1771 until its inauguration in 1799. Using the technical field books of surveyors Rausch and Länge and administrative correspondences, we will understand how – to use Deluc's words – this project was 'born on paper'. Far from being restricted to the planning phase, geometry also proved crucial to check the progress of operations and fine-tune the tunnel. In that context, the visits of Jean-André Deluc provide an original case study about the interactions between a natural scientist and engineers. Although this part of his scientific activity has never been analyzed, he sojourned in the Harz three times in 1776, 1778, and 1786, both during the planning phase and after digging operations had begun. Performing several sets of barometric observations to refine his physical formula, he engaged in fruitful exchanges with subterranean surveyors and reflected on the relationship between their respective methods. Finally, this encounter can be understood in the broader context of eighteenth-century scientific tourism. Deluc was indeed one of countless contemporaries visiting the Harz's mining cities, asking to enter the new tunnel or neighbouring pits. For a short period of time, mining acquired an almost public character, as numerous authors described the construction site of the Deep-George tunnel and the administration itself promoted it actively. This example highlights one last time the paradoxical position of subterranean geometry: just as scientists marvelled about it, and when its efficiency was clear for everyone to see, it would ultimately recede into the background.

---

[7] See Frängsmyr, Heilbron, and Rider, *The Quantifying Spirit in the Eighteenth Century* (1990).

## Thinking on Paper

On 26 July 1777, Captain-general von Reden, introduced by a cannon salute, surrounded by mining flags and musicians, held the most important speech of his career.[8] As the local mines 'went deeper by the day', the quantity of ground water had substantially increased. This had led 'many ore veins' being completely drowned. Water ponds, wheels, and pumps: all existing solutions were reaching their limits, becoming 'so expensive' to operate as to endanger the economy of the whole region.[9] This had led the mining administration in 1771 to inquire into a new, deeper gallery to drain ground water. It was the longest mine tunnel ever projected and would be, with its 284 m, almost twice as deep as the tunnels it replaced.[10]

Six years later, facing the entrance of the future tunnel, Reden concluded his speech as the first pickaxes were given. The Captain-general confidently forecast that this herculean work would be finished before the end of the century – and indeed it was. The Deep-George tunnel would not follow the sinuous ore veins, or even existing galleries, but had been planned in a modern fashion. It was bored through hard stone following mostly straight lines, in order to connect the mines directly with the lowest point in the region. The mining administration was confident enough to have the speech dedicated to King George III, printed, and widely circulated.[11]

Captain-general von Reden's confidence rested on solid ground. His administration had been working on the design and minutely exploring various alternatives. When discussing tunnel digging or engineering projects, the planning phase and its actors are generally neglected. Georg Andreas Steltzner (1725–1802), who had worked his way up from ore crusher to mining master, was in charge of these technical aspects.[12] He was in weekly, if not daily, contact with the two surveyors working on the project. Samuel Gottlieb Rausch had been appointed in Clausthal in 1742, and his son Carl August would soon succeed him. He had trained Johann Christian Heinrich Länge, who now worked for the twin city of Zellerfeld. When mentioned at all, the work of surveyors Länge

---

[8] The ceremony is extensively described in Gotthard, *Authentische Beschreibung von dem merkwürdigen Bau des Tiefen Georg Stollens am Oberharze* (1801), pp. 211–215.

[9] Von Reden, *Rede bei dem feyerlichen Anfange des tiefen Georg-Stollen-Baues unweit Grund* (1777), pp. 3–5. On the situation in the early 1770s, see Liessmann, *Historischer Bergbau im Harz* (2010), pp. 170–174. A thorough analysis of the technical water system of the Harz, known as *Wasserwirtschaft,* can be found in Schmidt, *Die Wasserwirtschaft des Oberharzer Bergbaus* (1989), pp. 30–37.

[10] Gotthard, *Authentische Beschreibung* (1801); Broel, 'Die markscheiderischen und bergmännischen Arbeiten beim Bau einiger Stollen und Schächte im Harzer Bergbau des Clausthaler Reviers' (1955), p. 39.

[11] Von Reden, *Rede bei dem feyerlichen Anfange des tiefen Georg-Stollen-Baues unweit Grund* (1777), pp. 10–11.

[12] On Steltzner, see Bartels, 'Der Harzer Oberbergmeister Georg Andreas Steltzner 1725–1802' (2015), pp. 275–288.

and Rausch on the Deep-George gallery is usually just a brief note before focusing on the digging work itself.[13] The justifiable certainty of Reden, however, meant that the critical part of the work had already been completed when digging operations began in the summer of 1777.

The triggering event for the whole project was a short report written in 1770 by Steltzner. The local mining master had greatly improved the water supply for pumps and water-wheels in the previous decade.[14] In his report, however, Steltzner described the 'highly necessary help' needed to keep 'the deepest mine workings' from flooding, and now frankly argued for a complete overhaul.[15] The report elicited lively reactions among the mining council, all of which were carefully protocolled. Captain-general von Reden suggested to consider a new drainage tunnel and pondered the pros and cons. 'A brand-new work' would be 'very lengthy and costly', he acknowledged, but the situation was getting desperate.[16] Luckily, routes linking the existing mines with the lowest points of the region – located near the city of Grund – might cross promising prospecting spots. The council prudently ordered a report:

Since it is necessary for such an important matter to be taken into careful consideration, and that all its circumstances be considered thoroughly: You will authorize that all records about the galleries [of the city of] Grund be inspected, and the surveyor Rausch will be charged of collecting all the information that could figure on local mining maps.[17]

Interestingly, the analysis that Rausch was asked to produce did not yet include new surveys. The existence of precise maps and reports allowed for a geometrical engineering expertise that the old surveyor could produce from his working cabinet. A few months later, his technical assessment was sent to mining master Steltzner, who added his own financial and mining expertise on the crucial question: 'whether the local mines could benefit so much from the construction of a deep tunnel as to warrant the gamble of such high costs?'[18] In a

---

[13] Bartels, *Vom frühneuzeitlichen Montangewerbe zur Bergbauindustrie* (1992), p. 389; Bartels, 'Der Harzer Oberbergmeister Georg Andreas Steltzner 1725–1802' (2015), p. 285: 'Für den Stollenbau waren bis 1774 noch keine Schritte zur praktischen Realisierung eingeleitet.'

[14] Ibid., pp. 279–280; Dennert, *Bergbau und Hüttenwesen im Harz* (1986).

[15] NLA HA, BaCl Hann. 84a, Nr. 9803, doc. 1a, *pro memoria*, 10 November 1770: 'höchst nöhtige hülffe denen tieffen Gruben'.

[16] Gotthard, *Authentische Beschreibung* (1801), p. 75; NLA HA, BaCl Hann. 84a, Nr. 9803, doc. 1b, meeting report, unpaginated: 'langwierig und viele Kosten erfodern'.

[17] NLA HA, BaCl Hann. 84a, Nr. 9803, doc. 1b, meeting report, unpaginated: 'Wie aber nothwendig, daß diese, so *importante* Sache in genügsame Überlegung genommen, und nach allen ihren Umstände gründlich erwogen werde : so wolten Sie veranlaßen, daß in der *Communion* die sämtlichen Acten von denen Gründerschen Stollens aufgesuchte würden, auch dem Marckscheidern *Rausch* hiemit aufgeben, das jenige, was davon etwa bey den hiesigen Rißen vorhanden seyn mögte, bey die Hand zu kriegen.'

[18] NLA HA, BaCl Hann. 84a, Nr. 9803, doc. 1b, meeting report, unpaginated: 'ob denen hiesigen Gruben durch die Herführung eines tiefen Stollens so viel Nutzen zu wachsen können, daß dagegen die großen anzuwendende Kosten zu *hazardir*en wären.'

region where mining had been going on for centuries, an extensive knowledge on the environment already existed. Several main routes came into question, corresponding to low points in neighbouring valleys, from which the mouth of a gallery could be dug and connected to the existing mine workings. Based on the surveying report, the mining master confirmed the feasibility of the project.

In 1772, however, Vice Captain-general August von Veltheim proposed a completely different solution. Instead of building a new tunnel, he suggested repurposing existing galleries. The project might not sound fashionable, Veltheim acknowledged, but required 'incomparably less costs and time than this [other] proposal'. His idea was to judiciously connect existing tunnels and to improve their efficiency using modern digging methods, but the practicability of his plan was doubtful. 'To be able to judge this to a certain extent', he asked another surveyor – Länge – to assess the relative positions of several galleries.[19] While Länge's preliminary observations were encouraging, the Clausthal surveyor – Rausch – soon discarded the possibility. 'He has presented to the mining office a map … on which the relation of one [gallery] to the other is presented', summed up Steltzner in his report, 'and the mining office has forwarded the map to me.' The mining master then used all the available evidence to assess the two alternative plans, using several mining maps and his own 'visual inspections' of the most important spots. He even 'collected the descriptive reports about nearby abandoned mine workings' and stored them in the mining office.[20] Steltzner finally concluded that the original idea of a new, deeper gallery was the most promising one.

During this first phase, most of the planning work had been carried out on paper, using old reports, existing field books and, most importantly, the growing stock of precise mining maps. Neither a general inspection nor extensive surveys had yet been organized, as used to be the case in earlier times. Surveyors relied on the vast trove of data gathered over the last century and were simply asked to perform measurements of a few limited areas, for which a greater accuracy was critical. This crucial remark shows the transformation that subterranean geometry had undergone. Going on site was expensive and, more importantly, direct observation could do little to decide between competing proposals when these were of a different nature and concerned distant places.

Once a consensus had been reached about the necessity of a new, then-unnamed drainage gallery, the second task was to ascertain its main direction,

[19] NLA HA, BaCl Hann. 84a, Nr. 9803, doc. 3, *pro memoria* of 18 December 1772: 'ungleich weniger Kosten und Zeit, als jener Vorschlag erfordert … Um davon einiger maaßen urtheilen zukönnen, so habe ich das Fallen dieses 13 Ltr Stollens von seinem Mundloche an, bis an die Einseitige Grentze, vom Comm: Markscheider ziehen laßen.'

[20] NLA HA, BaCl Hann. 84a, Nr. 9803, doc. 9, *Pro memoria*, 16 January 1774: 'Dieser hat Nr. 10. Trin. 1773 davon denn Riß in BergAmte vorgezeiget, in welchen die verhältniß eines gegen das andern bestimt ist, mir ist von BergAmt das Riß Zugestellet … Augenschein … von denen herum liegenden alten Wercken *discursive* Nachrichten eingezogen.'

the *Stollenlinie*. This was not a mere technical decision but entailed far-reaching economic and political considerations. An important question was to decide where exactly the entrance of the gallery – its *Mund* (mouth) – would be located. This obviously conditioned the depth of the gallery, but also influenced the direction it would follow. The lowest points were farthest away from the mines; they would thus require more time and costly investments in aeration shafts, machines, and workers. Each proposed route crossed several promising spots for further prospecting, each of which had to be carefully assessed. It was also crucial to decide which mines of the vast existing complex would be connected first, while trying to make the most of the existing infrastructure in order to lower the costs. In the end, the easiest part of the project was the naming of the gallery: since King George III of England and Hannover was open to the idea of funding the project, it was named after him the *Tiefer-Georg Stollen* (Deep-George Tunnel).[21]

In the summer of 1774, a full survey was then ordered to make a proper comparison between several possible pathways.[22] From this point on, we can track to the day – sometimes to the hour – the actions of the subterranean surveyors until the final completion of the project. Their measurements are precisely written down in dozens of field books (*Observationsbücher*). Analysing them reveals exactly how surveyors Länge and Rausch worked, while frequent annotations show how mining master Steltzner carefully checked their progress.[23] On 9 August, Länge began his work by taking a reference point close to the mouth of an abandoned gallery in Grund. Using his semicircle, compass, and surveying chain, he went through the hills and woods and arrived on 13 September at the entrance of the *Old-Blessing* mine, located in the main district (see Figure 7.2). From there, he spent six days carefully surveying down to the deepest point, where the projected tunnel would once be connected. Five weeks of continuous work finally enabled Länge to ascertain accurately the relative height of the two points.[24] A few weeks later, having connected his initial reference point in Grund to all other proposed entrances of the future tunnel, he drew his results on a map (see Figure 7.3). Together with the map, an extensive report summed up all possible entrances and pathways, with estimates for several parameters, including lengths, costs and – most importantly – the depth that each would add compared to existing tunnels. The result was checked

[21] See NLA HA, BaCl Hann. 84a, Nr. 9803, doc. 12, report of 26 May 1774 to the royal government and Gotthard, *Authentische Beschreibung* (1801), p. 30.

[22] NLA HA, BaCl Hann. 84a, Nr. 9803, doc. 16 and 17.

[23] In the *Bergarchiv* Clausthal, the section NLA HA Dep. 150 contains some 40 field books related to the digging of the *Deep-George Tunnel*, including some extensions in the early nineteenth century.

[24] NLA HA Dep. 150 Acc. 2018/700 Nr. 252, p. 1 for the beginning, p. 31 for the entrance in the *Old Blessing* pit, p. 36 for the last part of this first survey.

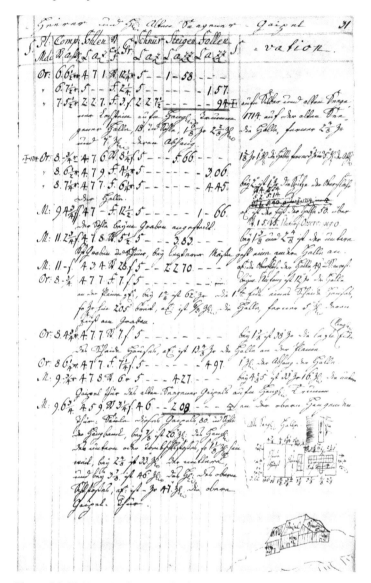

Figure 7.2  Field book of Johann Christian Heinrich Länge on 13 September 1774. The data table on the left is accompanied by textual observations providing details, cross-references, and even one correction by mining master Steltzner. In the bottom-right corner, a plan and sketch of the entrance of the *Old-Blessing* mine provide the exact dimension of the building, its machines, and the reference points used by the surveyor. Courtesy Bergarchiv Clausthal (NLA HA, BaCl Hann. 84a, Nr. 252, f. 31r. Photo: Thomas Morel).

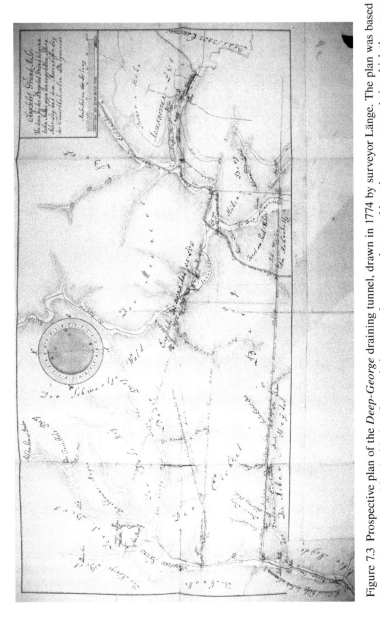

Figure 7.3 Prospective plan of the *Deep-George* draining tunnel, drawn in 1774 by surveyor Länge. The plan was based on a five-week survey combined with data from existing maps. It was supplemented by a long *pro memoria* in which the competing pathways for the future gallery were described, highlighting technical issues. Courtesy Bergarchiv Clausthal (NLA HA, BaCl Hann. 84a, Nr. 9803, unpaginated. Photo: Thomas Morel).

multiple times, most thoroughly by mining master Steltzner, who laconically added in the margins 'perused and found correct'.[25]

The large-size map, and most importantly the underlying data, offered solid quantitative evidence on the basis of which a decision could then be taken. At the end of that year, a one-week conference was organized between the political authorities of Hanover and the local mining administration.[26] The concept of the tunnel was officially just one aspect of a more general reform movement, which also included the opening of a mining class in Clausthal, in which future fore-men and engineers would be taught mathematics and physics. From now on, the project was in the hands of the Captain-general. As a wise cameralist trained at the University of Göttingen, von Reden ordered a new set of maps – this time less technical and more figurative – to be brought to London and he presented the mat-ter to King George III.[27] The main trade-off was between the length of the route and its difficulty: the deeper one went, the longer and costlier the operation would be. To demonstrate that the administration had made the most sensible choice, both financially sustainable and offering long-term security for the mines, several visualizations were produced. Along with all kinds of maps, numerous synthetic tables gathered together the main characteristics of the proposals.[28]

When Jean-André Deluc made his first visit to the Harz, the mining admin-istration was pondering one of the proposed routes. Delighted to be one 'of the party', Deluc admired not only the workmanship of the surveyors' instru-ments, but also the minuteness of their work. His description offers a glimpse of the multiple constraints involved in this kind of project:

The path that this tunnel will take is neither horizontal, nor in a straight line: & it is there again that the miner's skill can be witnessed. The gallery must be regularly inclined, so that the water runs off; & it must be inclined as little as possible, so as not to lose depth. Besides, it should follow certain contours, either to pass under valleys where shafts can be drilled, or to avoid parts that are too hard and take too long to drill, or parts that are too soft and require stamping. Sometimes again, if the surveyor sees near his path some gallery already drilled for other purposes, he detours in order to take advantage of the work already done.[29]

---

[25] NLA HA Dep. 150 Acc. 2018/700 Nr. 252, p. 102.

[26] See Bartels, *Vom frühneuzeitlichen Montangewerbe zur Bergbauindustrie* (1992), pp. 388–392.

[27] One copy of this map is preserved in Rissarchiv des LBEG, Clausthal, B832.

[28] See, for example, NLA HA, BaCl Hann. 84a, Nr. 9803, unpaginated, and NLA HA, BaCl Hann. 84a, Nr. 9805, unpaginated (reports of 8 August and 13 August 1776).

[29] Deluc, *Lettres physiques et morales sur l'histoire de la terre et de l'homme* (1779), vol. 4, pp. 623–624 : 'Cette route que devra tenir la *Galerie*, n'est ni horizontale, ni en droite ligne : & c'est là encore que se montre l'habileté des Mineurs. La galerie doit être régulièrement incli-née, pour que les eaux s'écoulent; & elle doit l'être le moins possible, pour ne pas perdre de la profondeur… Il faut de plus qu'elle suive certains contours; soit pour passer sous des Vallées où l'on puisse percer des puits; soit pour éviter des parties trop dures qui prendroient beaucoup trop de temps à percer, ou des parties trop molles qui exigeraient des étampages. Quelque fois encore, si le Géomètre voit près de sa route quelque *Galerie* toute percée pour d'autres usages, il se détourne, afin de profiter de ce travail déjà fait.'

The project went through a couple of years of hesitations, as George III's private counsellor and the Hanoverian administration negotiated with the Principality of Brunswick – who shared the political authority on parts of the Harz mines – about the specifics of the financing plan, a process in which maps again played a crucial role.[30] Meanwhile, surveyors continued their work, providing data on demand to feed the ongoing debates and considering numerous alternative solutions. Several rounds of negotiations happened in 1775 and 1776, and a final contract was signed in the spring of 1777. The digging of the Deep-George tunnel could begin.[31]

### Fine-Tuning the Tunnel

On 1 July 1777, the whole administration met at quarter past seven in front of the mining office in Clausthal and began a general visitation. From the existing mine workings, surveyors guided the group up to the proposed mouth of the Deep-George tunnel. Technical details about the exact position of aerating shafts were discussed and, most importantly, the location of the entrance was to be irrevocably fixed. 'During the visitation', a report stated, 'the surveyor Mr. Länge presented a map' that was used, together with 'the visual inspection', to reach a final decision.[32] Later this month, the mining council was informed that Länge had finished his last round of double-checking. The surveyor, reported mining master Steltzner, 'declares that he stand up by the correctness of his measurements up to a difference of a few inches'.[33] Six days later, Captain-general von Reden gave his speech, and dozens of miners soon began the digging operations. This chronology illustrates how the decision process of the enterprise largely rested on the surveyors' expertise.

Knowing the general route of the tunnel, however, was far from enough. In the late eighteenth century, boring 'by means of cutting and drilling' with blasting powder progressed, on average, less than one metre a week. At this pace, drilling the *Deep-George* tunnel from its mouth to the existing mine workings would have taken about two centuries – beginning from both sides would only halve the duration.[34] Von Reden claimed that he could achieve it

---

[30] NLA HA, BaCl Hann. 84a, Nr. 9805, Letter from 26 May 26 and 25 July 1776 from King George III. Same folder, a letter of surveyor Rausch lists the 'maps to send to Hanover' to inform the decision process (22 September 1776)

[31] The full text of the contract is printed in Gotthard, *Authentische Beschreibung* (1801), pp. 58–68.

[32] Ibid., p. 73: 'Bey der gedachten Befahrung producirte der damalige Communion Markscheider Herr Länge einen Special-Riß … auch durch den Augenschein.'

[33] NLA HA BaCl Hann. 84a Acc. 8 Nr. 2465/1, Extracts from Protocol, July 20,1777: 'da derselbe [Länge] *declari*ret, daß er bis auf eine Differenz von eine paar Zollen vor die Richtigkeit seines Zuges einstehe.'

[34] NLA HA BaCl Hann. 84a Acc. 8 Nr. 2465/1, meeting report from 2 Quartal Reminiscere, 1780, 'mittelst Schrämm und Bohren aufgefahren'.

in two decades by using several intermediary points, a task that was logistically and technically challenging. In order to plan and schedule excavations and counter-excavations, Rausch and Länge spent several months of full-time surveying work. Once again, the process is elegantly described by Jean-André Deluc. The Queen's reader, who had by then just arrived for his second geological journey in the Harz, wrote:

The day after we arrived in Clausthal, we visited … a mining city, close to the place where the new drainage tunnel they are working on will be coming out. We saw on our way several of the shafts being dug to speed up the works; & since it is one of the miner's works in which their skill is more evident, I consider it my duty to give a clearer idea to Her Majesty who knows this good people.[35]

Deluc went on to compare the miners to 'pilots who would have to meet at a point in the Atlantic Ocean, leaving from opposite coasts'. Without a compass and diligent observations, they would have little chance of meeting. Miners had it even harder, he continued, without 'the possibility to find one another in a wide horizon'. Nevertheless, and based on careful computations, 'six shafts are being dug, or have been, on the pathway of the tunnel. Once sunk to the necessary depth, two pairs of Miners will leave from the bottom of each one, heading in opposite directions on this planned route.'[36] To this aim, both surveyors conducted two full sets of observations and were strictly forbidden to 'correct the differences between themselves', but had to send the raw data to Steltzner.[37] The mining master thoroughly checked all the computations, condensing thousands of pages into four large folios. These tables were again reduced into a single one highlighting the differences between their geometrical measurements.[38]

The mining council then asked surveyors to account for their errors. This long work took place in their offices, comparing their respective field books to assess where computations mistakes had been made. In a second step, each section was considered separately, tracking down precisely where variations occurred. In the end, a couple of instances were left in which data could not be

[35] Deluc, *Lettres physiques et morales sur l'histoire de la terre et de l'homme* (1779), vol. 4, p. 621: 'Le lendemain de notre arrivée à Clausthal … nous allâmes visiter … l'une des *Villes de Mines*, & près du lieu où sortira la nouvelle *Galerie d'ecoulement* à laquelle on travaille. Nous vîmes sur notre chemin plusieurs des Puits qu'on perce pour accélérer l'ouvrage; & comme c'est là un des travaux des Mineurs où leur habileté se manifeste le plus, je me fais un devoir d'en donner une idée un peu distincte à V.M. qui connoît ces bonnes gens.'

[36] Ibid., vol. 4, pp. 622–623: 'Pilotes qui se seroient donné rendez-vous à un point de la Mer Atlantique en partant des Côtes opposées … la faculté de se découvrir dans un grand horizon … lorsqu'ils se seront enfoncés à la profondeur nécessaire, il partira deux couples de Mineurs du fond de chacun, qui se dirigeront en sens contraire sur cette route prévue.'

[37] NLA HA, BaCl Hann. 84a, Nr. 9825, unpaginated, meeting report 8 Quartal Luciae 1777: 'die vorfallenden Differenzien unter sich berichtigen sollen'.

[38] NLA HA, BaCl Hann. 84a, Nr. 9825, unpaginated, report dated 9 January 1778. For a similar review made by a Saxon official, see 40010 Bergamt Freiberg, Nr. 2790 Pabst von Ohain.

made to agree. Surveyor Länge then realized a few observations in the mines, using the signs previously carved in stones as fixed points, until nothing differed from more than a few inches. Observations at the surface were also necessary, but had to wait for the spring, since they could only be made 'during very calm weather, which is rare at this time of the year'.[39] The painstaking harmonization work was crucial, the mining master underlined, to know 'to which extent the *Deep* tunnel could then be reliably drilled'. Steltzner did not fully trust his surveyors and suspected Länge of altering his figures. In one report, he complained that 'he himself could only answer for the soundness of computations, not of the [surveying] work'. His fears ultimately proved unfounded: there would not be any major surprise until the last connection was reached twenty-two years later.

Once the road map had been set and the main direction fixed, surveyors still had plenty of work to ensure that the project was heading in the right direction. They were especially careful when connecting the various sections of the tunnel. As soon as routine observations indicated that an excavation was about to meet a counter-excavation, a local survey was performed to monitor the relative position of the two groups of miners. While theoretically superfluous, corrections were in practice needed for various reasons. Miners always deviated, even slightly, from projected courses. Surveyor Länge also complained about the 'incorrigible imperfection of the instruments and the deviation of the magnetic needle'. Even the slow drying of the paper on which the original large-scale maps had been drawn made them unreliable to ascertain directions with the wanted precision. It was thus necessary, after several years, to proceed to new observations 'and give the last instructions about the direction of the connection based on them'.[40]

The surveyors also supervised the operations of smoothing (*Nachhauen*) the sole of the gallery, to ensure that the slope was regular and would not interrupt the water flow.[41] A last daunting task was to connect the Deep-George tunnel with all related mines and prospecting sites. The first planning phase had only designed a tunnel relating one point in the mines to the lowest point in the valley. In the following two decades, surveyors then ascertained the relative height and position of all neighbouring mine workings.[42] Jean-André Deluc describes once again the multiple ramifications

---

[39] NLA HA, BaCl Hann. 84a, Nr. 9825, unpaginated, report dated 27 August 1778: 'bey recht stiller Lufft, welche aber bey jetziger Jahres Zeit selten ist'. During windy seasons, the plumb line and other instruments could rarely be used outside, see Schmidt, *Die Wasserwirtschaft des Oberharzer Bergbaus* (1989), p. 47.

[40] NLA HA, BaCl Hann. 84a, Nr. 9825, unpaginated, report of 10 January 1783: 'incorrigibelen Unvollkommenheit der Instrumente und der Abweichung der Magnet-Nadel … den Zug von neuen verrichten, und nach denselben die letzten Anweise Linien zum Durchschlag angeben'.

[41] See the complete file NLA HA, BaCl Hann. 84a, Nr. 9848. Depth differences of as little as a few inches are taken into consideration.

[42] See the data tables in NLA HA Dep. 150 Acc. 2018/700 Nr. 251.

of this engineering project: 'fifty other miners go from the mines to join this canal as it passes. For each mine should to a certain extent open a connection with this canal, in order to evacuate its [ground] water in it'.[43] Until the final inauguration, which happened in 1799, surveyors would bear a considerable responsibility.[44]

The Deep-George tunnel offers valuable lessons about the scope and nature of practical mathematics in the late eighteenth century. Its planning relied mostly on the observations and computations of a handful of surveyors. The mathematical tools tirelessly used by Rausch and Länge, while involving a large volume of data, were – all things considered – rather elementary. So too were the methods used by mining master Steltzner to check their results, and so were the corrections ultimately implemented. There were no theoretical quandaries or higher analysis, but this does not mean that surveying was an easy task. The real difficulty was to use stable methods, ensure accuracy, and scale up these seemingly easy methods. Subterranean geometry 'seems to be a minor art', noted surveyor Rösler in his *Mirror of Mining*, 'and the one who manages to correctly measure 6 or 10 angles thinks, and might well convince himself, that he is already a master. But if he is asked to perform a measurement of a hundred or more angles, it will never work.'[45]

The suspended compass, the semicircle, and the surveying chain produced impressive results, thanks in good part to the stable technical and cultural system in which they were operating. The culture of systematic survey had developed over generations and was in its very nature cumulative. A vast trove of data, regularly updated and improved by subterranean surveyors, was consigned in registers and maps. Measurements had a codified place in planning operations and informed the decision process. Quantitative data was presented to the mining council, ensuring that technical choices would be based on collective deliberations. Although its primary design was to solve specific technical problems, the mining administration was unexpectedly involved in a major scientific problem of the time: the laws of the atmosphere and the use of the barometer.

---

[43] Deluc, *Lettres physiques et morales sur l'histoire de la terre et de l'homme* (1779), vol. 4, pp. 624–625: 'Bien plus, cinquante autres Mineurs partent des Mines, pour aller joindre ce canal à son passage. Car il faut que chaque Mine ouvre des communications plus ou moins longue avec ce Canal, pour pouvoir y jetter ses eaux.'

[44] Surveyors had, for instance, to carefully plan the operations, for the draining tunnel was not yet functional, and each connection bore the risk of flooding the construction site. See NLA HA, BaCl Hann. 84a, Nr. 9828, EP 10 Quartal Crucis 1780, on connecting the whole Rosenhöfer complex to the tunnel.

[45] Rösler, *Speculum metallurgiae politissimum* (1700), p. 87: 'Dieses scheint zwar eine geringe Kunst zu seyn … und da es manchem in einem Zuge von 6. in 10. Winckeln zugetroffen / vermeinet er / und mag sich wohl einbilden / er sey schon Meister. Wann er aber einen Zug von 100. und mehr Winckeln verrichten soll / so will es denn nirgends zutreffen.'

## Jean-André Deluc Enters the 'Subterraneous Labyrinths'

When Jean-André Deluc (1727–1817) arrived in Clausthal in October 1776, he was a seasoned geologist with first-hand knowledge of many European mountains. Born in Geneva, Deluc had built his own barometers and thermometers from the early 1750s onwards, helped by the skills and tools of his father, a clock-maker. He had tirelessly toured the Alps in Switzerland, but also in Italy, trying to calibrate and standardize his observations. His general goal was to 'compare known heights with the lowering of the barometer observed at the same heights, to ascertain a general rule, from which one could in the future measure all accessible heights and know everywhere the density and absolute weight of air'.[46]

Barometric studies had been an important research topic ever since the experiences of Evangelista Torricelli (1608–1647) and Blaise Pascal (1623–1662) in the mid-seventeenth century.[47] Natural scientists working with barometers, however, had faced numerous difficulties. This was especially true for those trying to ascertain heights from differences in atmospheric pressure, a sub-discipline known as barometric hypsometry.[48] At first, scholars had tried to come up with a simple arithmetic relationship between height and pressure, hoping to deduce the former from the latter. They were mostly neglecting other factors such as temperature or humidity, and obtained mathematical laws whose computations did not correspond with actual observations. In the following decades, these discrepancies gradually convinced scientists to abandon a 'purely theoretical approach' in favour of empiricism; specifically, this meant taking into account meteorological observations.[49] Systematic campaigns in mountainous areas developed in the last third of the eighteenth century, when it became clear that existing formulas were not valid for great heights. The existing data was deficient, as both the Royal Society of London and the *Académie des Sciences* in Paris promoted standardization and calibration of barometers and thermometers.

Jean-André Deluc summed up many years of experiments – mostly in the Swiss Alps – in his *Recherches sur les modifications de l'atmosphère* (1772). This epoch-making work was soon adapted in several languages, ensuring a more homogeneous way of measuring.[50] Deluc's main formula to compute

[46] Deluc, *Recherches sur les modifications de l'atmosphère* (1772), vol. 2, p. 43.
[47] For a general overview, see Middleton, *The History of the Barometer* (1964), pp. 1–82, 172–260.
[48] A general analysis of barometric hypsometry can be found in Feldman, 'Applied Mathematics and the Quantification of Experimental Physics' (1985), pp. 127–195. On the barometer as a tool to measure mountains, see Cajori, 'History of determinations of the heights of mountains' (1929), pp. 482–514; Archinard, 'De Luc et la recherche barométrique' (1975), pp. 235–247.
[49] The expression is from Daniel Bernoulli (1700–1782), quoted in Feldman, 'Applied Mathematics and the Quantification of Experimental Physics' (1985), p. 140.
[50] See Hübner, *Jean André Deluc (1727–1817)* (2010), pp. 84–87.

height from atmospheric pressure did not rely on higher mathematics, for he was no mathematician. The *Recherches* approached the problem by gathering years of daily observations, trying to come up with a simple correction to account for variations of temperature. Its author sought a method relying on elementary mathematics only, rightly convinced that simplicity was the key to its adoption by the large public of *amateurs*.[51] The positive reception of his book endorsed his international recognition among the Republic of Letters. This success led Deluc, whose financial position in Geneva was somewhat precarious, to travel to London in 1773, where he was presented to King George III. He was quickly appointed as a Fellow of the Royal Society and, after demonstrating his barometer to Queen Charlotte, became her reader.

The new position allowed him to resume his scientific travels, provided he corresponded with the Queen, who was not allowed to do this kind of trip by herself.[52] He kept a diary describing in detail all matters of natural history and religion, subsequently published in the six-volume *Letters, Physical and Moral, upon the History of the Earth and of the Human Race, addressed to the Queen of Great-Britain* (1779–1780). Having secured an enviable position, Deluc began a new series of observations to test and refine his barometric rules. True to his empiricist credo, his main purpose was to try out his instruments and formula in new contexts. He set out to visit the deepest mines of Europe, to ascertain 'whether in those pits the condensations of air follow the same laws that they do out of them'.[53]

We can now understand Deluc's state of mind when he arrived in Clausthal in 1776. Despite decades of travels and countless barometric observations on mountains, he didn't know much about mining, let alone the dull craft of drilling drainage galleries such as the *Deep-George* tunnel. The natural scientist carried a home-made barometer, carefully transported in a wooden box also containing two thermometers, a water level, and a tripod.[54] Most importantly, he arrived with a simple idea: mines, he thought, were dug vertically, offering a perfect experimental setting. By observing two barometers, one at the surface and the other in the mine, one could compare two points at different heights and yet under 'the same vertical column of air'. These conditions offered a major improvement over the ascension of mountains, in which the atmosphere might vary significantly between the base and the top and distort

---

[51] Feldman, 'Applied Mathematics and the Quantification of Experimental Physics' (1985), pp. 176–177.

[52] On this aspect of Queen Charlotte's biography, see Orr, 'Queen Charlotte as Patron: Some Intellectual and Social Contexts' (2001), pp. 183–212, especially pp. 188–190 and 199–200 on her relationship with Deluc and the *Letters*.

[53] Deluc, 'Barometrical Observations on the Depth of the Mines in the Hartz' (1777), p. 402.

[54] Deluc describes his apparatus in detail in *Recherches sur les modifications de l'atmosphère* (1772), vol. 2, pp. 2–42.

the observation.[55] In addition to his two barometers, Deluc also needed to know the exact depth of the considered mine: 'knowing that the ore is drawn up in pales from the pits of the mines, I thought at first that it would be easy to measure their depths with a line, and I had accordingly provided myself with all the necessary implements for that purpose'. The assumption of the scientists, however, was soon disproved: 'when I arrived at Clausthal ... I found that those pits, being dug in the direction of the veins of ore, are too inclined to make such a mode of mensuration practicable'.[56]

This problem seemed insoluble. The natural experiment was aimed at computing a theoretical height difference, using two barometric observations and his formula. The result necessarily had to be compared with a concrete measurement in order to ascertain the validity of Deluc's mathematical rule. If the geometric measure was faulty, it could not serve as a reference and the whole point of going into mines disappeared. Deluc candidly described the situation in a letter to the Royal Society:

> At first this gave me great concern, because I had my experiments much at heart; but I was soon made easy by Baron Reden, Captain-general of the mines. 'You do not want these measures,' said he, 'since it is of much more consequence to us, that it can possibly be to you, to know exactly the depth of all the points of these mines. Without such knowledge, how could we direct ourselves in boring from one to the other?'[57]

Up to that point, Deluc did not even know of subterranean geometry. He would quickly learn that mining officials had 'registers which are kept of the depths of all these mines'. Von Reden led him 'in three mines in the environs of Clausthal; called the Dorothea, the Caroline, and the Benedict' (see Figure 7.1 above), in which he performed observations while a trusted official observed a similar barometer at the surface. The Captain-general then asked a mining clerk to prepare an extract from the surveyors' tables. The scientist could thus directly compare the two data sets: the geometrical and barometric measures turned out to be strikingly close. The depth of the Dorothea pit was 168.69 *toises* according to the barometer, and 169.53 *toises* according to the geometric survey (ca. 330 m).

Deluc was surprised, for he had expected that the 'exhalations of all kinds' would somehow 'affect the common laws of the air's elasticity'. Once again, miners could help: the prejudice came from his unawareness of the elaborated ventilation systems used in the mines. All things considered, the mines were

---

[55] Deluc only articulated clearly this explanation after his second visit to the Harz in 1778. See Deluc, 'A Second Paper Concerning Some Barometrical Measure in the Mines of the Hartz' (1779), pp. 492–493. In mountains, 'even the steepest' (p. 492), there is a considerable horizontal distance between any two observatory stations. In this respect, measurements made in the mines were much closer to the hypotheses 'supposed in the formula' (p. 493). Once again, the mine is here considered as a natural laboratory.

[56] Deluc, 'Barometrical Observations on the Depth of the Mines in the Hartz' (1777), pp. 406–407.

[57] Ibid., p. 407.

indeed a 'laboratory of nature' in which conditions were much more controlled and stable than outside.[58] Ventilation ensured an almost perfect 'homogeneity of the air', without any change of temperature and atmospheric pressure due to weather variations. This explained why the theoretical predictions of Deluc had turned out to be extremely close to the surveyors' measurements.[59]

His formula now empirically validated, Deluc wrote to the Royal Society, eager to inform his fellow scientists of the Republic of Letters. He had painstakingly recorded his own observations in detail, in accordance with the requirements of the society. In his inquiries about the depth of mines, the otherwise meticulous scientist had nevertheless made a risky assumption by assuming that the geometrical measurement given by the subterranean surveyors 'may be looked upon as the real heights'.[60] The endorsement of Baron von Reden notwithstanding, the data had been gathered by subterranean surveyors. These practitioners were unknown in the scholarly world, at a time when much of the actors' credits depended on their personal standing, the standing of the maker who had produced their instrument, and the public character of their actions.[61]

In experiences previously undertook in the Alps, Deluc had himself performed chain and level measurements, once 'levelling for thirty hours with a perch of 36 feet'. In that case, he had taken care to detail all his actions to give the reader 'the means to discover his own errors'.[62] Deluc's blind faith in the measures of subterranean surveyors was thus surprising, to say the least. Even assuming that these practitioners were both trustful and skilled, could an observation be so precise as to be regarded as 'the real heights'? Eighteenth-century scientists were well aware of instrumental errors, as they themselves struggled with their elaborated barometers. A few decades earlier, a Swiss fellow of Deluc, writing on *The Measure of Mountains*, had even decided to 'leave the instrument at home', claiming that 'geometrical instruments were for a part troublesome and for the other deceptive', giving 'unsatisfactory' results in mountainous areas.[63]

---

[58]  Ibid., p. 408. The expression 'laboratory of nature' was interestingly coined at the same time by Horace-Bénédict de Saussure in his *Voyages dans les Alpes* (1779), although it was a generic designation for mountainous areas, not mines. See Bigg, Aubin and Felsch, 'Introduction: The Laboratory of Nature – Science in the Mountains' (2009), pp. 316–317.

[59]  Deluc, 'Barometrical Observations on the Depth of the Mines in the Hartz' (1777), pp. 412–413; Deluc, 'A Second Paper Concerning Some Barometrical Measure in the Mines of the Hartz' (1779), pp. 492–493.

[60]  Deluc, 'Barometrical Observations on the Depth of the Mines in the Hartz' (1777), p. 409.

[61]  Sibum, 'Les gestes de la mesure. Joule, les pratiques de la brasserie et la science' (1998), p. 761, explains that 'in the late eighteenth century, the credit of experimental practices was mainly based on interactions carried out *in praesentia*'.

[62]  Close to his hometown of Geneva, Deluc had repeatedly surveyed the Mont Salève in order to 'compare known heights with the lowering of the barometer observed at the same heights, to ascertain a general rules' and described at length his geometrical methods. Jean-André Deluc, *Recherches sur les modifications de l'atmosphère* (Genève, 1772), vol. 2, pp. 43–47.

[63]  The scholar was Johann Jakob Scheuchzer (1672–1733), a leading naturalist of the time. See Scheuchzer, *Natur-Geschichte des Schweizerlandes* (1746), pp. 17–47, 'Von Abmessung der

Deluc was fully convinced by 'the daily experience of the truth' of these measures, but nevertheless felt that his readership would need more convincing arguments. He thus undertook a precise description of measuring techniques he had observed in the mines of Clausthal, warning his audience that they were 'made in so very singular a manner' that he also had doubted it in the first place. He described how a surveyor would suspend their compass and semi-circle from the 'wire' (surveying chain), how distances and height were computed from these observations, adding: 'it is likewise by this means that he goes over hills and vales, in order to determine points corresponding to his pits and galleries'.[64] This central argument, interestingly, had nothing to do with the accepted methodology of natural sciences and amounted to brute empiricism: Deluc trusted subterranean geometers simply because he had witnessed their efficiency. 'Is this a method that may safely be depended upon?' he rhetorically asked, aware that his previous description was unconvincing for his fellow scientists, adding:

The facts answers, and saves us the trouble of a long reasoning. A miner, solely upon the faith of his Geometer, and in the absolute obscurity of the entrails of the earth, undertakes a labour that is to cost him years in daily boring through a rock. Another miner sets out to meet him from some other mine, or from without. At the end of a determined measure, the Gnomes begin to hear each other, and at length they meet. I have observed some of these points of encounter in the galleries. It is sometimes difficult to perceive the small winding which has been necessary for their meeting end to end.[65]

This description shows the deep impression that these practitioners had had on him. Although it was published in the *Philosophical Transactions*, this paragraph sounds curiously unscientific. Deluc and contemporary natural philosophers prided themselves for dissecting every process, looking for false evidence and sources of errors. This was especially true for the research on barometric hypsometry, in which numerous factors had to be taken into account. The slightest variation in temperature, pressure, humidity, or an ill-calibrated instrument could deceive the most astute observer. Eighteenth-century natural science had set itself the goal of finding the true laws of nature. This process was sometimes labelled as 'applied mathematics', but the mathematicians trying to describe natural processes using equations often struggled in assessing what observations should be taken as true references.

Deluc's belief in the accuracy of underground geometry was directly grounded in his experience on the site of the *Deep-George* tunnel. While mine

Berg-Höhen', esp. p. 20: 'dieses Instrument zu Hause gelassen ... daß die Geometrischen Instrument theils beschwerlich, theils betrüglich, und die damit vorgenommene Meß-Arten an diesem Ort unzulänglich seyen'.
[64] Deluc, 'Barometrical Observations on the Depth of the Mines in the Hartz' (1777), pp. 423–424.
[65] Ibid., p. 424.

surveying surely was no exact science, its efficiency was tested with each project and comforted with each successful connection. Such accuracy was astonishing to an eighteenth-century scholar: surveyors worried about errors of a few inches when barometric observations led to incomparably greater differences.[66] The precision of the 'geometrical measures' and their description were, for Deluc, 'very interesting proofs of their exactness'.[67] Even as the mining administration, as shown above, sometimes questioned the accuracy of surveyors, Deluc argued that their data could safely be taken as 'real depths' since it had proven substantially more reliable and precise than his own barometer.

The reaction of the scientific community to his *Barometrical Observations on the Depth of the Mines in the Hartz*, published in 1777 in the *Philosophical Transactions*, was equally remarkable.[68] Deluc was at the time a polarizing figure; many aspects of his work were indeed highly debated, and occasionally rejected.[69] In this case, however, there was a general agreement about his decision of taking practitioners' measures for the real height. The *Monthly Review* of London even centred its analysis of Deluc's paper precisely on 'the geometrical measures that had before been taken by the miners'.[70] Although it was the key element supporting the superiority of Deluc's formula, and an easy target for criticisms in an active field of research, the data was taken up in the specialized literature without discussion.

## A Deep Connection

In the dangerous environment of mines, Deluc had to be constantly guided and accompanied. He also needed a skilled assistant to monitor the barometer set at the surface, in order to compare with his own underground observations. Mining clerks, one named Leyser on the first trip and another named Mayer on the second, offered themselves to make the coordinated observations 'every quarter on an hour'.[71] Deluc's account corroborates the mining archive in showing that officials were both curious and qualified, knowing how to operate elaborate instruments such as his portable siphon barometer. The natural scientist also made repeated barometric observations at the surface, as he followed Captain-general von Reden, surveyor Rausch, and other officials on the path of the future Deep-George tunnel. He toured several proposed entrances

---

[66] Deluc, 'A Second Paper Concerning Some Barometrical Measure in the Mines of the Hartz' (1779), p. 495.
[67] Ibid., p. 495.
[68] Deluc, 'Barometrical Observations on the Depth of the Mines in the Hartz' (1777), pp. 401–449.
[69] See, for example, Sigrist, 'Scientific Standards in the 1780s: A Controversy over Hygrometers' (2011), pp. 148–183.
[70] The *Monthly Review, or, Literary Journal*, vol. 58, 1778, pp. 455–456.
[71] Deluc, 'A Second Paper Concerning Some Barometrical Measure in the Mines of the Hartz' (1779), p. 486.

and gathered barometric observations 'since their relative depth to the mines has been geometrically measured with great care'.[72] The measures of surveyors Länge and Rausch, described in the above sections, thus directly contributed to the improvement of eighteenth-century theories of the barometer.[73] The *Letters* frequently mention planning work related to the Deep-George tunnel. It was obviously a time-consuming issue for mining administration at the time, and yet its highest members readily participated in Deluc's own barometric endeavours.

Deluc's exploration in the Harz and his encounter with the Deep-George drainage tunnel is a singularly well-documented contact point between two different worlds. While their interests and methods diverged on several points, both were interested in issues of instrumentation, measuring depth, and ascertaining the relative position of points. For Jean-André Deluc, this concern was framed in a theory of the atmosphere: he strove for generality and had found in the Harz a perfect setting to test his universal law. The barometer was meant to refine a much-needed formula for further explorations. In a letter to the Queen, he claimed that – provided one could calibrate barometers accurately – '[he] could have used it to calculate the heights of many mountains, & other places in which [he] would make natural history observations'.[74] Mining officials operated in a more local context already well known to them. They sought to obtain a higher degree of precision required by their technical operations.

Although they worked in different conceptual frameworks and pursued separate goals, Deluc and the Clausthal officials were interested in each other's work. The Harz surveyors never considered replacing their methods by the new barometer. After all, even in the best situation, Deluc's measurements were routinely off by several fathoms, a difference that did not matter when ascertaining a mountain's height, but was way too large to safely dig a draining tunnel. They nevertheless welcomed the barometer as 'an easy way to check geometrical operations', to ensure that one has avoided major mistakes during 'a moment of inattention'.[75] Deluc recalled with pride that a suspiciously large discrepancy between his observations and the surveyor's tables had helped identify a miscalculation on their side: 'and they found that in writing the

---

[72] Deluc, *Lettres physiques et morales sur l'histoire de la terre et de l'homme* (1779), vol. 3, p. 403: 'leur abaissement, relativement aux Mines, a été mesuré géométriquement avec beaucoup de soin, & que c'étoit là que je me proposois d'observer le *Baromètre*'. See also p. 388 for his description of preliminary operations of a new 'draining gallery for these mines'.

[73] Ibid., p. 403. The future draining tunnel is mentioned on pp. 321, 387–388, 395–396, and 401–404, as well as in the fourth volume, pp. 560, 621–625.

[74] Ibid., vol. 4, p. 94. Another famous example is Alexander von Humboldt's use of the instrument during his scientific trip to South America in 1799–1804.

[75] Ibid., vol. 4, p. 404. Indeed, J.A. Scheidhauer (see Chapter 6) made barometric experiments in 1771, as indicated in his *Nachlass* (Scheidhauer XVII 300e, *Versuch, welches über den Stand des Barometers in verschiedenen Höhen am 17 Juni 1771*).

depths of the different parts of the mine, the sum of which ought to give the whole depth, one of these parts had been twice put down'.[76] Mining officials were thus open to innovation, even as they operated in a technical system in which the barometer could at best play a complementary role.[77] Studying the history of barometric hypsometry, this instrument momentarily turns usual assumptions about early modern science and technology upside down. Natural scientists had to rely on arbitrary rules of thumb while calibrating their instruments, correcting their observation step by step to match the data of practitioners. Subterranean surveyors, on the other hand, had already established a stable methodological system. They were in the position of making reliable, precise predictions about the future, not in the form of general laws, but in the design of an engineering project. In the late eighteenth century, they were able to plan precisely the direction of an underground tunnel, just as an astronomer could precisely compute the path of a comet.

Jean-André Deluc is representative of his milieu inasmuch as many scientists went to the mountains – and most of the ones who did used a Deluc barometer. What is less usual about him is that he did not, unlike some other scientists, feel superior to miners.[78] He stressed their mutual complementarity and took pains to describe what his interlocutors actually did. His insistence in presenting their geometrical measure as 'true values' ensured their acceptance in scientific societies. Instead of assuming a discontinuity between natural scientists and engineers, Deluc saw a clear kinship in their approaches. Furthermore, he presented their work as demonstrating the usefulness of sciences:

One sees here that geometry and physics are no vain sciences. It is by the rules of the former, applied as simply as possible, & by one of the magnet's properties, that the subterranean surveyor directs the miner, when, starting from the bottom of his shafts or galleries, & making his way in the bowels of the earth by dint of powder, he seeks to connect with another gallery or in some Shafts … & he does all this so accurately that he arrives at the exact point he was looking for, though it may have been several thousand fathoms away when he started his work.[79]

---

[76] Deluc, 'A Second Paper Concerning Some Barometrical Measure in the Mines of the Hartz' (1779), p. 491.

[77] Barometric and geometrical measurements could also be complementary in larger-scale operations, for instance comparing the *Deep-George* tunnel in Clausthal with the height of Hanover, Göttingen or even the North Sea. This was drawn up in a remarkable table in Gotthard, *Authentische Beschreibung* (1801), pp. 278–280.

[78] Deluc's attitude is best compared with Eberhard Zimmermann's stance that surveyors could have 'saved much efforts' by using a barometer, but nevertheless dismissed it simply because 'the miner has that in common with the farmer, that he despises everything his father did not know'. Zimmermann, *Beobachtungen auf einer Harzreise* (1775), p. 45.

[79] Deluc, *Lettres physiques et morales sur l'histoire de la terre et de l'homme* (1779), pp. 307–308: 'On voit là que la géométrie & la physique ne sont pas des sciences vaines. C'est par les règles de la première, appliquées le plus simplement possible, & par une des propriétés de l'Aimant, que le Géomètre souterrein dirige le Mineur, quand, partant du fond de ses puits ou de ses

How can we make sense of this encounter? Deluc's adventures show once again how inadequate linear narratives about the dynamic of sciences and crafts can sometimes be. It would certainly seem that the natural scientist was in that case not informing practitioners, or even formalizing their intuitions, but lagged way behind. This inversion, however, would be hardly more fruitful, for the very categories are biased. Admittedly, Jean-André Deluc uses 'we' for the community of scholars and 'them' for the miners in his correspondence with the Royal Society, but the mining officials were far from being ignorant of the current scientific issues. Captain-general von Reden, for instance, was a baron who had studied in the most prestigious universities. In the neighbouring city of Saint-Andreasberg, Deluc was received by another official whose mineral collection was reportedly used by 'all the naturalists who visit these mountains'.[80] Vice Captain-general Von Veltheim freely spoke of 'mathematics, physics, chemistry, and natural history' as the 'fundamental sciences of a miner' that would help 'uncover the laws of nature'.[81]

A stark dichotomy between scholars and miners is inadequate to understand the exchanges happening at the end of the eighteenth century. When members of the Republic of Letters explored the mountains, the persons they met were not illiterate miners. They collaborated with skilled engineers, higher officials, and more generally with 'hybrid experts', to use a term coined by Ursula Klein.[82] These shared the curiosity of scientists but acted in a specific technical and economic environment, and thus had a different set of issues and goals in mind. Over the course of their long discussions about mines and natural history, Deluc and von Reden indeed became friends. The Swiss natural scientist inquired in letters about the progress of the Deep-George tunnel, reporting at one point: 'I hear from Mr. Baron von Reden that two sections of the tunnel have already met perfectly.'[83] Von Reden later visited Deluc in England, and

---

galeries, & se faisant chemin dans le sein de la Montagne à force de poudre, il cherche à percer dans quelqu'autre galerie ou dans quelque Puits ... & il execute tout cela avec tant de justesse, qu'il arrive au point exact qu'il cherchoit, quoiqu'il en fut peut-être à plusieurs milliers de toises quand il a commencé à s'y diriger.

[80] Ibid., vol. 4, p. 469: 'les Naturalistes qui visitent ces montagnes'. Deluc even implied that this collection had a scientific purpose, adding 'rassemblés dans leur plus grande perfection, tous ces accidens rare des *Filons de plomb & argent*, qui indiquent des Causes sécondaires, & qui peuvent aider à les découvrir'.

[81] Bartels, *Vom frühneuzeitlichen Montangewerbe zur Bergbauindustrie* (1992), pp. 384–386.

[82] Klein, *Nützliches Wissen* (2016), pp. 87–104. See also p. 9 for a critique of an account of the Industrial Revolution presenting entrepreneurs and engineers as 'tinkerers and amateurs' (*Tüftler und Bastler*).

[83] Deluc, *Lettres physiques et morales sur l'histoire de la terre et de l'homme* (1779), vol. 4, p. 624: 'J'apprends par Mr. le Baron *de Reden* que deux de ces portions de la Galerie se sont déjà parfaitement recontrées.' On p. 550, Deluc writes of the 'sweet influence of friendship' he had for von Reden as a 'not the least advantage I gained from this trip'.

was led by him to visit Watt and Boulton's factory, eager to discuss the potential of steam machines for the Harz mines.[84]

In the late eighteenth century, subterranean geometry had become strikingly efficient and, above all, had its own dynamic. Mining techniques were not yet fully ruled by scientific precepts or laws of nature, but formed a distinct and at least equally efficient set of practices. We can finally come back to Lewis Mumford's seminal question raised in the Introduction: 'Did the mine acclimate us to the views of science? Did science in turn prepare us to accept the products and the environment of the mine?'[85] Taken literally, we now understand what is missing from this formulation: it takes the mining administration and its advanced technical culture out of the equation. It sounds as if scientists were entering a new *terra incognita* or had observed stubborn workers endlessly reproducing a fixed set of practices. And yet Mumford was unarguably right in another sense: mines were popular as a scientific model, a writing topic, and above all as a travel destination.

### Travelling to the Harz

The digging of a new draining tunnel – even the longest in the world – lacks the spectacular aspect of many natural phenomena, if only because it takes place underground. The construction site of the *Tiefer-Georg* tunnel nevertheless became a fashionable destination and attracted an important number of visitors. During his first trip to the Harz, Deluc had already noted that, its isolated situation notwithstanding, 'curiosity often brings foreigners to Clausthal'.[86] The underground had always inspired myths and legends. Among scholars, and later natural scientists, old quarrels about the generation of metals were replaced by new debates about the formation and structure of the earth. In a broader context, the eighteenth century saw a development of a culture of travelling and inquiring about the natural world. The publishing success of travel literature in turn gave the bourgeoisie a taste for voyages, at a time when roads and circulation were improving.[87]

In this specific case, the appeal of the new tunnel was enhanced by an active self-fashioning of the local administration. As mines were now administered by the German states, they had become a sort of 'public space'. The inaugural speech by Captain-general von Reden was published as construction began in 1777. When the completion happened in 1799, as forecast two decades before,

---

[84] Hübner, *Jean André Deluc 1727–1817* (2010), pp. 124–125.
[85] Mumford, *Technics and Civilization* (2010), p. 70.
[86] Deluc, *Lettres physiques et morales sur l'histoire de la terre et de l'homme* (1779), vol. 3, p. 187.
[87] On the evolving attitude towards travel in the seventeenth and eighteenth century, see Bourguet and Licoppe, 'Voyages, mesures et instruments : une nouvelle expérience du monde au Siècle des lumières' (1997), pp. 1124–1129, talking about 'l'attrait des sommets et des pays perdus'. See also Bourguet, Licoppe and Sibum, *Instruments, Travel and Science* (2002); Von Zimmermann, *Wissenschaftliches Reisen – reisende Wissenschaftler* (2003).

the tunnel fulfilled his promises exactly as expected: the once endangered mining district was fully relieved from ground water. Moreover, considerable sums were spared as many pumping machines became unnecessary. The success of the operation was actively advertised by the administration. Thousands attended the official ceremony held at the entrance of the tunnel, as the royal anthem was sung: 'God save great George our King'.[88] More significantly, all the technical notes gathered by mining master Steltzner were lent to a writer, Johann Christian Gotthard (1756–1813), who readily acknowledged that he was 'no miner himself'. He arranged a book for a larger audience that appeared in 1801, the *Authentic Description of the Remarkable Construction of the Deep-George Tunnel*.[89] The original notes were a tedious text and would not have been understandable outside the circle of professionals. Rewriting a technical text for a larger audience was an ingenious way to ensure publicity about this project; this, however, led to the exclusion of most events related to subterranean geometry.[90]

In the last quarter of the eighteenth century, a large body of literature dealt not only with the natural history of mountains, but more specifically about tunnel digging and the Harz mines. A few years before Deluc published his *Letters to the Queen of Great-Britain*, Leonard Euler had written his *Letters to a German Princess* (1768), mentioning the compass and subterranean geometry.[91] More specifically, German journals kept the general public updated by publishing news about the progress of the digging work of the Deep-George tunnel.[92] University professors had found a fantastic object for their studies: Eberhard von Zimmermann, who taught mathematics and natural history in Brunswick, published his *Observations on a Harz Journey* in 1775. He was a harsh critic of miners' methods, even though he too relied on surveyor's Rausch measures to test his own barometer.[93] Christoph Gatterer, lecturer at the nearby University of Göttingen, organized a 'preparatory *collegium* for the benefit of students' in which he described how to make the most of a trip to the Harz. His teaching was eventually published as a five-volume travel guide, the *Instructions to Profitably Visit the Harz Mountains and Other Mines* (1785–1792).[94] Students

---

[88] Gotthard, *Authentische Beschreibung* (1801), p. 250.

[89] Ibid., introduction: 'weil ich selbst nicht Bergmann bin'.

[90] Kolb, 'Gotthard's Authentische Beschreibung von dem merkwürdigen Bau des Tiefen Georg' (1999), pp. 9–12.

[91] As a reader to the Queen, Deluc himself recommended Charlotte to read Euler's *Lettres à une princesse d'Allemagne*, together with the *Lettres* of the Abbé Nollet and philosophical works by Adam Smith or Willem Jacob 's Gravesande.

[92] See August Ludwig Schlözer's published correspondence (*Briefwechsel*), vol. 2, Heft XI, p. 323, Vol. 5, 1779, Heft XXVI, pp. 69–78 and Heft XXX, p. 386; Alexander Wilhelm Köhler's *Bergmännisches Journal*, vol. 4, Stück 3, pp. 216–236, then 236–238.

[93] Zimmermann, *Beobachtungen auf einer Harzreise* (1775), pp. 33–45.

[94] Gatterer, *Anleitung den Harz und andere Bergwerke mit Nuzen zu bereisen* (1785), vol. 1, p. vi: 'ein solches Vorbereitungs-Collegium über den Harz zu lesen, und ein kleines Handbuch darüber zu schreiben'.

from the newly founded academies of Freiberg and Schemnitz were of course frequent visitors of the local mines and were especially keen to visit the Deep-George tunnel.[95] Deluc himself described the mines as a place of interest, not merely for his own barometric and geological studies, but also as 'an object that belongs as much to the *oeconomy* as to natural history'.[96]

The testimony of Georg Heinrich Hollenberg, a civil engineer from Osnabrück visiting the Harz, illustrates how commonplace these visits had become. 'In order to see the local mines', he wrote, 'one notifies the mining master, who then has a miner call you off early in the morning.' Mining master Steltzner then 'usually brings foreigners to the *Burgstädter* mine and the *Dorothea*' (see Figure 7.1).[97] Distinguished guests such as Deluc had the privilege to see one of the 'various collections of minerals', the most prestigious being von Reden's, but Hollenberg was politely refused. He did not take offence, knowing that 'foreigners' and 'dilettantes ... come here too often that one could spend more time on you than is usually intended'.[98] Despite his fondness for mechanics, Hollenberg was not even allowed to visit the 'model collection' of Steltzner and was this time bitterly disappointed. Fortunately, he found a 'subaltern' ready to pursue the tour, 'who had the unexpected kindness to show me the new *George* tunnel ... this astonishing work is one of the most important and remarkable to be seen here'.[99] Wilhelm Ferdinand Müller, an even more modest visitor than Hollenberg, was curious of the tunnel as well, but could not find a guide. He had to walk by himself to the mouth of the tunnel and 'went in for a good hour, accompanied by two miners'.[100]

The Harz mines had become a public display of mining know-how and technology, so popular that the major pits introduced *Fremdenbücher* (guests' books), to be signed at the entrance by 'foreigners' to ensure that nobody went missing.[101] The guest book of the *Deep-George* tunnel counts hundreds of

---

[95] See the testimony of Freiesleben in Freiesleben, *Bemerkungen über den Harz* (1795), pp. 295–313.
[96] Deluc, *Lettres physiques et morales sur l'histoire de la terre et de l'homme* (1779), vol. 3, p. 396: 'objet qui appartient autant à l'Oeconomie qu'à l'Histoire naturelle'.
[97] Hollenberg, *Bemerkungen über verschiedene Gegenstände auf einer Reise durch einige Deutsche Provinzen* (1782), p. 61: 'um die hiesigen Bergwerke zu sehen, meldet man sich bey dem Bergmeister, welcher einen sodann durch einen Bergmann des Morgens frühzeitig abrufen läßt ... ein Fremder gewöhnlich auf den Burgstädter Zug nach der Dorothee'.
[98] Ibid., pp. 62–63: 'verschiedene Sammlungen von Mineralien ... Die Fremdem kommen hier zu häufig, daß man auf einen mehr Zeit verwenden könnte, als gewöhlich dafür bestimmet ist.'
[99] Ibid., p. 66: 'die unerwartete Gefälligkeit, mir den neuen Georgstollen zu zeigen ... Dieses erstaunende Werk ist eines der wichtigsten und merkwürdigsten, die hier zu sehen sind.'
[100] Müller, *Meine Streifereyen in den Harz und in einige seiner umliegenden Gegenden* (1801), p. 100: 'ich ging eine gute Stunde weit hinein, von zwey Bergleuten begleitet', and more generally pp. 94–101 on the tunnel.
[101] See, for example, NLA HA BaCl Hann. 84a Nr. 10203, *Fremdenbuch des Tiefen Georg-Stollen 1779–1856*. Similar registers existed in Saxony, and likely in most mining regions. See SächsStA–F, 40012 Bergamt Johanngeorgenstadt, Nr. 664, *Besichtigung, Befahrung und Erkundung des hiesigen Bergbaus durch Ausländer 1730–1788*.

names, whose diversity illustrates the fame of the project. The son of King George III, Prince Frederick, visited the site and enjoyed a private 'subterranean fireworks', as did Karl-August, Grand-Duke of Saxe.[102] Entrepreneur James Watt and technologist Johann Beckmann visited the Harz mines at the turn of the nineteenth century, along with French, Norwegian, and Dutch groups of up to twenty people. Not only naturalists, but also merchants and their wives, having read about it in the newspapers, asked to visit. Mountains and mines logically found an echo in the *Sturm und Drang*, a powerful artistic movement that developed in the late eighteenth century German-speaking world. J.W. von Goethe published his *Winter Journey in the Harz* based on two visits he had made there in the late 1770s.[103]

Besides a few technical descriptions, however, most of the existing accounts do not mention subterranean geometry. Even official publications briefly hailed the accuracy of measurements before turning to more impressive descriptions. Surveys were abstract objects which, unlike water-wheels, mining machines and fireworks hardly made for thrilling readings. The very efficiency of surveyors paradoxically made their work invisible: mundane eighteenth-century travellers were looking for 'sublime sensations' and had little interest for the cold rationality of geometry.[104] In hailing the efficiency of surveyors, Jean-Andre Deluc himself had written that it was 'difficult to perceive the small winding which has been necessary for [the] meeting' of various tunnel sections.[105] There was a general, if tacit, agreement 'about the value of subterranean geometry, which appeared in such bright light during the execution of this undertaking', according to a contemporary.[106] The interest quickly faded as its efficiency was then taken for granted. Planning a drainage tunnel used to be an important issue and a major technical gamble. In earlier times, there are numerous examples of printed sermons, to be read in a whole district every Sunday 'until further notice', that is, until completion of the tunnel.[107] Thanks to the accurate work of Länge, Rausch, and their peers, forecasting and planning tools had now reached a point when suspense or anxiety had disappeared.

---

[102] Gotthard, *Authentische Beschreibung* (1801), pp. 231–232.
[103] Scientific artistic concerns could be mixed, as in Dannenberg's scientific poetry *Der Harz: ein Gedicht in sieben Gesängen* (1782), in which each chant was illustrated by a scientific plate. The *Deep-George* tunnel and its mastermind Steltzner are described on pp. 81–82.
[104] Bigg, Aubin and Felsch, 'Introduction: The Laboratory of Nature – Science in the Mountains' (2009), p. 311.
[105] Deluc, 'Barometrical Observations on the Depth of the Mines in the Hartz' (1777), p. 424.
[106] Freiesleben, *Bemerkungen über den Harz* (1795), p. 296: 'über den Werth der Markscheidekunst, der bey der Ausführung dieses Unternehmens in so hellem Lichte erschien'.
[107] Landesarchiv Sachsen-Anhalt, Aa Nr. 426, contains examples from 1671, 1685, and 1716. See the printed document of 1716 explicitly praying for a 'new and costly deep tunnel started many years ago, on which relies the welfare of the whole region.

Their work then became transparent. Even Deluc, who used to be so enthusiastic about the prowess of surveyors, came to see this as routine. During his third trip to Clausthal – in 1786 – subterranean geometry was not on his agenda anymore. He only mentions the Deep-George tunnel in passing, stating: 'I found the work not yet completed, though it was carrying on at various points, by means of shafts, which had been sunk on the line of their intended course.'[108] What Deluc had once seen as the main reason for a scientific excursion had, less than a decade later, receded into the background.

\*

>You stand with heart unplumbed
>Mysteriously revealed
>Above the marveling world
>J.W. von Goethe, *Winter Journey in the Harz*, 1777[109]

When J.W. von Goethe wrote his *Winter Journey in the Harz* – one of his most famous poems – the region seemed wild and mysterious to visitors ('but who is it stands aloof? His path is lost in the brake,' Goethe further asked). In spite of appearances, however, centuries of extracting operations had deeply transformed the region into a 'landscape with numbers'. The hills, valleys, and veins had been quantified by generations of surveyors, as nature was methodically subjected to mine extraction.[110] The careful design of the *Deep-George* tunnel illustrates the transformation of mining sciences, and most prominently subterranean geometry.[111] One usually thinks of practical mathematics and modern engineering as transformative in the aftermath of the Industrial Revolution. In the Harz, however, Deluc encountered plenty of examples of efforts largely based on mathematics, from the careful calculations used to design water-column engines to the design of draining tunnels.

These projects were made possible, despite their large scale and their huge price tag, by two factors: a coherent administrative policy and the accuracy of mathematics. A long tradition of engineering and an extensive knowledge of local districts were combined with improvements in theoretical training. The system of mining maps was being constantly standardized and refined, in order to reach new levels of precision. The development of subterranean geometry, which was accelerating due to the foundation of *Bergakademien*, formed an integral part of the technical system of the mines. Barometers were included, without revolution or major upheaval, to a growing stock of methods

---

[108] Deluc, *Geological Travels in Some Parts of France, Switzerland, and Germany* (1813), p. 354.
[109] 'Harzreise im Winter', translated by Christopher Middleton, Suhrkamp/Insel Publishers, Boston, (1983) pp. 66–71.
[110] Bourguet, 'Landscape with Numbers' (2002).
[111] See Bartels, 'Der Harzer Oberbergmeister Georg Andreas Steltzner 1725–1802' (2015), p. 287.

and resources, whose reliability paved the way for even greater investments and technical innovation.[112] The continuation of a mathematical culture that had developed in the sixteenth century would culminate half a century later, when the mine theodolite once again pushed the boundaries of accuracy.

When Deluc arrived in the Harz, the seasoned mountaineer was surprised by a context that was new to him. The mines turned out to be a fertile field for experiments, a laboratory of nature where he was assisted, and often even guided, by skilled engineers. Conversing with Captain-general von Reden, mining master Steltzner and surveyor Rausch, Deluc soon realized how complementary their purposes were. Around a cluster of common issues related to depth, standardization, and instrumentation, fruitful and respectful exchanges developed. The natural scientist was surprisingly frank about his own prejudices and the limits of his barometer, while showing a genuine curiosity about the 'subterraneous Geometers'.[113] According to the scientific standard of his readership, he should have at least tried to replicate their measurements, or witness their operations and inquire about their instruments.[114] However, the experience of their repeated successes, and the long-standing reputation of mining offices were enough to persuade him and those who read his reports in the *Philosophical Transactions*. The popular appeal of mines, of which Deluc's publications are but one of many examples, drew countless *amateurs* on the construction site of the future tunnel. This site offered a perfect venue to observe the usefulness of underground surveying, and yet its very efficiency ultimately rendered it inconspicuous.

---

[112] Bartels, *Vom frühneuzeitlichen Montangewerbe zur Bergbauindustrie* (1992), pp. 33–35, 385–387.
[113] Deluc, 'Barometrical Observations on the Depth of the Mines in the Hartz' (1777), p. 422.
[114] Sibum, 'Les gestes de la mesure' (1998), pp. 751–752, 761, 771; Fleetwood, 'No Former Travellers Having Attained Such a Height on the Earth's Surface' (2018), p. 17.

# Conclusion

Confidence in the usefulness and accuracy of mathematics is not a timeless truth. In the mining states of the Holy Roman Empire, it emerged at the turn of the sixteenth century with the realization that many seemingly unrelated issues could successfully be tackled by supplementing the mindful hand and the experience of the senses with a compass, a surveying chain, and a water level.[1] In medieval times, German-speaking miners once relied on customs and 'anthropometric measures' to set the boundaries of their concessions – rudimentary procedures that slowly matured into a constructive geometry.[2]

The exceptionally high status of mining mathematics was therefore not a given, static fact, but the outcome of a multifarious social process. In the long term, the new values of accuracy and equity came to be associated with measuring tools and surveying procedures. Geometrical methods gradually became the most convincing tool to ensure fairness in setting concession limits and solving increasingly complex technical problems. During the silver rush of the late fifteenth century, the rules governing the delimitation of concessions gradually turned into a full-fledged discipline, performed in public and practised by sworn-in officials – subterranean geometry, also known in the dialect of miners as 'the art of setting limits' or *Markscheidekunst*. From then on, legal texts and mining sermons emphasized, the right way to behave was to respect mathematically guided decisions, or to challenge them using other measurements. The geological structure of mining districts, in conjunction with the political and religious settings of the German states, had given birth to an original culture of practical mathematics.

This book has traced how the *geometria subterranea* acquired a remarkable reputation and spread from the booming mining cities to the courts of early modern rulers, where it encountered the world of the learned. Surveying methods and their associated value system circulated among the mining states

---

[1] On the concept of 'mindful hand', see Roberts, Schaffer, and Dear, *The Mindful Hand* (2007); Halleux, *Le savoir de la main* (2009).

[2] 'Anthropometric measures' refers to the use of a man's own body parts, be it foot, finger or outstretched arms to apprehend the world, as defined in Kula, *Measures and Men* (1986), pp. 24–28, ch. 5, 'Man as the Measure of All Things'.

of Saxony, Bohemia and beyond, while the *Markscheider* became courtiers and served as versatile experts for the military or for the hydraulic endeavours of Elector August of Saxony, Duke Julius of Brunswick-Lüneburg, and their peers. In the princely courts, this culture of mathematics born in the mining pits met the 'valuation of precision' cherished by early modern Prince-practitioners, to quote Bruce T. Moran.[3] Collaborations between surveyors and scholars or mechanical artisans were leveraged by local rulers into a geometry of power that expanded to all part of their realms, as described in Chapter 3.[4] These numerous contacts, that occasionally led to close collaborations, were highly significant but ultimately short-lived. Courtly interests soon diverged as the Holy Roman Empire spiralled into the Thirty Years War.

In the aftermath of the war, the artisanal mathematics of mine surveyors proved remarkably resilient, adapting to the new technical conditions of deep-level mining and gunpowder blasting. The mathematization of the underground even accelerated as a new generation of surveyors, spearheaded by Balthasar Rösler and Adam Schneider, proved able to combine traditional practices with recent theoretical developments. With the incremental introduction of data table and mining maps, the seventeenth-century witnessed a new level of mathematization and increased control over mountainous landscapes. The purpose of the discipline itself evolved as early successes opened up new possibilities, emboldening both surveyors and mining administrations. This highly situated and eminently vernacular knowledge was then written down in a corpus of manuscripts which, in turn, came to form the backbone of a codified teaching system. An increasing precision allowed for the emergence of standardized mining maps, suddenly making the underground visible, and hazardous enterprises manageable.

In the second part of the book, I analysed how the nascent nation-states of Saxony, Prussia, and Hanover seized on the new mining geometry by acquiring collections of maps and instruments, amalgamating them in their reorganized mining offices. By regulating the dissemination of quantitative information, administrations learned to tame their landscapes and to monitor their territories more closely than ever.[5] They merged once scattered concessions into vast mining districts that were ruled and policed like subterranean cities. In the Holy Roman Empire, arithmetic and geometry were then extended to forest management, agriculture, and commerce, trying

---

[3] Moran, 'Princes, Machines and the Valuation of Precision in the 16th Century' (1977), pp. 209–228.

[4] Korey, *The Geometry of Power* (2007); Dolz, *Genau messen, Herrschaft verorten* (2010).

[5] On the emergence of a modern concept of landscape in the sixteenth century, see Blackbourn, *The Conquest of Nature* (2006); Rosenberg, 'The Measure of Man and Landscape in the Renaissance and Scientific Revolution' (2009). On the quantification of the environment, see Bourguet, 'Landscape with Numbers' (2002).

to replicate the successful policies developed in the mines.[6] This 'land culture technology', to use a concept coined by Ernst Pitz, led to a 'rationalism that, beyond its economic success, was of general importance for the development of Western sciences'.[7] The dual use of practical mathematics – at once technical and political – culminated with the foundation of the first mining schools and academies in the eighteenth century, as described in Chapter 6. There, an industrious rationality developed outside of the traditional university system as a synthesis between the secular experience of mining administrations and the reform impetus of the cameralist movement. Spurred by a handful of mining officials, attempts were made to develop advanced mathematical theories on the basis of concrete problems. In the nineteenth century, it would then pave the way for the German polytechnic institutes, which in turn influenced a good part of modern engineering and technical education.[8]

The scientific influence of the seemingly minor 'art of setting limits' extended to large portions of natural history, blurring the modern boundaries erected between academic disciplines. None other than Abraham Gottlob Werner (1749–1817), arguably the leading geologist of his time, engaged with the topic in his *Treatise on the External Characters of Fossils* (1774, English translation in 1805).[9] In this work, which offered an influential theory for classifying minerals, he claimed that 'it is to the Saxon philosophers and miners that we are indebted' for the development of earth sciences. 'The different theories' proposed by natural scientists were not just based on thin air or abstract reasoning, but on the life's work of generations of surveyors and craftsmen who, Werner continued, 'provided all the information we possess on this subject'. The geologist then took the time to name them, mentioning successively surveyors Rösler and Voigtel, as well as Professor Charpentier and mining master Scheidhauer, all of whom we encountered in the course of this book.[10] At that time, artisanal and unpublished quantifying methods could still exert an influence in specific areas of natural sciences, as shown both by Werner's mineralogical theories and Deluc's barometric studies. Underground surveying was a discipline tied to mining regions and to their specific problems, and yet it was episodically relevant in other contexts, producing rich and revealing encounters.

---

[6] On the related concept of 'knowledge management', see Popplow, 'Knowledge Management to Exploit Agrarian Resources as Part of Late-eighteenth-century Cultures of Innovation' (2012), pp. 413–433.

[7] Pitz, *Landeskulturtechnik, Markscheide- und Vermessungswesen* (1967), p. 8: 'dieser Rationalismus ist doch über den wirtschaftlichen Erfolg hinaus von Bedeutung für die Entwicklung der abendländischen Wissenschaften gewesen'.

[8] See Morel, *Mathématiques et politiques scientifiques en Saxe 1765–1851* (2013); Klein, *Technoscience in History* (2020).

[9] On Abraham Gottlob Werner, see Albrecht and Ladwig, *Abraham Gottlob Werner and the Foundation of the Geological Sciences* (2002).

[10] Werner, *New Theory of the Formation of Veins* (1809), p. 47.

*

When analysing the growing use of mathematics in the early modern period and how it shaped the way people considered both nature's laws and the world's order, the elephant in the room is the Scientific Revolution. And yet, this book has carefully avoided dealing directly with the questions of if and how 'the active life of early modern Europe' had been responsible for the rise of the new sciences.[11] Ever since Edgard Zilsel engaged with the topic during his WWII exile, the relationships between scholars and craftsmen have formed an important debate, whose ultimate goal was to assess their respective contributions to the development of modern sciences. Remarkable studies have underlined how mines were among a handful of privileged venues in which the interactions between the learned and the skilled could be observed.[12] Indeed, subterranean geometry developed (just as accounting, navigation, fortifications and many disciplines did) at a time when scholars were increasingly relying on mathematical laws, equations, and formulas to interpret the natural world. In the long run, it seems obvious that sciences and crafts have acted as mutually reinforcing forces. Still, direct connections and straightforward explanations, let alone coherent overarching narratives, so far elude the most astute analyses.[13]

Taking a step aside, the present series of case studies has traced the rise of mining mathematics in civic society and human affairs, of which science was but one element. In other words, I am convinced that it might be a more fruitful approach to focus on the artisans and engineers themselves, without teleological afterthoughts. Using a deliberately wide range of sources, where one would not normally expect to find mathematics, we have witnessed how geometric and arithmetical thinking gradually made its way into the religion, technology, and culture of mining states. In the sixteenth century, high numeracy and literacy rates ensured a public consensus about concession settings and an optimal circulation of methods. Surveyors skilfully adapted the existing compasses, blending their instrumental know-how with an extensive knowledge of the mines' physical and legal environment. Far from being restricted to counting and measuring, their expertise encompassed mineralogy, local topography, as well as the existing rules and customs.

---

[11] Cohen, *The Scientific Revolution: A Historiographical Inquiry* (1994), pp. 321–367, here p. 321.
[12] In recent years, there have been important contributions by Pamela H. Smith, Eric Ash, Warren A. Dym, and Hjalmar Fors, to name but a few. About Zilsel's thesis, see Zilsel, *The Social Origins of Modern Science* (2003); Cormack, Walton, and Schuster, *Mathematical Practitioners and the Transformation of Natural Knowledge in Early Modern Europe* (2017).
[13] This point has been summed up exhaustively in Cohen, *The Scientific Revolution*, pp. 491–505. Attempts to link the rise of new sciences with one specific group of practices, for instance 'the reckoning up of value on behalf of the mercantile interests', has generally found no wide agreement from historians, see Hadden, *On the Shoulders of Merchants* (1994), p. xiv.

In mining districts, a concrete mathematization of nature was made not only theoretically possible: it was factually accepted and efficiently used in a daily setting. This development was largely unmediated by the academic culture of its time which, as the first chapter has shown, should be interpreted with caution. Nor was my argument in this book to subvert the usual narrative by claiming that underground geometry had a revolutionary influence. In fact, while the ability of subterranean surveyors to find their way underground both astonished and perplexed scholars, it is fair to say that this did not elicit new concepts or lead to groundbreaking theoretical results. Their realizations simply served as tangible examples of what relatively elementary mathematical tools could achieve.

It would thus be far-stretched to claim that the *Markscheidekunst* – a word that is today unknown even to most early modern historians – contributed to the 'mathematical way in the scientific revolution', closely studied by Peter Dear and other colleagues.[14] Indeed, the subterranean geometers were as focused as contemporary craftsmen on concrete, down-to-earth issues, plainly satisfied 'to leave philosophizing to others and to concentrate instead on the practical improvements they could make with this mathematical knowledge and these instruments'.[15] Conversely, the theoretical breakthroughs of the early modern period, from astronomy and algebra to mechanics, rarely found their way into the daily practices of common men. What could be deduced from theory was then barely commensurable with what could be achieved in practice. When Georgius Agricola described to his fellow humanists the surveying of shafts and galleries, his presentation of actual practices was flawed, just as his deductions based on the *practica geometriæ* were too simplistic to be used in the mines. As Sophie Roux has argued, 'the gap between the alleged program of mathematizing nature and its effective realization' was much too broad to be bridged any time soon.[16]

This history of subterranean geometry told in this book is nevertheless a success story of the slow and gradual mathematization of the material world. It documents the growing influence of new values – accuracy and quantifica-tion – on society at large, of which science was then but one part. Mundane and pragmatic uses of geometric rules and instruments were highly significant at the time, albeit in a different context from that of the learned. Taking into account early modern practitioners helps in supplementing studies of schol-arly discourses by focusing on the very 'local idiosyncrasies' that usually go unstudied.[17] If the present book shows anything, it is that the quantification of nature and search for accurate procedures were not confined to learned circles.

---

[14] Dear, *Discipline and Experience* (1995); Goldenbaum, 'The Geometrical Method as a New Standard of Truth, Based on the Mathematization of Nature' (2017), pp. 274–307.
[15] Cormack, 'Mathematics for Sale' (2017), p. 85.
[16] Roux, 'Forms of Mathematization, 14th–17th Centuries' (2010), p. 321.
[17] Dear, *Discipline and Experience* (1995), p. 5.

Discourses on the usefulness and efficiency of mathematics were ubiquitous, from religious sermons to collections of jurisprudence. If the early modern period indeed saw the establishment of universally valid laws by small communities of high-minded individuals, these breakthroughs happened not in a static world but within rapidly changing societies. Acknowledging the existence of artisanal numerate traditions and studying their own dynamics is crucial, as shown by the case study about Jean-André Deluc presented in the last chapter. In order to refine his general formula for ascertaining heights with a barometer, Deluc toured Europe and calibrated his general law using, among other resources, the highly situated knowledge of subterranean surveyors.

Shifting our focus from the rise of the new sciences to a broader history of popular mathematical knowledge, as proposed here, offers fresh research perspectives. Up to the end of the eighteenth century, analysing what professionals, artisans, and engineers were actually doing is a most fruitful lens through which to analyse the growing influence of the mathematical arts. Besides courts, universities, and academies, there were several places in which people tackled new problems using arithmetic and geometry. This was particularly true in the Holy Roman Empire, with its myriad of local powers and often arcane regulations, all of which encouraged the development of a 'trade mathematics' (*Berufsmathematik*), to build on Ivo Schneider's promising concept.[18]

The versatility of quantifying expertise has been highlighted in Chapter 3: the dynasties of the Öders and the Rieses show how the measuring and counting methods developed in the mines could seamlessly be adapted to new issues. Theirs sons became gaugers, *Rechenmeister*, surveyors, or instrument-makers in Saxony and beyond, at a time when university mathematics was generally seen as an unprofitable discipline (*brotlose Kunst*). In recent years, brilliant studies have shown how a vernacular epistemology of quantification developed in other countries as well. Just as Mathesius, Spangenberg, or Eichholtz were preaching geometry in the mining states, 'the reformed minister Peter Plancius taught nautical science to seamen from the pulpit of the Oudezyds Kapel in Amsterdam', while 'Thomas Hood gave public lectures on mathematics and astronomy in London for mariners, craftsmen and soldiers'.[19] Dutch bookshops were eagerly selling mathematical tables and booklets for navigation, while the French *artistes* contributed to the development of mechanical arts;

---

[18] On Ivo Schneider's *Berufsmathematik* 'that was bound to the conditions of the market, of supply and demand, and was partially organized in guilds, like a craft', see Schneider, *Johannes Faulhaber 1580–1635* (1993), introduction. Gunthild Peters has recently studied how the regulation on wine-gauging shaped groups of professional in various German states: Peters, *Zwei Gulden vom Fuder* (2018), pp. 104–152.

[19] Hooykaas, *Religion and the Rise of Modern Science* (1972), p. 91. On the training of Dutch seamen, see also Schotte, *Sailing School* (2019), pp. 41–47; on instrument-makers' shops in London, and informal gatherings promoting practical mathematics, see Cormack, 'Mathematics for Sale' (2017).

practitioners even played an active role in promoting the 'mathematical sciences of the state' within the Swedish mining administration.[20]

The 'art of setting limits' epitomizes an order of knowledge typical of the early modern period. Subterranean geometry is worth studying precisely because it allows us to understand how practical mathematics could be grounded in, and evolve along, the operative knowledge of its time.[21] The attempts of early modern practitioners to quantify nature were certainly local and modest in their means. They may well have been often inconclusive in their results and not readily generalizable. Still, unexpected connections nevertheless arose here and there between technical problems and abstract theories. These punctual relationships warrant further analyses and this much more modest goal might be easier to achieve. The number of available studies on practical mathematics is far from matching both the number of practitioners and their significance. More importantly, the worldly uses of mathematics tend to be mentioned mostly in direct relation with contemporary achievements in the sciences. Depending on the case, such practices appear as spurs, explanatory factors, or even simply as contextualizing elements for supposedly more crucial epistemological transformations. This perspective distorts our understanding of the silent rise of a mathematical culture which, as this book has hopefully shown, can only be properly understood in its own terms.[22]

*

Underground geometry was primarily a vernacular expertise, first transmitted orally and then recorded in manuscripts using the *Bergmannsprache*. This dialect was as unintelligible to laymen as the specialized languages of philosophers, lawyers, or physicians. Usually seen as a sign of maturity, and even a criterion of scientificity for learned disciplines, idioms tend to be perceived as insular or esoteric when practised by craftsmen. In both cases, however, it is a matter of being as precise as possible in communicating with one's peers.[23] This prejudice already led Agricola to craft, in his *De re metallica,* a whole new Latin vocabulary to describe the underground world. However, his move was based on a fundamental misunderstanding: Latin, or even colloquial German, could not serve as a medium for an operational knowledge

---

[20] On Dutch bookshops, see Schotte, *Sailing School* (2019), pp. 48–62; on the French *artistes*, see Bertucci, *Artisanal Enlightenment* (2017); on craftsmen and *mechanicus* in Sweden, see Orrje, *Mechanicus: Performing an Early Modern Persona* (2015), here quoted from p. 97.
[21] Bennett, 'Practical Geometry and Operative Knowledge' (1998), pp. 195–222.
[22] Morel, 'Mathematics and Technological Change' (2023).
[23] On the influence of language among scholars, related to the renewal of the quadrivium and the mathematization of nature, see Reiss, *Knowledge, Discovery and Imagination in Early Modern Europe* (1997); Hoppe, 'Die Vernetzung der mathematisch ausgerichteten Anwendungsgebiete mit den Fächern des Quadriviums in der frühen Neuzeit' (1996).

so intrinsically linked to specific places and people, whose living dialect was in constant development. Scholars and rulers similarly complained that mine surveying was an esoteric art. By this, they criticized not only the use of an obscure dialect but also the fact that this knowledge was not put into print, which led them to the conclusion that it was intentionally hidden.

Such criticisms of artisans, craftsmen, and technicians have long been unquestioningly accepted and passed on in a historiography which, more often than not, presents them in a folkloric or archaic perspective. Decentring our view and looking at the customs, chronicles, and archives, what one witnesses here is a well-codified discipline. Local accounts and customs unarguably indicate that surveys were publicly displayed, discussed, and challenged. Craftsmen relied on their dialect and system of manuscripts mainly because their authority and their careers, which developed within secretive administrations, depended less on literary activity than on their concrete technical achievements. It is clear that the scattered and seemingly uncoordinated developments of quantitative arts and crafts has obscured their general significance. The absence of a printed tradition, however, should be interpreted with nuance and not be mistaken as a sign of backwardness; likewise, a teaching system based on companionship is not necessarily static. In the Holy Roman Empire, but also in Scandinavia and Central Europe, the *geometria subterranea* enjoyed a wide, albeit intermittent, recognition. This mathematical art developed in a handful of regions, from which surveyors and mining masters travelled and could be found in many early modern states. Advanced teaching systems that used contracts, examinations, and certificates largely paved the way for the development of modern technical schools, as the history of the Freiberg mining academy illustrates.

The most obvious parallel is that of land reclaiming operations carried out all over Europe by Dutch *landmeters* (surveyors). From the English Fens to the French marshes, they applied their measuring expertise and engineering techniques, while local adaptations occasionally spurred further improvements.[24] Military architecture, mercantile arithmetic, or navigation, and other disciplines that have been studied in more detail followed similar paths. Virtually no European region was left untouched by these enterprises of land improvement, and everywhere mathematics was on display, taming nature in specific contexts but according to broadly similar patterns. Vernacular expertise travelled with its practitioners, as the competing nations-states sought to improve their realms and enviously looked at what rivals and neighbours had already achieved. It is highly paradoxical to note how widespread quantifying, counting, and measuring activities were, and still appear unconnected with the

---

[24] See Willmoth, *Sir Jonas Moore* (1993), pp. 88–120; Streefkerk, Werner, and Wieringa, *Perfect gemeten* (1994); Ash, *The Draining of the Fens* (2017).

improving status of contemporary mathematics.[25] Hidden in plain sight, these empirical practices appear as underground mathematics, only perceived when they fit within, or are mentioned by, scholarly discourses.

From the Renaissance up to the onset of industrialization, scholars and natural scientists were fascinated by the multifarious character and esoteric language of subterranean geometry. In the sixteenth century, the mining world prompted inquiries from curious humanists, from Georgius Agricola to Erasmus Reinhold. If we have shown that their published works were much more complex than a mere description of practices, the interest that these books aroused in learned circles attest of a significant curiosity. More substantial influences developed when subterranean surveyors collaborated with scholars and instrument-makers at the court of Dresden, at the turn of the seventeenth century. Subterranean surveyors bemused the most famous mathematicians, from Leibniz – who annotated his personal copy of Voigtel's handbook – to Euler who described their instruments in his *Letters to a German Princess*. Even in the Age of Enlightenment, no simple or clear-cut hierarchy existed between the learned and the skilled. Both groups shared common issues around measurement and precision, but tackled them with a variety of means to achieve different ends. Most importantly, a good measure of respect and genuine mutual curiosity existed.

In the early modern period, numbers, geometric figures, maps, and tables were not solely used for scientific purposes but developed in society at large.[26] In this respect, reckoning masters, artists and inventors, navigators and instrument-makers, gaugers and underground surveyors all contributed to the advancement of knowledge. In their own way, they were instrumental in a nascent mathematization of nature. If one wishes to better grasp the social changes that mathematics was undergoing at the time, one has to study these practitioners in their context. This entails understanding the concrete issues they faced and appraising their methods by their own standards. Instead of rigorous hypotheses, one is confronted with fluctuating tariffs and a myriad of weights and measures. Instead of abstract demonstrations, one suddenly encounters technical conundrums that were solved with imperfect information, itself acquired with artisanal instruments. Their limitations, frequent failures, and uncoordinated progresses notwithstanding, contemporaries witnessed how the new mathematical tools affected their daily lives, for better or worse. Among them, the most perceptive managed to articulate it explicitly. Let us one

[25] The recent study of Margaret Schotte on navigation has sought to highlight the international circulation of knowledge and to approach this history in a more global (albeit still restricted to Europe) fashion. See Schotte, *Sailing School* (2019).

[26] See Kula, *Measures and Men* (1986); Folkerts, Knobloch, and Reich, *Maß, Zahl und Gewicht* (1989).

last time quote a Lutheran minister, Cyriacus Spangenberg, from the mining city of Eisleben. In a sermon held in 1574 for his regular audience of miners, Spangenberg casually underlined how ubiquitous mathematics were:

A Christian may thus well be a reckoning master, a composer, an organist, a painter, an astronomer, a subterranean surveyor, a weigher, a wine gauger, an assayer, etc., because they deal with things in the way God orders everything, and there is no craft, duty, or handling that could be practised without counting, measuring, and weighing.[27]

The eminently local and social character of early modern subterranean geometry might seem at odds with the usual description of mathematics as abstract and universal.[28] Looking at the actual practices, however, one sees how surveying and counting seamlessly blended with mining customs, geology, and religion. The composite character of this mathematical culture makes it sometimes hard to recognize, as it mixed not only various kind of knowledge, but also crafts and skills. Its traces are scattered and have to be tracked in lists of problems, maps, administrative reports, sermons, and rulings. Moreover, the influence of underground surveyor grew largely apart from universities and scientific academies, with sporadic if meaningful contacts with the learned world, and yet it exerted an immense influence in the mining states of the Holy Roman Empire. In the early modern period, the mathematization of nature could take many intermediary forms, and theory was certainly not the only way to advance it. Subterranean geometry illustrates how the daily toil of industrious, yet often overlooked practitioners formed another pathway to mathematical rationality.

[27] Spangenberg, *Die XIX. Predigt Von Doctore Martino Luthero* (1574), pp. 45–46: 'ein Christ wol kan ein Rechenmeister / Componist / Organist / ein Maler / ein Astronomus / ein Marscheider / ein Wagmeister / ein Visirer / ein Probirer / etc. seyn / denn sie gehen damit umb / darnach Gott alles ordnet / unnd ist zwar kein Handwerck / kein Ampt noch Hantierung / das one Zelen / Messen und Wägen … könde geübet oder getrieben werden'.
[28] Roberts, Schaffer, and Dear, *The Mindful Hand* (2007).

# Bibliography

**Primary Sources**

*Manuscripts*

*Banská Štiavnica (Slovenský banský archív)*
SUBA, HKG I, *Beschreibung des Nachlasses von Georg Zacharias Angerstein (17.11.1761)*.

*Bochum (Deutsche Bergbau-Museum)*
Sign. 875, Beer, Johann Gabriel, *Geometrie Subteranea oder Marckscheide-Kunst* (manuscript, 1739).

*Clausthal (Niedersächsisches Landesarchiv – Bergarchiv Clausthal)*
NLA HA, BaCl Hann, Historische Nachrichten über die Harzverhältnisse 84a, 7(1), Nr. 7/5, *Rund-Erlass des Herzogs Julius, den Landmesser Gottfried Mascopius bei Erfüllung des ihm erteilten Auftrages, vom ganzen Fürstentum eine 'Landtaffel' (Landkarte) anzufertigen, tatkräftig zu unterstützen (16.01.1572)*.
NLA HA, BaCl Hann. 84a 12, Nr. 12/29, *Schreiben des Markscheiders Wolff Seydel an den Ober-Zehntner, seinen Vetter Zacharias Schneyder, das Markscheiden lernen zu lassen (01.06.1677)*.
NLA HA, BaCl Hann. 84a Acc. 8, Nr. 2465/1, *Nachlass Steltzner zum Tiefer Georg Stollen*.
NLA HA, BaCl Hann. 84a, Nr. 6684, *Grubenvermessungen – Clausthaler Revier (1659–1726)*.
NLA HA, BaCl Hann. 84a, Nr. 6692/1, *Acta betr: Markscheider-Observations-Bücher 1695 seq.*
NLA HA, BaCl Hann. 84a, Nr. 6698, *Markscheiderregistraturen und -inventarien, Zellerfelder Revier (1711–1809)*.
NLA HA, BaCl Hann. 84a, Nr. 6699, *Markscheiderregistraturen und -inventarien – St. Andreasberger Revier (1739–1819)*.
NLA HA, BaCl Hann. 84a, Nr. 9803, *Tiefer Georg Stollen, speziell: Die wegen Herantreibung desselben entworfene Pläne und Vorverhandlungen (1770–1776)*.
NLA HA, BaCl Hann. 84a, Nr. 9805, *Tiefer Georg Stollen, speziell: Die wegen Herantreibung desselben entworfene Pläne und Vorverhandlungen (1776–1781)*.

253

NLA HA, BaCl Hann. 84a, Nr. 9825, *Tiefer Georg Stollen, speziell: Die wegen Herantreibung desselben geleisteten Markscheidearbeiten (1774–1797).*

NLA HA, BaCl Hann. 84a, Nr. 9828, *Tiefer Georg Stollen, speziell: Durchtreibung desselben durch den Rosenhöfer und Burgstätter Zug (1776–1802).*

NLA HA, BaCl Hann. 84a, Nr. 9848, *Nachschießen und Verwahrung der Sohle, Förste und Wange desselbem, Grund und Profilriß Strossenbau unter der 13. Strecke (1777–1800).*

NLA HA, BaCl Hann. 84a, Nr. 10203, *Fremdenbuch des Tiefen Georg-Stollen (1779–1856).*

NLA HA Dep. 150 K. Acc 2018/700, Nr. 3, *Bergrisse Daniel Flach.*

NLA HA Dep. 150 Acc. 2018/700, Nr. 251, *Berechnungen derer Gegen-Sohlen des Tiefen-Georg Stollen.*

NLA HA Dep. 150 Acc. 2018/700, Nr. 252, *1te Observation vom Tiefer Georg Stollen. Die Vorschlaege übern Hüllfe Gottes und Isaacks Tanner- ingleichen übern Laubhütter Stollen Gang nach dem Thurm Rosenhoeffer Zug bettrefend (1774–1775).*

NLA BaCl, Acc. 8, Nr. 300, *Bewilligung von Markscheider-Gebühren (1765–1768).*

### Dresden (Sächsisches Staatsarchiv/Hauptstaatsarchiv Dresden)

SächsStA–D, 10005 Hof- und Zentralverwaltung (Wittenberger Archiv), Nr. Loc. 4322/03, *Schneeberg I, Allgemeine Sachen/Vermischte Sachen/Schreiben (Kurfürst Ernsts und Herzog Albrechts) an Hans Setener (?), Markscheider zu Graupen, über die ihm aufgetragene Verrichtung einer Markscheidung auf dem Schneeberg. Dresden, Dienstag nach Quasimodogeniti 77 (1477).*

SächsStA–D, 10024 Geheimer Rat (Geheimes Archiv), Nr. 08341/06, *Schwarzenbergische Handlung wegen Bereitung, Vergleich, Verrainung, Abteilung und Abmessung der Herrschaft und des Amtes Schwarzenberg, (1549–1552).*

SächsStA–D,10024 Geheimer Rat (Geheimes Archiv), Nr. 08342/01, *Markscheiderische Abmessung des Amtes Schwarzenberg (1550).*

SächsStA–D,10024 Geheimer Rat (Geheimes Archiv), Nr. 09762/03, *Mathematica, Mechanica et Geographica (1532–1726).*

SächsStA–D, 10024 Geheimer Rat (Geheimes Archiv), Nr. Loc. 09762/05, *Schriften an und von dem mit der Landesvermessung beauftragten Markscheider Matthias Öder (u.a. von Jöstel M.), unter anderem die Vermessung des Kriegsstücks im Amt Frauenstein betreffend (1602–1604).*

SächsStA–D, 10026 Geheimes Kabinett, Nr. Loc. 01254/06, *Hans August Nienborgs Bestellung zum Landfeldmesser, nachgehends zum Oberlandfeldmesser auch Markscheider (1701).*

SächsStA–D, 10036 Finanzarchiv, Loc. 32833 Rep. LII, f. 100, *Ersetzung der Markscheider Stellen (1733–1815).*

### Dresden (Sächsische Landesbibliothek – Staats- und Universitätsbibliothek)

SLUB Mscr. C.1, *Collectanea astronomica (Abraham Ries, Melchier Jöstel)* (sixteenth century).

SLUB Mscr. C.3, *Abraham Ries Miscellanea mathematica* (1584–1612).

SLUB Mscr. C.5, *Abraham Ries Sammelhandschrift* (1596).

SLUB Mscr. C.81, *Abraham Ries Sammelhandschrift (?)*.
SLUB Mscr. C.433, *Abraham Ries, Arithmomachia. Endliche erclerung Churfurstlicher Sexischer Arithmomachiae. Die tafel Arithmomachiae d. i. der Zanck streith krig vnd Kampf zwischen den geraden vnd vngeraden tzalen (1562)*.
SLUB, Kartensammlung, A13534, *Hiob Magdeburg, Duringische und Meisnische Landtaffel (Meißen, 1566)*.

### Freiberg (Sächsisches Staatsarchiv/Bergarchiv Freiberg)

SächsStA–F, 40001 Oberbergamt Freiberg, *Nr. 608, Markscheidearbeiten an und Vortrieb der Saidenbacher Rösche (1599–1603)*.
SächsStA–F, 40001 Oberbergamt Freiberg, Nr. 1362, *Gesuch um eine Leitung für das Schul- und Bethaus Bräunsdorf sowie für die anzulegende Bergwerksakademie, gestellt von M. Christian Ehrenfried Seifert, 1726*.
SächsStA–F, 40001 Oberbergamt Freiberg, Nr. 2310, *Landesherrliche Befehle über die Wiederbesetzung verschiedener Dienststellen (1568–1773)*.
SächsStA–F, 40001 Oberbergamt Freiberg, Nr. 2479, *Ausmessung und Berechnung der im Bergamtsrevier Marienberg befindlichen Wassersäulenmaschinen (1780–1802)*.
SächsStA–F, 40001 Oberbergamt Freiberg, Nr. 2805, *Genaue Bestimmung des beim Bergbau üblichen Lachtermaßes sowie Anschaffung eines Normalmaßes (1772–1830)*.
SächsStA–F, 40001 Oberbergamt Freiberg, Nr. 3477, *Fasciculus von Befehlen, Instructione und Lehr-Contracten, über die Markscheide-Kunst (1551–1593)*.
SächsStA–F, 40001 Oberbergamt Freiberg, Nr. 3578, *Gedrucktes Regulativ über das Volumen von Förderbehältnissen (1788)*.
SächsStA–F, 40001 Oberbergamt Freiberg, Nr. 3631, *Anfertigung und Hinterlegung von Grubenrissen im Bergamt*.
SächsStA–F, 40006 Bergamt Altenberg, Nr. 1480, *Appellation von Gottfried Wilhelm Grellmann, Markscheider in Altenberg (1753–1757)*.
SächsStA–F, 40010 Bergamt Freiberg, Nr. 326, *Auf Beschert Glück 1698 erfolgtes Erbbereiten und die dabei nicht richtig gesetzten Lochsteine (1698–1721)*.
SächsStA–F, 40012 Bergamt Johanngeorgenstadt, Nr. 664, *Besichtigung, Befahrung und Erkundung des hiesigen Bergbaus durch Ausländer (1730–1788)*.
SächsStA–F, 40012 Bergamt Johanngeorgenstadt, Nr. 949, *Acta über Herrn Bergmeister Johann Andreas Scheidhauern zu Freyberg (1765)*.
SächsStA–F, 40013 Bergamt Marienberg, Nr. 1438, *Gewältigung der Tiefbaue unter der Sohle des Weiße Taube Stolln bei den drei Lautaer Berggebäuden Drei Weiber Fundgrube, Vater Abraham Fundgrube und Antritt Fundgrube (1771–1783)*.
SächsStA–F, 40015 Bergamt Schneeberg, Nr. 1309, *Vermessbuch 1597–1632*.
SächsStA–F, 40015 Bergamt Schneeberg, Nr. 1448, *Gangstreitigkeit zwischen den Gewerken des Silbergebäudes Obere Katharina Fundgrube gegen das Zwittergebäude Neujahr Fundgrube im Raschauer Gemeindewald*.
SächsStA–F, 40040 Fiskalische Risse zum Erzbergbau, G 5003, *Freiberga Subterranea cum Ditionibis Exteris eo pertinentibus metalli feris – oder Unterirdisches Freiberg mit den dahin gehörigen auswärtigen Bergrevieren (1693–1695)*.
SächsStA–F, 40040, Fiskalische Risse zum Erzbergbau, G 5004, *Freiberga Subterranea cum Ditionibis Exteris eo pertinentibus metalli feris – oder Unterirdisches Freiberg mit den dahin gehörigen auswärtigen Bergrevieren (18. Jh.)*.
SächsStA–F, 40040 Fiskalische Risse zum Erzbergbau, G 5005, *Freiberga Subterranea cum Ditionibis Exteris eo pertinentibus metalli feris – oder Unterirdisches Freiberg mit den dahin gehörigen auswärtigen Bergrevieren (Konzept 2, 1686–1693)*.

SächsStA–F, 40040 Fiskalische Risse zum Erzbergbau, G 5006, *Freiberga Subterranea cum Ditionibis Exteris eo pertinentibus metalli feris – oder Unterirdisches Freiberg mit den dahin gehörigen auswärtigen Bergrevieren (Konzept 1, 1686–1693).*

SächsStA–F, 40040 Fiskalische Risse zum Erzbergbau, H3585, *Margaretha und Himmelskrone Fundgrube bei St. Michaelis (24 November 1698).*

SächsStA–F, 40047 Winkelbücher, Nr. 1429, *Mensurata Geodaesia subterranaeae et Geome. Bene observata et diligenter colligata a Paulo Christophero Beidlero (1692–1703).*

SächsStA–F, 40168 Grubenakten Marienberg, Nr. 512, *Instruktion des Berg- und Amtshauptmanns Abraham von Schönberg über die Bestellung des Schichtmeisters Adam Schneider zum Markscheider im Obergebirge.*

SächsStA–F, 40168 Grubenakten des Bergreviers Marienberg, Nr. 909, *Vater Abraham Fundgrube am Stadtberg bei Marienberg (1767–1802).*

### Freiberg (Archiv der TU Bergakademie Freiberg)

UAF, OBA 5, *Stipendiengesuche (1738–1750).*

UAF, OBA 6, *Stipendiengesuche (1750–1761).*

UAF, OBA 10, *Acta, Verbesserung der Akademie betreffend (1794–1795).*

UAF, OBA 181, *Aufnahmegesuch Selbstzähler (?–?).*

UAF, OBA 182, *Aufnahmegesuch Selbstzähler (1754–1778).*

UAF, OBA 236, *Acta, Wie bey der Berg-Academie wegen derer Stipendiaten und Anhörung der Vorlesungen zutreffenden Einrichtungen und dieserhalb zu erstattenden Haupt-Anzeigen betreffend (1765–1769).*

UAF, OBA 237, *Acta, die Errichtung einer Bergakademie allhier in Freyberg betreffend (1769–1777).*

UAF, OBA 242, *Acta, die bey der Berg-Academie wegen derer Stipendiaten und Anhörung der Vorlesungen zutreffenden Einrichtungen und was dem sonst anhängig betreffend (1780–1782).*

UAF, OBA 244, *Acta, die bey der Berg-Academie wegen derer Stipendiaten und Anhörung der Vorlesungen zutreffenden Einrichtungen und was dem sonst anhängig betreffend (1784).*

UAF, OBA 246, *Acta, die bey der Berg-Academie wegen derer Stipendiaten und Anhörung der Vorlesungen zutreffenden Einrichtungen und was dem sonst anhängig betreffend (1786).*

### Freiberg (Wissenschaftlicher Altbestand der TU Bergakademie Freiberg)

TU BAF – UB XVII 12, *Geometria subterranea oder Marckscheide Kunst, das ist Meß-Kunst unter dem Erden (1708–1729).*

TU BAF – UB XVII 15, *Gründliche und aus erfahrenheit stammende Marckscheidekunst oder Gründlicher unterricht von Marckscheiden und Meßen ober und unter Erden (copy ca. 1750).*

TU BAF – UB XVII 14, *Geometriae pars, Vom Marckscheiden und Feldmessenn unnter der erden darinnen geleret wird wie pA metall- unnd mineralischen bergwercken geometrice verfahrenn (ca. 1700).*

TU BAF – UB XVII 18, *Neu Marckscheide-Buch: darinnen begriffen die Tabulae Sinuum samt derer Gebrauch nebst Beschreibung des Gruben-Heng und Zulege Compasses, item der Waßer-Wage (1669–?).*

TU BAF – UB XVII 333, *Gründliche und aus erfahrenheit Stammete Marckscheide-Kunst oder Gründlicher Unterricht von Marckscheiden (1693)*.

TU BAF – UB XVII 636 *Gründliche und aus erfahrenheit Stammende Marckscheide-Kunst, oder Gründlicher Unterricht von Marckscheiden (ca. 1700)*.

TU BAF – UB, Nachlass 115 Johann Andreas Scheidhauer:
o 300d, *Studiorum Mathematicorum*.

o 330e, *Tabulae decimales, Studiorum Mathematicorum, Studia Mathematica varia et Exercitationes scientine ..., Studiorum Mathematicorum elementa arithmeticae, Versuch welcher über den Stand des Barometers in verschiedenen Höhen am 17.06.1771 angestellt worden (enthält u.a. Abtheilung eines Kunstgestänges mit Zeichnungen), verschiedene Abhandlungen zur Mathematik u.a. Geodäsie 1757, Geometrie 1757*.

o 300m, *Beyträge zur Markscheidekunst in einzelnen Abhandlungen*.

### Gotha (Forschungs- und Landesbibliothek)
Signatur Chart A 972, *Gründlicher Unterricht in Bergbau und fürnemlich in der Marckscheider Kunst (1718)*.

### Innsbruck (Universitäts- und Landesbibliothek Tirol)
Cod. 1187, *Johann Andreas Scheidhauer, Beiträge zur Markscheidekunst. Sammlung von 81 Aufgaben zum Markscheidewesen mit Einführung (1775–1776)*.

### Rennes (Bibliothèque départementale d'Ille-et-Vilaine)
C 1485, *Procès verbal de la mine du Pont Péan fait en execution de l'Arrêt du Conseil du 23 juin 1761*.

### Wernigerode (Landesarchiv Sachsen-Anhalt)
Aa Nr. 426, *Acta, die Ablesung des Bergk-Gebeths, von denen Cantzeln wegen der Berg-Wercke dieser löblichen Graffschaft Mannßfeldt, undt deren Dancksagung, vor geschehenen glücklichen Durchschlag in neuen Stollen, betreffend (1671–1717)*.

F8, Bb Nr. 16, *Vereidigung und Anweisung des neuen eislebischen Bergvogts Peter Gerhardt Enthält auch: Befahrung der Baderzeche und der Lichtlöcher. – Kassenvisitazion und Vergleich der Achtwochenzettel mit der Jahresrechnung (1613–1614)*.

### Wolfenbüttel (Niedersächsisches Landesarchiv)
NLA WO, 2 Alt, Nr. 19675, *Abgelehntes Gesuch des Schloss- und Garnisonkantors Johann Adam Heydecke in Blankenburg, seinen Sohn zum Markscheider auszubilden (1750)*.

NLA WO, 33 Alt, Nr. 414, *Markscheider am Kommunion-Unterharz: Annahme, Besoldung, Markscheidergebühren, Anfertigung von Rissen Enthält (1681–1806)*.

NLA WO, 112 Alt, Nr. 1777, *Die Ausbildung des L. C. Ilse aus Hüttenrode zum Markscheider und dessen Tätigkeit (1766–1788)*.

### Printed Primary Sources

Agricola, Georgius. *Georgii Agricolae medici Bermannus, sive de re metallica* (Basel: Froben1, 1530).

Agricola, Georgius. *Libri quinque De mensuris et ponderibus* (Paris: Wechel, 1533).

Agricola, Georgius. *Georgii Agricolae de re metallica libri XII* (Basel: Frobenius, 1556).

Agricola, Georgius. *Vom Bergkwerck: xij. Bücher* (Basel: Frobenius, 1557).

Agricola, Georgius. *De natural fossilium* (Basel: Frobenius, 1558).

Agricola, Georgius. *Schriften über Maße und Gewichte (Metrologie)*. Edited by Georg Fraustadt and Walter Weber (Berlin: VEB Deutscher Verlag der Wissenschaften, 1959).

Agricola, Georgius, Robert Halleux, and Albert Yans. *Bermannus le mineur: un dialogue sur les mines* (Paris: Les Belles Lettres, 1990).

Agricola, Georgius, Herbert Clark Hoover, and Lou Henry Hoover. *De Re Metallica: Translated from the First Latin Edition of 1556 with Biographical Introduction, Annotations and Appendices* (London: The Mining Magazine, 1912).

Albinus, Petrus. *Auszug der Eltisten und fürnembsten Historien, des uralten streitbarn und beruffenen Volcks der Sachsen* (Dresden: Bergen, 1598).

Anonymous. *Das ABC der Bergwercks-Wissenschaften. In Frage und Antwort abgefasset zum Unterricht für die Jugend* (Freiberg: Reinhold, 1747).

Beck, Dominikus. *Briefe eines Reisenden von \*\*\* an seinen guten Freund zu \*\*\* über verschiedene Gegenstände der Naturlehre und Mathematik* (Salzburg: Erbinn, 1781).

*Bergk-Ordenung / des Durchlauchtigsten / Hochgebornen Fürsten und Herrn / Christianen / Hertzogen zu Sachssen* (Dresden: Hieronymus Schütz, 1589).

*Bergkordnung des Freyen Königlichen Bergkwercks Sanct Joachimsthal* (Zwickau: Wolfgang Meyerpeck, 1548).

Bernhardi, Gotthelf Benjamin. *Drey Fragen über die Berggerichtsbarkeit im Königreich Sachsen* (Freiberg: Craz, 1808).

Beyer, Adolph. *Otia Metallica. Oder bergmännische Neben-Stunden*, 2 vols. (Schneeberg: Fulden, 1748–1751).

Beyer, August. *Gründlicher Unterricht von Berg-Bau, nach Anleitung der Marckscheider-Kunst* (Schneeberg: Fulden, 1749).

Biringuccio, Vannoccio. *The Pirotechnia of Vannoccio Biringuccio: The Classic Sixteenth-Century Treatise on Metals and Metallurgy* (Mineola: Dover, 2005).

Brown, Edward. *A Brief Account of Some Travels in Divers Parts of Europe* (London: Tooke, 1673).

Brunn, Lucas. *Euclidis elementa practica, Oder, Ausszug aller Problematum und Handarbeiten auss den 15. Büchern Euclidis* (Nuremberg: Halbmayer, 1625).

Busse, Friedrich Gottlieb von. *Betrachtung der Winterschmidt- und Höll'schen Wassersäulenmaschine nebst Vorschlägen zu ihrer Verbesserung und gelegentlichen Erörterungen über Mechanik und Hydraulik* (Freiberg: Craz & Gerlach, 1804).

Christian, Meltzer von Wolckenstein. *Historia Schneebergensis Renovata: das ist, Erneuerte Stadt- und Berg-Chronica* (Schneeberg: Fulden, 1716).

Christian, Wolff. *Mathematisches Lexicon, darinnen auch die Schriften, wo jede Materie zu finden, angeführet werden* (Leipzig: Gleditsch, 1716).

Cochlaeus, Johannes, Karl Langosch, and Volker Reinhardt. *Kurze Beschreibung Germaniens = Brevis Germanie descriptio (1512)* (Darmstadt: WBG, 2010).

Dannenberg, Erich Christian Heinrich. *Der Harz: ein Gedicht in sieben Gesängen* (Göttingen: Bossiegel, 1782).

Deluc, Jean-André. *Recherches sur les modifications de l'atmosphère*, 2 vols. (Geneva, 1772).

Deluc, Jean-André. 'Barometrical Observations on the Depth of the Mines in the Hartz'. *Philosophical Transactions of the Royal Society of London* 67 (1777): 401–449. https://doi.org/10.1098/rstl.1777.0023.

Deluc, Jean-André. 'A Second Paper Concerning Some Barometrical Measure in the Mines of the Hartz'. *Philosophical Transactions of the Royal Society of London* 69 (1779): 485–504. https://doi.org/10.1098/rstl.1779.0032.

Deluc, Jean-André. *Lettres physiques et morales sur l'histoire de la terre et de l'homme. Addressees a la reine de la Grande Bretagne, par J.A. de Luc* (The Hague and Paris: De Tune et Duchesne, 1779).

Deluc, Jean-André. *Geological Travels in Some Parts of France, Switzerland, and Germany* (London: Rivington, 1813).

Daubuisson, Jean-François. *Des mines de Freiberg en Saxe et de leur exploitation*, 2 vols. (Leipzig: Wolf, 1802).

Duhamel, Jean-Pierre François. *Géométrie souterraine, élémentaire, théorique et pratique* (Paris: Imprimerie Royale, 1787).

Eichholtz, Peter. *Geistliches Bergwerck, Das ist: Andächtiger, lieblicher und beweglicher Betrachtungen* (Goslar: Duncker, 1655).

Euclid. *The Thirteen Books of Euclid's Elements Translated from the Text of Heiberg with Introduction and Commentary.* Translated by Sir Thomas Heath. 3 vols. (Cambridge: Cambridge University Press, 1908).

Euler, Leonhard. *Lettres à une princesse d'Allemagne: Sur divers sujets de physique & de philosophie* (St Petersburg: Académie Impériale des Sciences, 1768)

Fehling, Carl Heinrich Jacob. *Die Kleidungen derer hohen und niedren Berg Officiers, Berg Beamdten und Berg Arbeiter, wie solche in dem Bergmännischen Aufzug in Plaischen Grund, ohnweit Dresden gegangen am 26. Sept. 1719*, Staatliche Kunstsammlungen Dresden, Kupferstich-Kabinett, Inv.-Nr. C 677.

Flaco, Siculo. *De agrorum conditionibus & constitutionibus limitum, Siculi Flacci lib. I. Iulii Frontini lib. I. Aggeni Vrbici lib. II. Hygeni Gromatici lib. II. Variorum auctorum ordines finitionum. De iugeribus metiundis. Finium regundorum. Lex Manilia. Coloniarum pop. Romani descriptio. Terminorum inscriptiones & formae. De generibus lineamentorum. De mensuris & ponderibus: Omnia figuris illustrata* (Paris: Turnebum, 1554).

Freiesleben, Johann Carl. *Bemerkungen über den Harz. Erster Theil: Bergmännische Bemerkungen* (Leipzig: Schäfer, 1795).

Frisius, Gemma. *Libellus de locorum describendorum ratione & de eorum distantijs inueniendis nunquam ante hac visus* (Antwerp: Grapheus, 1533).

Gatterer, Christoph Wilhelm Jakob. *Anleitung den Harz und andere Bergwerke mit Nuzen zu bereisen.* 5 vols. (Göttingen: Vandenhoeck, 1785).

Gotthard, Johann Christian. *Authentische Beschreibung von dem merkwürdigen Bau des Tiefen Georg Stollens am Oberharze* (Wernigerode: Struck, 1801).

Grosgebauer, Martin. *Vom Feldmessen: Eigentlicher und kurtzer unterricht, wie dasselbe auffs förderlichst und gewiesest durch die Triangel oder dreyeckichte Figuren zu verrichten* (Schmalkalden: Schmück, 1596).

Gruber, Johann Sebastian. *Neue und gründliche mathematische Friedens- und Kriegs-Schule* (Nürnberg: Riegel, 1697).

Grübler, Johann Samuel. *Ehre der Freybergischen Todten-Grüffte nebst kurtzen Lebens-Beschreibungen der meisten dasigen Patriciorum und Geschlechter.* 2 vols. (Leipzig: Lanckisch, 1731).

Henning, Calvör. *Programma de historia recentiori Hercyniæ superioris mechanica* (Clausthal, 1726).

Henning, Calvör. *Acta Historico-Chronologico-Mechanica circa metallurgiam in Hercynia superiori, Oder, Historisch-chronologische Nachricht und theoretische und practische Beschreibung des Maschinenwesens, und der Hülfsmittel bey dem Bergbau auf dem Oberharze* (Braunschweig: Fürstlische Waysenhaus-Buchhandlung, 1763).

Herttwig, Christoph. *Neues und vollkommenes Berg-Buch, Bestehend in sehr vielen und raren Berg-Händeln und Bergwercks-Gebräuchen* (Dresden: Zimmermann, 1734).

Hollenberg, Georg Heinrich. *Bemerkungen über verschiedene Gegenstände auf einer Reise durch einige Deutsche Provinzen, in Briefen* (Stendal: Franzen & Große, 1782).

Hübner, Johann. *Curieuses Natur-Kunst-Gewerk und Handlungs-Lexicon* (Leipzig: Gleditsch, 1712).

Hugh of Saint-Victor. *Practical Geometry [Practica Geometriæ].* Translated by Frederick A. Homann (Milwaukee: Marquette University Press, 1991).

Jacobsson, Johann Karl Gottfried. *Technologisches Wörterbuch, oder alphabetische Erklärung aller nützlichen mechanischen Künste, Manufakturen, Fabriken und Handwerker* (Berlin: Nicolai, 1781).

Jenisch, Paul. *Chronicon Annaebergense continuatum* (Annaberg: Hasper, 1812).

Jugel, Johann Gottfried. *Gründlicher und deutlicher Begriff von dem gantzen Berg-Bau-Schmeltz-Wesen und Marckscheiden, in drey Haupt-Theile eingentheilet* (Berlin: Rüdiger, 1744).

Jugel, Johann Gottfried. *Geometria subterranea, oder Unterirdische Messkunst der Berg- und Grubengebäude, insgemein die Markscheidekunst genannt, etc* (Leipzig: Kraus, 1773).

Kästner, Abraham Gotthelf. *Anmerkungen über die Markscheidekunst, nebst einer Abhandlung von Höhenmessungen durch das Barometer* (Göttingen: Vandenhoeck, 1775).

Kirchmaier, Georg Caspar. *Institutiones metallicae, das ist Wahr- und klarer Unterricht vom Edlen Bergwerck* (Schrödter, 1687).

Kirchmaier, Georg Caspar. *Hoffnung besserer Zeiten, durch das Edle Bergwerck, von Grund, und aus der Erden zuerwarten* (Wittenberg: Meyer & Zimmermann, 1698).

Köhler, Alexander Wilhelm, *Bergmännisches Journal* (Freiberg: Craz, 1791), volume 4.

Kreysig, Georg Christoph. *Nachlese zum Buchdrucker-Jubilaeo in Ober-Sachsen* (Dresden: Kraus, 1741).

Lehmann, Christian. *Christian Lehmanns Historischer Schauplatz derer natürlichen Merckwürdigkeiten in dem Meißnischen Ober-Ertzgebirge* (Leipzig: Lankisch, 1699).

Lehmann, Johann Gottlob. *Kurze Einleitung in einige Theile der Bergwerks-Wissenschaft* (Berlin: Nicolai, 1751).

Lempe, Johann Friedrich. *Gründliche Anleitung zur Markscheidekunst* (Leipzig: Crusius, 1782).

Lempe, Johann Friedrich. 'Auflösung einer Aufgabe aus der Markscheidekunst'. *Leipziger Magazin zur Naturkunde, Mathematik und Oekonomie*, 3(2) (1783): 177–188.

Lempe, Johann Friedrich. *Bergmännisches Rechenbuch* (Freiberg: Barthel, 1787).

Lempe, Johann Friedrich. *Magazin der Bergbaukunde*, 13 vols. (Dresden: Walther, 1785–1799).

Löhneysen, Georg Engelhard. *Bericht vom Bergwerck, wie man dieselben bawen und in guten wolstande bringen sol* (Zellerfeld, 1617).

Lünig, Johann Christian. *Fortgesetzter Codex Augusteus oder neuvermehrtes corpus juris Saxonici*, 3 vols. (Leipzig: Heinsius, 1772).

Marperger, Paul Jacob. *Das neu-eröffnete Berg-Werck* (Hamburg: Schiller, 1704).

Mathesius, Johannes. *Sarepta oder Bergpostill: Sampt der Jochimßthalischen kurtzen Chroniken* (Nuremberg: Berg & Newber, 1562).

Meltzer, Christian. *Bergkläufftige Beschreibung der Bergk-Stadt Schneebergk* (Schneeberg: Pfüzner, 1684).

Moeller, Andreas. *Theatrum Freibergense chronicum; Beschreibung der alten löblichen Berghauptstadt Freyberg in Meissen* (Freiberg: Beuther, 1653).

Müller, Wilhelm Ferdinand. *Meine Streifereyen in den Harz und in einige seiner umliegenden Gegenden* (Weimar: Gädicke, 1801).

Münster, Sebastian. *Cosmographei. Oder Beschreibung aller Länder, Herschafften, fürnemsten Stetten, Geschichten, Gebreuche, Hantierung etc* (Basel: Heinrich Petri, 1550).

Oppel, Friedrich Wilhelm von. *Anhang der Anleitung zur Markscheidekunst* (Dresden: Walther, 1752).

Oppel, Friedrich Wilhelm von, and Johann Gottlieb Kern. *Bericht von Bergbau* (Leipzig: Crusius, 1772).

Paltz, Johannes von. *Die Himmlische Fundgrube* (Leipzig: Konrad Kachelofen, 1490).

Peithner, Johann Taddäus Anton. *Erste Gründe der Bergwerkswissenschaften* (Prague: Johann Joseph Clauser, 1769).

Petri, Friedrich. *Der Teutschen Weissheit: Das ist: Außerlesen kurtze, sinnreiche, lehrhaffte vnd sittige Sprüche vnd Sprichwörter*, 2 vols. (Hamburg: Phillip von Ohr, 1605).

Pfinzing, Paul. *Methodus Geometrica, das ist Kurtzer wolgegründter unnd auszführlicher Tractat von der Feldtrechnung und Messung* (Nuremberg : Fuhrmann, 1598).

Ramus, Petrus. *Scholarum mathematicarum libri unus et triginta* (Basel: Episcopius, 1569)

Reden, Claus Friedrich von. *Rede bei dem feyerlichen Anfange des tiefen Georg-Stollen-Baues unweit Grund* (Clausthal: Wendeborn, 1777).

Reinhold, Erasmus. *Gründlicher vnd Warer Bericht. Vom Feldmessen, Sampt allem, was dem anhengig.* (Erfurt: Bawman, 1574).

Richter, Daniel Adam. *Umständliche aus zuverläßigen Nachrichten zusammengetragene Chronica der im Meißnischen Ober-Ertz-Gebürge gelegenen Königliche Churfürstliche Sächßischen freyen Berg-Stadt St. Annaberg* (Annaberg: Friese, 1746).

Riese, Adam, and Stefan Deschauer. *Das 1. Rechenbuch von Adam Ries: mit einer Kurzbiographie, einer Inhaltsanalyse, bibliographischen Angaben, einer Ubersicht uber die Fachsprache und einem Metrologischen Anhang* (Munich: Institut fur Geschichte der Naturwissenschaften, 1992).

Rösler, Balthasar. *Speculum metallurgiae politissimum, oder, Hell-polierter Berg-Bau-Spiegel* (Dresden: Winckler, 1700).

Rost, Johann Leonhard. *Mathematischer Lust- und Nutz-Garten* (Nuremberg: Weigel, 1724).

Rudhart, Hans. *Antzeigung des nauenn Breyheruffen Bergwerks Sanct Joachimsthal* (Leipzig: Thamer, 1523).

Rudolphi, Friderich. *Gotha diplomatica oder ausführliche historische Beschreibung des Fürstenthums Sachsen-Gotha* (Gensch, 1717).

Saussure, Horace-Bénédict de. *Voyages dans les Alpes, précédés d'un Essai sur l'histoire naturelle des environs de Genève* (Neuchatel: Fauche, 1779)

Scheuchzer, Johann Jacob. *Natur-Geschichte des Schweizerlandes* (Zürich: Gessner, 1746).

Schlözer, August Ludwig. *Briefwechsel meist historischen und politischen Inhalts* (Göttingen: Vandenhoeck, 1779)

Schmid, Wolffgang. *Das erst buch der Geometria: Ein kurste unterweisung was vn warauff Geometria gegründet sey* (Nuremberg: Johann Petreius, 1539).

Schmidt, Franz Anton. *Chronologisch-systematische Sammlung der Berggesetze der österreichischen Monarchie. Fünfter Band, vom Jahre 1656 bis 1708* (Vienna: Hof- und Staats-Aerarial-Druckerey, 1835).

Schönberg, Abraham. *Ausführliche Berg-Information* (Leipzig & Zwickau: Fleischer & Büschel, 1693).

Schreiter, Johannes. *Decimae metallicae oder Zehen Bergpredigten* (Leipzig: Jacob Apel, 1615).

Schwenter, Daniel. *Simonis Stevini Kurtzer doch gründlicher Bericht von Calculation der Tabularum Sinuum, Tangentium und Secantium* (Nuremberg: Halbmayer, 1628)

Seyffert, Christian Ehrenfried. *Bibliotheca Metallica, oder Bergmännischer Bücher-Vorrath* (Leipzig: Zunkel, 1728).

Span, Sebastian. *Sechshundert Bergk-Urthel: Schied vnd Weisunge bey vorgefallenen Bergkwercks Differentien vnterschiedener Orten* (Zwickau: Melchior Göpner, 1636).

Spangenberg, Cyriacus. *Die XIX. Predigt Von Doctore Martino Luthero, wie er so ein getreuwer Marscheider auff vnsers Herrn Gottes Berge gewesen* (Frankfurt am Main: Nicolaus Bassee, 1574).

Stöffler, Johannes. *Von künstlicher Abmessung aller grösse, ebene oder nidere in die lenge, höhe, breite unnd tiefe* (Frankfurt: Christoph Egenolph, 1512).

Sturm, Leonhard Christoph. *Geographia mathematica* (Frankfurt an der Oder: Zeidler, 1705).

Sturm, Leonhard Christoph. *Vier kurtze Abhandlungen: 1. von der geometrischen Verzeichnung der regulären Vielecke 2. Von dem Gebrauch des Proportional-Circuls 3. Von der Trigonometria plana 4. Von der Marckscheide-Kunst: als ein Anhang dem kurzen Begriff des gesamten Mathesis beizufügen.* (Frankfurt an der Oder: Schrey und Hartmann, 1710).

Tacitus, Publius Cornelius. *On Germany.* Translated by Thomas Gordon (New York: Collier & Son, 1910).

Trebra, Friedrich Wilhelm Heinrich von. *Erfahrungen vom Innern der Gebirge* (Dessau: Verlagskasse für Gelehrte und Künstler, 1785).

Trebra, Friedrich Wilhelm Heinrich von. *Bergmeister-Leben und Wirken in Marienberg 1767–1779* (Freiberg: Craz & Gerlach, 1818).

Trebra, Friedrich Wilhelm Heinrich von. Der Bericht von F.W.H. von Trebra über den sächsischen Bergbau zwischen 1766 und 1815. In *Akten und Berichte vom sächsischen Bergbau*, vol. 9. Edited by Lothar Riedel (Kleinvoigsberg: Jens-Kugler-Verlag, 2005).

*Ursprung vnd Ordnungen der Bergwerge inn Königreich Böheim* (Leipzig: Henning Gross, 1616).

Voigtel, Nicolaus. *Geometria subterranea, oder Marckscheide-Kunst* (Eisleben: Johann Diebeln, 1686).

Voigtel, Nicolaus. *Perfectionirte Geometria subterranea, Oder vollkommene Marck-Scheide Fortifications-Fester-Gebäude-Feldmessen- und Wasser-Leitungs-Kunst* (Leipzig: In Verlegung der Autoris, 1692).

Voigtel, Nicolaus. *Vermehrte Geometria Subterranea oder Markscheidekunt* (Eisleben: Georg Andreas Leg, 1713).

Weidler, Johann Friedrich. *Institutiones mathematicae decem et sex purae mixtaeque matheseos disciplinas complexae* (Wittenberg: Hannaver, 1718).

Weidler, Johann Friedrich, and Niklas Fuchsthaler. *Anleitung zur unterirdischen Meß- oder Markscheidekunst* (Vienna: Trattner, 1765).

Weigel, Christoph. *Abbildung und Beschreibung derer sämtlichen Berg-Wercks-Beamten und Bedienten nach ihrem gewöhnlichen Rang und Ordnung im behörigen Berg-Habit* (Nuremberg: Weigel, 1721).

Werner, Abraham Gottlob. *New Theory of the Formation of Veins: With Its Application to the Art of Working Mines*. Translated by Charles Anderson (Edinburgh: Constable, 1809).

Will, Georg Andreas. *Briefe über eine Reise nach Sachsen* (Altdorf: Monath, 1785)

Wolff, Christian. *Vollständiges Mathematisches Lexicon* (Leipzig: Gleditsch, 1734).

Zeidler, Christian Salomon. *Erbvermessen der Silber-Gruben Catharina/Aufm Fasten-Berg/in Johann Georgen Stadt* (Schneeberg: Fulden, 1714).

Zeiher, Johann Ernst. *De studio mathematico eruditis non satis commendando* (Wittenberg: Charisius, 1784).

Zimmermann, Carl Friedrich. *Ober-Sächsische Berg-Academie, in welcher die Bergwercks-Wissenschaften nach ihren Grund-Wahrheiten untersucher, und nach ihrem Zusammenhange entworfen werden* (Dresden & Leipzig: Hekel, 1746).

Zimmermann, Eberhard August Wilhelm von. *Beobachtungen auf einer Harzreise: nebst einem Versuche, die Höhe des Brockens durch das Barometer zu bestimmen* (Braunschweig: Fürstlische Waysenhaus-Buchhandlung, 1775).

## Secondary Sources

Adlung, Stephan. *Markscheiderische Tafeln und Inschriften im sächsischen Erzbergbau*, Akten und Berichte vom sächsischen Bergbau, vol. 22 (Kleinvoigtsberg: Jens-Kugler-Verlag, 1999).

Ageron, Pierre, 'Mathématiques de la guerre souterraine : Bélidor dans l'Empire Ottoman'. In *Les Mathématiques et le Réel: expériences, instruments, investigation*. Edited by Evelyne Barbin, Dominique Bénard, and Guillaume Moussard (Rennes: Presses Universitaires de Rennes, 2018): pp. 211–224.

Alberti, Hans Joachim. *'Entwicklung des bergmännischen Rißwesens'*, unpublished Diplomarbeit (Bergakademie Freiberg, 1927).

Albrecht, Helmuth, and Roland Ladwig, eds. *Abraham Gottlob Werner and the Foundation of the Geological Sciences: Selected Papers of the International Werner Symposium in Freiberg 19th to 24th September 1999* (Freiberg: Technische Universität Bergakademie, 2002).

Andronov, Svetoslav, Dietmar Baum, Helmut Hartmann et al. *Der Elsterfloßgraben Geschichte und Gestalt eines technischen Denkmals* (Naumburg: Burgenlandkreis, 2005).

Archinard, Margarida. 'De Luc et la recherche barométrique'. *Gesnerus: Swiss Journal of the History of Medicine and Sciences* 32(3–4) (1975): pp. 235–247. https://doi .org/10.5169/SEALS-521075.

Ash, Eric H. *Power, Knowledge, and Expertise in Elizabethan England* (Baltimore: Johns Hopkins University Press, 2004).

Ash, Eric H. *Expertise: Practical Knowledge and the Early Modern State* (Chicago: University of Chicago Press, 2010).

Ash, Eric H. *The Draining of the Fens: Projectors, Popular Politics, and State Building in Early Modern England* (Baltimore: Johns Hopkins University Press, 2017).

Asmussen, Tina. 'The Kux as a Site of Mediation: Economic Practices and Material Desires in the Early Modern German Mining Industry'. In *Sites of Mediation*. Edited by Susanna Burghartz, Lucas Burkart, and Christine Göttler (Leiden: Brill, 2016): pp. 159–182.

Axworthy, Angela. *Le mathématicien renaissant et son savoir: le statut des mathématiques selon Oronce Fine* (Paris: Classiques Garnier, 2016).

Bachmann, Manfred. *Der silberne Boden: Kunst und Bergbau in Sachsen* (Stuttgart & Leipzig: Deutsche Verlag-Anst, 1990).

Baron, Roger. 'Sur l'introduction en Occident des termes "geometria theorica et practica"' *Revue d'histoire des sciences* 8(4) (1955): pp. 298–302.

Bartels, Christoph. *Vom frühneuzeitlichen Montangewerbe zur Bergbauindustrie : Erzbergbau im Oberharz 1635–1866* (Bochum: Deutsches Bergbau-Museum, 1992).

Bartels, Christoph. 'Der Harzer Oberbergmeister Georg Andreas Steltzner (1725–1802) und die Montanwissenschaften in der zweiten Hälfte des 18. und am Beginn des 19. Jahrhunderts'. In *Staat, Bergbau und Bergakademie*. Edited by Hartmut Schleiff and Peter Konečný. Vierteljahrschrift für Sozial- und Wirtschaftsgeschichte 223 (Stuttgart: Franz Steiner, 2015): pp. 275–288.

Bartels, Christoph, Beatus Frey, and Andreas Bingener. *Das Schwazer Bergbuch. II. Band: Der Bochumer Entwurf und die Endfassung von 1556, Textkritische Editionen* (Bochum: Deutsches Bergbau-Museum, 2006).

Bartusch, Paul. *Die Annaberger Lateinschule zur Zeit der ersten Blüte der Stadt und ihrer Schule* (Annaberg: Graser, 1897).

Baumgärtel, Hans. 'Die 200-Jahrfeier der Bergakademie. Probleme und Aufgaben'. In *Die Bergakademie und ihre Verantwortung für die Pflege wertvoller nationaler Traditionen: Vorträge des XIII. Berg- und Hüttenmännischen Tages 24. bis 27. Mai 1961 in Freiberg*. Freiberger Forschungsheften, D 42 (Berlin: Akademie Verlag, 1962): pp. 77–86.

Baumgärtel, Hans. *Bergbau und Absolutismus: der sächsische Bergbau in der zweiten Hälfte des 18. Jahrhunderts und Massnahmen zu seiner Verbesserung nach dem Siebenjährigen Kriege* (Leipzig: Deutscher Verlag für Grundstoffindustrie, 1963).

Baumgärtel, Hans. 'Von Bergbüchlein zur Bergakademie: Zur Entstehung der Bergbauwissenschaften zwischen 1500 und 1765/1770'. In *Freiberger Forschunshefte*, vol. D50 (Berlin: Akademie Verlag, 1965).

Bennett, James Arthur. 'Practical Geometry and Operative Knowledge'. *Configurations* 6(2) (1998): pp. 195–222. https://doi.org/10.1353/con.1998.0010.

Benoit, Paul. *La mine de Pampailly: XV^e–XVIII^e siècles, Brussieu-Rhône* (Lyon: Service régional de l'archéologie de Rhône-Alpes, 1997).

Beretta, Marco. 'Humanism and Chemistry: The Spread of Georgius Agricola's Metallurgical Writings'. *Nuncius* 12(1) (1997): pp. 17–47. https://doi.org/10.1163/182539197X00023.

Berhardt, Wilhelm. *Philipp Melanchthon als Mathematiker und Physiker* (Wittenberg: Verlag für Heimatskunde, 1865).

Berkel, Klaas van. 'Stevin and the Mathematical Practitioners, 1580–1620'. In *A History of Science in the Netherlands: Survey, Themes and Reference*. Edited by Klaas van Berkel, Albert van Helden, and Lodewijk Palm (Leiden: Brill, 1999): pp. 13–36.

Berndorff, Lothar. '"Und da habe ich müssen mach ihrer sprach reden". Einsichten in die lutherischen Bergmannspredigten des Cyriacus Spengenberg'. In *Martin Luther und der Bergbau im Mansfelder Land: Aufsätze zur Ausstellung*. Edited by Rosemarie Knape(Leipzig: Stiftung Luthergedenkstätten in Sachsen-Anhalt, 2000): pp. 189–203.

Bertucci, Paola. *Artisanal Enlightenment: Science and Mechanical Arts in Old Regime France* (New Haven: Yale University Press, 2017).

Bigg, Charlotte. 'Diagrams'. In *A Companion to the History of Science* (Chichester: Wiley, 2016): pp. 557–571. https://doi.org/10.1002/9781118620762.ch39.

Bigg, Charlotte, David Aubin, and Philipp Felsch. 'Introduction: The Laboratory of Nature – Science in the Mountains'. *Science in Context* 22(3) (2009): pp. 311–321. https://doi.org/10.1017/S0269889709990020.

Birembaut, Arthur. 'L'enseignement de la minéralogie et des techniques minières'. In *Enseignement et diffusion des Sciences en France au XVIIIè siècle*. Edited by René Taton (Paris: Hermann, 1964): pp. 365–418.

Blackbourn, David. *The Conquest of Nature: Water, Landscape and the Making of Modern Germany* (London: Cape, 2006).

Blair, Ann M. *Too Much to Know: Managing Scholarly Information before the Modern Age* (New Haven & London: Yale University Press, 2010).

Blaschke, Karlheinz. *Sachsen im Zeitalter der Reformation, Schriften des Vereins für Reformationsgeschichte* (Gütersloh: Gütersloher Verlagshaus, 1970).

Boettcher, Susan R. 'Martin Luthers Leben in Predigten: Cyriakus Spangenberg und Johannes Mathesius'. In *Martin Luther und der Bergbau im Mansfelder Land: Aufsätze zur Ausstellung* (Leipzig: Stiftung Luthergedenkstätten in Sachsen-Anhalt, 2000): pp. 163–188.

Bogsch, Walter. *Der Marienberger Bergbau seit der zweiten Hälfte des 16. Jahrhunderts: 4 Studien* (Cologne: Böhlau, 1966).

Bönisch, Fritz. 'The Geometrical Accuracy of 16th and 17th Century Topographical Surveys'. *Imago Mundi* 21 (1967): pp. 62–69.

Bönisch, Fritz. *Die erste kursächsische Landesaufnahme: ausgeführt von Matthias Öder und Balthasar Zimmermann von 1586 bis in die Anfangszeit des Dreißigjährigen Krieges* (Leipzig: Sächsische Akademie der Wissenschaften, 2002).

Bornhardt, Wilhelm. *Geschichte des Rammelsberger Bergbaues von seiner Aufnahme bis zur Neuzeit* (Berlin: Preussischer Geologischer Landesanstalt, 1931).

Bourguet, Marie Noëlle, Christian Licoppe, and H. Otto Sibum, eds. *Instruments, Travel and Science: Itineraries of Precision from the Seventeenth to the Twentieth Century* (London: Routledge, 2002).

Bourguet, Marie-Noëlle. 'Lanscape with Numbers. Natural History, Travel and Instruments in the Late Eighteenth and Early Nineteenth Centuries'. In *Instruments, Travel and Science* (London: Routledge, 2002): pp. 96–125.

Bourguet, Marie-Noëlle, and Christian Licoppe. 'Voyages, mesures et instruments : une nouvelle expérience du monde au Siècle des lumières'. *Annales* 52(5) (1997): pp. 1115–1151. https://doi.org/10.3406/ahess.1997.279622.

Braunstein, Philippe. 'Les statuts miniers de l'Europe médiévale'. *Comptes rendus des séances de l'Académie des Inscriptions et Belles-Lettres* 136(1) (1992): pp. 35–56.

Breithaupt, August. *Die Bergstadt Freiberg im Königreich Sachsen* (Freiberg: Craz & Gerlach, 1847).

Brianta, Donata. 'Education and Training in the Mining Industry, 1750–1860: European Models and the Italian Case'. *Annals of Science* 57(3) (2000): pp. 267–300. https://doi.org/10.1080/00033790050074165.

Briggs, Henry. 'The Development of Mine Surveying Methods from Early Times to 1850'. *Transactions of the Institute of Mine Surveyors*, VI (1925): pp. 19–32, 77–83, 120–124.

Broel, Theodor. 'Die markscheiderischen und bergmännischen Arbeiten beim Bau einiger Stollen und Schächte im Harzer Bergbau des Clausthaler Reviers. Ein Beitrag
zur Geschichte des Markscheidewesens'. *Mitteilungen aus dem Markscheidewesen*
62(2) (1955): pp. 37–55.

Brough, Bennett Hooper. *A Treatise on Mine-Surveying* (London: Griffin, 1913).

Brown, Christopher Boyd. *Singing the Gospel: Lutheran Hymns and the Success of the
Reformation* (Harvard: Harvard University Press, 2009).

Brown, Kendall W. *A History of Mining in Latin America: From the Colonial Era to the
Present* (Albuquerque: University of New Mexico Press, 2012).

Bürger, Stefan. 'Die Annaberger St. Annenkirche. Die Besonderheiten ihrer Architektur
als Zeichen für kulturellen Wandel'. In *Das Erzgebirge im 16. Jahrhundert.
Gestaltwandel einer Kulturlandschaft im Reformationszeitalter*. Edited by Martina
Schattkowsky (Leipziger Universitätsverlag, 2013): pp. 353–376.

Büttner, Jochen. 'Shooting with Ink'. In *The Structures of Practical Knowledge* (Cham:
Springer, 2017): pp. 115–166.

Cajori, Florian. 'History of Determinations of the Heights of Mountains'. *Isis, A Journal
of the History of Science Society* 12(3) (1929): pp. 482–514.

Clauss, Herbert, and Siegfried Kube. *Freier Berg und vermessenes Erbe:
Untersuchungen zur Frühgeschichte des Freiberger Bergbaus und zur Entwicklung
des Erbbereitens* (Berlin: Akademie Verlag, 1957).

Cohen, H. Floris. *The Scientific Revolution: A Historiographical Inquiry* (Chicago:
University of Chicago Press, 1994).

Cooter, Roger, and Stephen Pumfrey. 'Separate Spheres and Public Places: Reflections
on the History of Science Popularization and Science in Popular Culture'. *History of
Science* 32(3) (1994): pp. 237–267. https://doi.org/10.1177/007327539403200301.

Cormack, Lesley B. *Charting an Empire: Geography at the English Universities,
1580–1620* (Chicago: University of Chicago Press, 1997).

Cormack, Lesley B. 'Mathematics for Sale: Mathematical Practitioners, Instrument
Makers, and Communities of Scholars in Sixteenth–Century London'. In
*Mathematical Practitioners and the Transformation of Natural Knowledge
in Early Modern Europe*. Edited by Lesley B. Cormack, Steven A. Walton,
and John A. Schuster (Cham: Springer, 2017): pp. 69–85. https://doi
.org/10.1007/978-3-319-49430-2_4.

Cormack, Lesley B., Steven A. Walton, and John A. Schuster, eds. *Mathematical
Practitioners and the Transformation of Natural Knowledge in Early Modern
Europe*. Studies in History and Philosophy of Science 45 (Cham: Springer,
2017).

Danna, Raffaele. 'Figuring Out: The Spread of Hindu-Arabic Numerals in the European
Tradition of Practical Mathematics (13th–16th Centuries)'. *Nuncius* 36(1) (2021):
pp. 5–48. https://doi.org/10.1163/18253911-bja10004.

Darmstaedter, Ernst. *Berg-, Probir-und Kunstbüchlein* (Munich: Münchner Drucke, 1926).

De Munck, Bert, and Antonella Romano. *Knowledge and the Early Modern City. A
History of Entanglements Knowledge Societies in History* (London & New York:
Routledge, 2020).

De Vries, Jan. *Economy of Europe in an Age of Crisis, 1600–1750* (Cambridge & New
York: Cambridge University Press, 1976).

Dear, Peter. *Discipline and Experience: The Mathematical Way in the Scientific
Revolution* (Chicago: University of Chicago Press, 1995).

Demeulenaere-Douyère, Christiane, and David Sturdy, eds. *L'enquête du Régent, 1716–1718: sciences, techniques et politique dans la France pré-industrielle* (Turnhout, Belgique: Brepols, 2008).

Dennert, Herbert. *Bergbau und Hüttenwesen im Harz: vom 16.–19. Jh. dargestellt in Lebensbildern führender Persönlichkeiten* (Clausthal-Zellerfeld: Pieper, 1986).

Denzel, Markus A. 'Die Bedeutung der Rechenmeister für die Professionalisierung in der oberdeutschen Kaufmannschaft des 15./16. Jahrhunderts'. In *Verfasser und Herausgeber mathematischer Texte der frühen Neuzeit, Schriften des Adam-Ries-Bundes*, vol. 14 (Freiberg: TU Bergakademie Freiberg, 2002): pp. 23–30.

Déprez-Masson, Marie-Claude. 'Richesse minières et traités techniques : le *De re metallica* de Georg Bauer Agricola'. In *Normes et pouvoir à la fin du Moyen Âge* (Montreal: Céres, 1989).

Déprez-Masson, Marie-Claude. *Technique, mot et image: le De re metallica d'Agricola* (Turnhout: Brepols, 2006).

Dietrich, Richard. *Untersuchungen zum Frühkapitalismus im mitteldeutschen Erzbergbau und Metallhandel* (Hildesheim: Olms, 1991).

Dolz, Wolfram. *Genau messen, Herrschaft verorten: das Reisgemach von Kurfürst August, ein Zentrum der Geodäsie und Kartographie* (Berlin & Munich: Staatliche Kunstsammlungen Dresden und Deutscher Kunstverlag, 2010).

Dolz, Wolfram, and Klaus Schillinger. 'Markscheideinstrumente in den sächsischen kurfürstlichen Sammlungen im Spiegel handschriftlicher Inventare und Kataloge'. *Berichte der Geologischen Bundesanstalt* 35 (1996): pp. 93–97.

Drissen, Alfred. *Das Sprachgut des Markscheiders* (Recklinghausen: Schürmann und Klagges, 1939).

Dubourg Glatigny, Pascal, and Hélène Vérin. *Réduire en art: la technologie de la Renaissance aux Lumières* (Paris: Maison des sciences de l'homme, 2008).

Dumasy-Rabineau, Juliette. 'La vue, la preuve et le droit : les vues figurées de la fin du Moyen Âge'. *Revue historique*, 668(4) (2013): pp. 805–831.

Duris, Pascal. 'Faire feu de tout bois. Ou l'historien de la biologie au travail'. In *Méthode et Histoire. Quelle histoire font les historiens des sciences et des techniques?* (Paris: Classiques Garnier, 2013): pp. 233–243.

Dym, Warren Alexander. *Divining Science Treasure Hunting and Earth Science in Early Modern Germany* (Leiden & Boston: Brill, 2011).

Eamon, William. *Science and the Secrets of Nature: Books of Secrets in Medieval and Early Modern Culture* (Princeton: Princeton University Press, 1994).

Eisenstein, Elizabeth L. *The Printing Press as an Agent of Change* (Cambridge: Cambridge University Press, 1980).

Epple, Moritz, Tinne Hoff Kjeldsen, and Reinhard Siegmund-Schultze. 'From "Mixed" to "Applied" Mathematics: Tracing an Important Dimension of Mathematics and Its History'. *Oberwolfach Reports* 10(1) (2013): pp. 657–733. https://doi .org/10.4171/OWR/2013/12.

Ermisch, Hubert. *Das sächsische Bergrecht des Mittelalters* (Leipzig: Giesecke & Devrient, 1887).

Ernsting, Bernd, ed. *Georgius Agricola, Bergwelten 1494–1994, Veröffentlichungen aus dem Deutschen Bergbau-Museum Bochum 55* (Essen: Edition Glückauf, 1994).

Feldman, Theodore S. 'Applied Mathematics and the Quantification of Experimental Physics: The Example of Barometric Hypsometry'. *Historical Studies in the Physical Sciences* 15(2) (1985): pp. 127–195. https://doi.org/10.2307/27757551.

Felten, Sebastian. 'Mining Culture, Labour, and the State in Early Modern Saxony'. *Renaissance Studies* 34(1) (2020): pp. 119–148. https://doi.org/10.1111/rest.12583.

Fessner, Michael, and Christoph Bartels. 'Von der Krise am Ende des 16. Jahrhunderts zum deutschen Bergbau im Zeitalter des Merkantilismus'. In *Geschichte des deutschen Bergbaus*, vol. 1 (Münster: Aschenforff Verlag, 2012): pp. 453–590.

Flachenecker, Helmut. 'Zwischen Universalität und Spezialisierung: Agricola als Humanist'. In *Georgius Agricola. Bergwelten 1494–1994*. Edited by Bernd Ernsting (Essen: Glückauf Verlag, 1994): pp. 101–103.

Fors, Hjalmar. *The Limits of Matter: Chemistry, Mining, and Enlightenment* (Chicago: University of Chicago Press, 2015).

Fleetwood, Lachlan. '"No Former Travellers Having Attained Such a Height on the Earth's Surface": Instruments, Inscriptions, and Bodies in the Himalaya, 1800–1830'. *History of Science*, 56(1) (2018): pp. 3–34. https://doi.org/10.1177/0073275317732254.

Flood, Raymond. *The History of Mathematical Tables: From Sumer to Spreadsheets* (Oxford: Oxford University Press, 2003).

Folkerts, Menso. 'Die Mathematik im sächsisch-thüringischen Raum im 15. und 16. Jahrhundert'. In *Kaufmanns- Rechenbücher und mathematische Schriften der frühen Neuzeit, Schriften des Adam-Ries-Bundes*, vol. 22 (Freiberg: TU Bergakademie Freiberg, 2011): pp. 1–22.

Folkerts, Menso, Eberhard Knobloch, and Karin Reich. *Maß, Zahl und Gewicht: Mathematik als Schlüssel zu Weltverständnis und Weltbeherrschung*, Ausstellungskataloge der Herzog August Bibliothek (Weinheim: Acta Humaniora, 1989).

Fors, Hjalmar, and Jacob Orrje. 'Describing the World and Shaping the Self: Knowledge-Gathering, Mobility and Spatial Control at the Swedish Bureau of Mines'. In *Transnational Cultures of Expertise*. Edited by Lothar Schilling and Jakob Vogel (Berlin, Boston: De Gruyter, 2019): pp. 107–128. https://doi.org/10.1515/9783110553734-007.

Frängsmyr, Tore, J. L. Heilbron, and Robin E. Rider. *The Quantifying Spirit in the 18th Century* (Berkeley: University of California Press, 1990).

Fritzsch, Karl Ewald. 'Die Schönbergporträts der Bergakademie Freiberg'. *Bergakademie* 14 (1962): pp. 311–317.

Garçon, Anne-Françoise. 'Réduire la mine en science … ? : Anatomie des *De re metallica* d'Agricola (1528–1556)'. In *Réduire en art: La technologie de la Renaissance aux Lumières*. Edited by Hélène Vérin and Pascal Dubourg Glatigny (Paris: Éditions de la Maison des sciences de l'homme, 2008): pp. 317–336.

Gätzschmann, Moritz Ferdinand. 'Zur Geschichte der Wassersäulenmaschinen in Sachsen'. *Berg- und Hüttenmännische Zeitung* 32 (1–2) (1873): pp: 1–5; 13–16.

Gautier Dalché, Patrick, and Armelle Querrien. 'Mesure du sol et géométrie au Moyen Âge'. *Archives d'histoire doctrinale et littéraire du Moyen Âge* 82(1) (2015): pp. 97–139.

Gebhardt, Rainer, ed. *Die Annaberger Brotordnung von Adam Ries: kommentierte und bearbeitete Faksimileausgabe der 1533 erstellten und 1536 gedruckten Brotordnung* (Annaberg-Buchholz: Adam-Ries-Bund, 2004).

Gebhardt, Rainer, ed. *Zur Wirkungsgeschichte der Brotordnung von Adam Ries, Schriften des Adam-Ries-Bundes Annaberg-Buchholz*, vol. 18 (Freiberg: TU Bergakademie Freiberg, 2006).

Gille, Bertrand. *Les ingénieurs de la Renaissance* (Paris: Hermann, 1964).

Gingerich, Owen. 'The Role of Erasmus Reinhold and the Prutenic Tables in the Dissemination of the Copernican Theory'. In *Colloquia Copernicana*2 (1973): pp. 43–62; 123–126.

Gleue, Axel W. *Wie kam das Wasser auf die Burg?: vom Brunnenbau auf Höhenburgen und Bergvesten* (Regensburg: Schnell & Steiner, 2008).

Goldenbaum, Ursula. 'The Geometrical Method as a New Standard of Truth, Based on the Mathematization of Nature'. In *The Language of Nature: Reassessing the Mathematization of Natural Philosophy in the Seventeenth Century*. Edited by Geoffrey Gorham, Benjamin Hill, Edward Slowik, and C. Kenneth Waters (Minneapolis: University of Minnesota Press, 2017): pp. 274–307.

Goldstein, Catherine. 'Les fractions décimales : un art d'ingénieur ?' In *Penser la technique autrement XVIᵉ-XXIᵉ siècle. En hommage à l'œuvre d'Hélène Vérin*, edited by Robert Carvais, Anne-Françoise Garçon, and André Grelon (Paris: Classique Garnier, 2017): pp. 185–203. https://hal.archives-ouvertes.fr/hal-00734932.

Graulau, Jeannette. *The Underground Wealth of Nations: On the Capitalist Origins of Silver Mining, A.D. 1150–1450* (New Haven & London: Yale University Press, 2019).

Haasbroek, N. D. *Gemma Frisius, Tycho Brahe and Snellius and their triangulations* (Delft, Netherlands: Rijkscommissie voor Geodesie, 1968).

Habashi, Fathi. 'Christlieb Ehregott Gellert and His Metallurgic Chymistry'. *Bulletin for the History of Chemistry* 24 (1999): 32–39.

Hadden, Richard W. *On the Shoulders of Merchants: Exchange and the Mathematical Conception of Nature in Early Modern Europe* (Albany: State University of New York Press, 1994).

Hake, Hardanus, and Heinrich Denker. *Die Bergchronik des Hardanus Hake, Pastors zu Wildemann* (Schaan/Liechtenstein: Sändig, 1981).

Hall, Bert. 'Der meister sol auch kennen schreiben und lesen : Writings about Technology ca. 1400–ca. 1600 and Their Cultural Implications'. In *Early Technologies, Invited Lectures on the Middle East at the University of Texas at Austin* (1979): pp. 47–58.

Halleux, Robert. *Le savoir de la main. Savants et artisans dans l'Europe pré-industrielle* (Paris: Armand Colin, 2009).

Hannaway, Owen. 'Georgius Agricola as Humanist'. *Journal of the History of Ideas* 53(4) (1992): pp. 553–560. https://doi.org/10.2307/2709936.

Hannaway, Owen. 'Reading the Pictures: The Context of Georgius Agricola's Woodcuts'. *Nuncius* 12(1) (1997): pp. 49–66.

Heilbron, John L., and René Sigrist, eds. *Jean-Andre Deluc: Historian of Earth and Man* (Geneve: Slatkine, 2011).

Herrmann, Walther. 'Bergbau und Kultur : Beiträge zur Geschichte des Frieberger Bergbaus und der Bergakademie'. In *Freiberger Forschunshefte, Kultur und Technik*, vol. D2 (Berlin: Akademie Verlag, 1953).

Herrmann, Walther. 'Ein bergmännisches Gedicht aus der Barockzeit Freiberg'. *Sächsische Heimatblätter* 17 (1971): pp. 207–212.

Hilaire-Pérez, Liliane. 'Transferts technologiques, droit et territoire: le cas francoanglais au XVIIIe siècle'. *Revue d'histoire moderne contemporaine* 44(4) (1997): pp. 547–579.

Hilaire-Pérez, Liliane, Stéphane Blond, and Michèle Virol, eds. *Les ingénieurs, des intermédiaires ? Transmission des connaissances et coopération chez les ingénieurs (Europe, XVe–XVIIIe siècle)* (Toulouse: Presses Universitaires du Midi, 2022).

Hillegeist, Hans-Heinrich. 'Auswanderungen Oberharzer Bergleute nach Kongsberg/Norwegen im 17. und 18. Jahrhundert'. In *Technologietransfer und Auswanderungen im Umfeld des Harzer Montanwesens* (Lukas Verlag: Berlin, 2001): pp. 9–48.

Hooykaas, Reijer. *Religion and the Rise of Modern Science* (Edinburgh: Scottish Academic Press, 1972).

Hooykaas, Reyer. *Humanisme, science et réforme: Pierre de La Ramée (1515–1572)* (Leiden: Brill, 1958).

Hoppe, Brigitte. 'Die Vernetzung der mathematisch ausgerichteten Anwendungsgebiete mit den Fächern des Quadriviums in der frühen Neuzeit'. In *Der 'mathematicus'. Zur Entwicklung und Bedeutung einer neuen Berufsgruppe in der Zeit Gehard Mercators, Duisburger Mercator-Studien* 4. Edited by Hantsche Irmgard (Bochum: Brockmeyer, 1996): pp. 1–33.

Horst, Thomas. 'Kartographie und Grundstückseigentum in der Frühen Neuzeit'. *ZfV – Zeitschrift für Geodäsie, Geoinformation und Landmanagement* 6 (2014): pp. 369–376. https://doi.org/10.12902/zfv-0035-2014.

Horst, Thomas. 'Alpine Grenzen auf frühneuzeitlichen Manuskriptkarten unter besonderer Berücksichtigung der Augenscheinkarten'. In *Grenzen – Frontières Geschichte der Alpen – Histoire des Alpes – Storia delle Alpi* 23 (Zurich: Chronos, 2018): pp. 49–70.

Horst, Ulrich. 'Leibniz und der Bergbau'. *Der Anschnitt* 18(5) (1966): pp. 36–51.

Hübner, Marita. *Jean André Deluc (1727–1817): Protestantische Kultur und moderne Naturforschung* (Göttingen: Vandenhoeck & Ruprecht, 2010).

Iwańczak, Józef. *Die Kartenmacher: Nürnberg als Zentrum der Kartographie im Zeitalter der Renaissance* (Darmstadt: WBG, 2009).

Jardine, Boris. 'Instruments of Statecraft: Humphrey Cole, Elizabethan Economic Policy and the Rise of Practical Mathematics'. *Annals of Science* 75(4) (2018): pp. 304–329. https://doi.org/10.1080/00033790.2018.1528510.

Jeannin, Pierre. *Les marchands au XVIe siècle* (Paris: Le Seuil, 1957).

Jeannin, Pierre. *Marchands d'Europe: pratiques et savoirs à l'époque moderne* (Paris: Editions Rue d'Ulm-ENS, 2002).

Jobst, Wolfgang, and Walter Schellhas. *Abraham von Schönberg, Leben und Werk. Die Wiederbelebung des Erzgebirgischen Bergbaus nach dem dreißigjährigen Krieg durch Oberghauptmann Abraham von Schönberg*. Freiberger Forschunshefte, vol. D 198 (Leipzig & Stuttgart: Verlag für Grundstoffindustrie, 1994).

Johnston, Stephen. *'Making Mathematical Practice: Gentlemen, Practitioners and Artisans in Elizabethan England'* (PhD, University of Cambridge, 1994).

Kaden, Herbert. *Das sächsische Bergschulwesen: Entstehung, Entwicklung, Epilog* (Cologne: Böhlau, 2012).

Kamenicky, Miroslav. *Banícke školstvo na Slovensku do založenia Baníckej akadémie v Banskej Štiavnici* (Bratislava: Slovak Academic Press, 2006).

Kasiarova, Elena. *Bergbau- und Hüttenwesenvergangenheit der Slovakei in kartographischen Quellen* (Kosice: Banska Agentura, 2010).

Kasper, Hanns-Heinz. 'Das Rechnungswesen im kursächsischen Hüttenwesen im 16. Jahrhundert'. In *Adam Ries und seine Zeit* (Berlin, Akademie der Wissenschaften der DDR, 1984): pp. 67–73.

Kaufhold, Karl Heinrich, and Wilfried Reininghaus. *Stadt und Bergbau* (Böhlau Verlag: Köln & Weimar, 2004).

Kaunzner, Wolfgang. 'Johannes Widmann, Cossist und Verfasser des ersten großen deutschen Rechenbuches'. In *Rechenmeister und Cossisten der frühen Neuzeit, Schriften des Adam-Ries-Bundes* vol. 7 (Freiberg: TU Bergakademie, 1996): pp. 37–51.

Keller, Alex. 'Renaissance Theaters of Machines'. *Technology and Culture* 19(3) (1978): pp. 495–508. https://doi.org/10.2307/3103380.

Kessler-Slotta, Elisabeth. 'Die Illustrationen in Agricolas *De Re Metallica*', *Der Anschnitt*, XLVI(2–3) (1994): pp. 55–64.

Kirchhoff, Alfred. 'Matthias Öder grosses Kartenwerk über Kursachsen aus der Zeit um 1600'. *Neues Archiv Für Sächsische Geschichte* 11 (1890): pp. 319–332.

Kirnbauer, Frank. 'Die Entwicklung des Grubenrißwesens in Österreich'. *Blätter Für Technikgeschichte* 24 (1962): pp. 60–129.

Klein, Ursula. 'Ein Bergrat, zwei Minister und sechs Lehrende. Versuche der Gründung einer Bergakademie in Berlin um 1770'. *NTM Zeitschrift für Geschichte der Wissenschaften, Technik und Medizin* 18(4) (2010): pp. 437–468.

Klein, Ursula. *Nützliches Wissen. Die Erfindung der Technikwissenschaften* (Göttingen: Wallstein, 2016).

Klein, Ursula. 'Hybrid Experts'. In *The Structures of Practical Knowledge*. Edited by Matteo Valleriani (Cham: Springer, 2017): pp. 287–306. https://doi.org/10.1007/978-3-319-45671-3_11.

Klein, Ursula. *Technoscience in History: Prussia, 1750–1850* (Cambridge, MA: MIT Press, 2020).

Klinger, Kerrin, and Thomas Morel. 'Was ist praktisch am mathematischen Wissen? Die Positionen des Bergmeisters J. A. Scheidhauer und des Baumeisters C. F. Steiner in der Zeit um 1800'. *NTM Zeitschrift für Geschichte der Wissenschaften, Technik und Medizin* 26(3) (2018): pp. 267–299. https://doi.org/10.1007/s00048-018-0197-8.

Knape, Rosemarie, ed. *Martin Luther und der Bergbau im Mansfelder Land: Aufsätze zur Ausstellung* (Leipzig: Stiftung Luthergedenkstätten in Sachsen-Anhalt, 2000).

Knape, Rosemarie, ed. *Martin Luther und der Bergbau im Mansfelder Land: Rundgang durch die Ausstellung* (Lutherstadt Eisleben: Stiftung Luthergedenkstätten in Sachsen-Anhalt, 2000).

Knothe, Christian. 'Die Dresdener Forstzeichenbücher und Waldrisse-Vermessungstechnische Dokumente des 16. Jahrhunderts'. *Mitteilungsblatt DVW Sachsen* 8(1) (1998): pp. 6–17.

Koch, Manfred. *Geschichte und Entwicklung des bergmännischen Schrifttums, Schriftenreihe Bergbau-Aufbereitung* (Goslar: Hübener, 1963).

Koeppe, Wolfram, ed. *Making Marvels: Science and Splendor at the Courts of Europe* (New York: Metropolitan Museum of Art, 2019).

Kolb, Hans Emil. 'Gotthard's Authentische Beschreibung von dem Merkwürdigen Bau des Tiefen Georg-Stollens'. In *200 Jahre Tiefer Georg-Stollen* (Arbeitsgemeinschaft Harzer Montangeschichte, 1999).

Konečný, Peter. 'Die Montanistische Ausbildung in Der Habsburgermonarchie, 1763–1848'. In *Staat, Bergbau und Bergakademie: Montanexperten im 18. und frühen 19. Jahrhundert* (Stuttgart: Steiner, 2013): p. 95–124.

Konečný, Peter. 'Kameralisten, Bildungsreformer und aufstrebende Bergbeamte: Johann Thaddäus Anton Peithner und sein Vorschlag zur Gründung einer Bergakademie im Habsburgerreich'. *Montánna história* 8–9 (2015): pp. 66–89.

Korey, Michael. *The Geometry of Power: Mathematical Instruments and Princely Mechanical Devices from around 1600 in the Mathematisch-Physikalischer Salon* (Berlin: Deutsche Kunstverlag, 2007).

Koser, Reinhold. *Über eine Sammlung von Leibniz-Handschriften im Staatsarchiv zu Hannover, Sitzungsberichte der Preußischen Akademie der Wissenschaften* (Berlin: De Gruyter, 1902).

Kratzsch, Klaus. *Bergstädte des Erzgebirges: Städtebau und Kunst zur Zeit der Reformation* (Munich: Schnell und Steiner, 1972).

Krumm, Markus. 'Visualisierung als Problem. Konflikte zwischen Vermessungsexperten am Haller Salzberg vor Erfindung der Grubenkarte 1531'. *Forum Hall in Tirol. Neues zur Geschichte der Stadt* 3 (2012): pp. 294–313.

Kühn, Heidi. 'Die Mathematik im deutschen Hochschulwesen des 18. Jahrhunderts (unter besonderer Berücksichtigung der Verhältnisse an der Leipziger Universität)' (PhD, Karl-Marx-Universität Leipzig, 1988).

Kühne, Andreas. 'Augustin Hirschvogel und sein Beitrag zur praktischen Mathematik'. In *Verfasser und Herausgeber mathematischer Texte der frühen Neuzeit, Schriften des Adam-Ries-Bundes*, vol. 14 (Freiberg: TU Bergakademie, 2002): pp. 237–251.

Kula, Witold. *Measures and Men* (Princeton: Princeton University Press, 1986).

Küpker, Markus. 'Manufacturing'. In *The Oxford Handbook of Early Modern European History, 1350–1750*, vol. 1 (Oxford: Oxford University Press, 2015): pp. 509–542.

Laube, Adolf. *Studien über den erzgebirgischen Silberbergbau von 1470 bis 1546* (Berlin: Akademie Verlag, 1974).

Lefèvre, Wolfgang. 'Picturing the World of Mining in the Renaissance: The Schwazer Bergbuch (1556)'. *Preprint of the Max-Planck-Institut für Wissenschaftsgeschichte* 407 (Berlin, 2010).

Lefèvre, Wolfgang. 'Architectural Knowledge'. In *The Structures of Practical Knowledge* (Cham: Springer, 2017): pp. 247–270.

L'Huillier, Hervé. 'Practical Geometry in the Middle Ages and the Renaissance'. In *Companion Encyclopedia of the History and Philosophy of the Mathematical Sciences*. Edited by Ivor Grattan-Guinness, vol. 1 (Baltimore: Johns Hopkins University Press, 2003): pp. 185–191.

Liessmann, Wilfried. *Historischer Bergbau im Harz* (Berlin: Springer, 2010).

Lindgren, Uta. 'Astronomische und geodätische Instrumente zur Zeit Peter und Philipp Apians'. In *Philipp Apian und die Kartographie der Renaissance* (Weißenhorn: Konrad, 1989): pp. 43–65.

Lindgren, Uta. 'Land Surveys, Instruments, and Practitioners in the Renaissance'. In *The History of Cartography* vol. 3(1) (Chicago: University of Chicago Press, 2007): pp. 477–508.

Lindgren, Uta. 'Maß, Zahl und Gewicht im alpinen Montanwesen um 1500'. In *Kosmos und Zahl Boethius* (Stuttgart: Franz Steiner Verlag, 2008): pp. 347–364.

Lindqvist, Svante. *Technology on Trial: The Introduction of Steam Power Technology into Sweden, 1715–1736* (Uppsala & Stockholm: Alqvist & Wiksell, 1984).

Lommatzsch, Herbert. 'Petrus Eichholtz und sein "Geistliches Bergwerk"'. *Der Anschnitt* 15(6) (1963): pp. 5–10.

Long, Pamela O. 'The Openness of Knowledge: An Ideal and Its Context in 16th-Century Writings on Mining and Metallurgy'. *Technology and Culture* 32(2) (1991): pp. 318–355. https://doi.org/10.2307/3105713.

Long, Pamela O. *Openness, Secrecy, Authorship: Technical Arts and the Culture of Knowledge from Antiquity to the Renaissance* (Baltimore: Johns Hopkins University Press, 2001).

Long, Pamela O. *Artisan/Practitioners and the Rise of the New Sciences, 1400–1600* (Corvallis: Oregon State University Press, 2011).

Löscher, Hermann. *Das erzgebirgische Bergrecht des 15. und 16. Jahrhunderts* (Berlin: Akademie Verlag, 2003).

Lothar Suhling. *Aufschließen, Gewinnen und Fördern: Geschichte des Bergbaus* (Munich: Deutsches Museum, 1983).

Lüschen, Hans. *Die Namen der Steine: das Mineralreich im Spiegel der Sprache* (Thun: Ott, 1968).

Lutz, Robert Hermann. *Wer war der gemeine Mann?* (Munich: Oldenbourg, 1979).

Marr, Alexander. 'Copying, Commonplaces, and Technical Knowledge: The Architect-Engineer as Reader'. In *The Artist as Reader: On Education and Non-Education of Early Modern Artists* (Leiden: Brill, 2013): pp. 421–446.

Matthes, Erich. *Das Erste Bergbuch von St. Joachimsthal 1518–1520*, Die Fundgrube. Eine Sammlung genealogischen *Materials* vol. 33 (Regensburg: Korb'sches Sippenarchiv, 1965).

McLean, Matthew. *The Cosmographia of Sebastian Münster: Describing the World in the Reformation,* St Andrews Studies in Reformations History (Burlington: Ashgate, 2013).

Meixner, Heinz, Walter Schellhas, Peter Schmidt. *Balthasar Rösler: Persönlichkeit und Wirken für den Bergbau des 17. Jahrhunderts: Kommentarband zum Faksimiledruck 'Hell-polierter Berg-Bau-Spiegel'* (Leipzig: Deutscher Verlag für Grundstoffindustrie, 1980).

Merrill, Elizabeth M. 'Pocket-Size Architectural Notebooks and the Codification of Practical Knowledge'. In *The Structures of Practical Knowledge* (Cham: Springer, 2017): pp. 21–54.

Métin, Frédéric. 'La fortification géométrique de Jean Errard et l'école française de fortification (1550–1650)' (PhD, Université de Nantes, 2016).

Michel, H. 'Boussoles de Mines des XVIe et XVIIe siècles'. *Ciel et Terre* 72 (1956): pp. 617–631.

Middleton, W. E. Knowles. *The History of the Barometer* (Baltimore: Johns Hopkins University Press, 1964).

Mihalovits, Johann. 'Die Gründung der ersten Lehranstalt für technische Bergbeamte in Ungarn'. In *Historia eruditionis superioris rerum metallicarum et saltuariarum in Hungaria 1735–1935*, vol. 1 (Sopron: Universitatis Regiae Hungaricae, 1938): pp. 4–31.

Moran, Bruce T. 'Princes, Machines and the Valuation of Precision in the 16th Century'. *Sudhoffs Archiv* 61(3) (1977): pp. 209–228.

Moran, Bruce T. 'German Prince-Practitioners: Aspects in the Development of Courtly Science, Technology, and Procedures in the Renaissance'. *Technology and Culture* 22(2) (1981): 253–274. https://doi.org/10.2307/3104900.

Morel, Thomas. 'Mathématiques et politiques scientifiques en Saxe (1765–1851). Institutions, acteurs et enseignements' (PhD, Université de Bordeaux, 2013).

Morel, Thomas. 'Le microcosme de la géométrie souterraine : échanges et transmissions en mathématiques pratiques'. *Philosophia Scientiae* 19(2) (2015): pp. 17–36. https://doi.org/doi.org/10.4000/philosophiascientiae.1089.

Morel, Thomas. 'Circulating Mining Knowledge from Freiberg to Almaden: The Life and Career of J.M.Hoppensack (1741–1815)' (Workshop, Paris, July 2016).

Morel, Thomas. 'Usefulness and Practicability of Mathematics: The German Mining Academies in the 18th Century'. In *Preprint of the Max-Planck-Institut für Wissenschaftsgeschichte 481* (Berlin, 2016): pp. 49–62.

Morel, Thomas. 'Bringing Euclid into the Mines. Classical Sources and Vernacular Knowledge in the Development of Subterranean Geometry'. In *Translating Early Modern Science*, Intersections 51 (Brill: Leiden, 2017): pp. 154–181.

Morel, Thomas. 'Subterranean Geometry and Its Instruments: About Practical Geometry in the Early Modern Period'. In *Oberwolfach Reports 58* (Oberwolfach: Mathematisches Forschungsinstitut Oberwolfach, 2017): pp. 80–83.

Morel, Thomas. 'Five Lives of a Geometria subterranea (1708–1785). Authorship and Knowledge Circulation in Practical Mathematics'. *Revue d'histoire des mathématiques* 24(2) (2018): pp. 207–258.

Morel, Thomas. 'Mathematics and Technological Change: The Silent Rise of Practical Mathematics'. In *Bloomsbury Cultural History of Mathematics*, vol. 3 (London: Bloomsbury, 2023).

Morel, Thomas, Giuditta Parolini, and Cesare Pastorino, eds. *The Making of Useful Knowledge, Preprint of the Max Planck Institute for the History of Science 481* (Berlin, 2016).

Morera, Raphael. 'Maîtriser l'eau et gagner des terres au XVIe siècle. La poldérisation au prisme du *Tractaet van Dyckagie* d'Andries Vierlingh'. *Artefact* 4 (2016): pp. 149–160.

Mukerji, Chandra. *Impossible Engineering: Technology and Territoriality on the Canal Du Midi* (Princeton: Princeton University Press, 2009).

Mumford, Lewis. *Technics and Civilization* (Chicago: University of Chicago Press, 2010).

Munck, Bert de, Steven Laurence Kaplan, and Hugo Soly. *Learning on the Shop Floor: Historical Perspectives on Apprenticeship* (New York: Berghahn Books, 2007).

Nasifoglu, Yelda. 'Reading by Drawing: The Changing Nature of Mathematical Diagrams in Seventeenth-Century England'. In *Reading Mathematics in Early Modern Europe* (New York: Routledge, 2020).

Nehm, Walter. 'Die ersten Ansätze des Markscheidewesens auf dem Rammelsberg'. *Mitteilungen aus dem Markscheidewesen* 44 (1933): pp. 79–88.

Nehm, Walter. 'Georg Öder und seine markscheiderische Tätigkeit auf dem Rammelsberg'. *Neues Archiv für sächsische Geschichte* 55 (1934): pp. 64–72.

Nehm, Walter. 'Materialen zur Geschichte des sächsischen Markscheidewesen A: Über die Anfänge der Rißlichen Darstellungen- B: Die Zeit Balthasar Rößlers und Nikolaus Voigtels'. *Mitteilungen aus dem Markscheidewesen* 58 (1951): pp. 18–23; 124–137.

Nuria, Valverde Pérez. 'Underground Knowledge: Mining, Mapping and Law in Eighteenth-Century Nueva España'. In *The Globalization of Knowledge in the Iberian Colonial World*. Edited by Helge Wendt (Berlin: Edition Open Access, 2016): pp. 227–258.

Orr, Clarissa Campbell. 'Queen Charlotte as Patron: Some Intellectual and Social Contexts'. *The Court Historian* 6(3) (2001): pp. 183–212. https://doi.org/10.1179/cou.2001.6.3.001.

Orrje, Jacob. 'Mechanicus: Performing an Early Modern Persona' (PhD, Uppsala Universitet, 2015).

Peters, Gunthild. *Zwei Gulden vom Fuder. Mathematik der Fassmessung und praktisches Visierwissen im 15. Jahrhundert*, Boethius 69 (Stuttgart: Steiner, 2018).

Pfläging, Kurt and Westfalia (Lünen). *Bergbuch Massa Marittima: 1225–1335 = Constitutum comunis et populi civitatis Massae* (Westfalia: Lünen, 1977).

Pitz, Ernst. *Landeskulturtechnik, Markscheide- und Vermessungswesen im Herzogtum Braunschweig bis zum Ende des 18. Jahrhunderts* (Göttingen: Vandenhoeck & Ruprecht, 1967).

Popplow, Marcus. 'Knowledge Management to Exploit Agrarian Resources as Part of Late-Eighteenth-Century Cultures of Innovation: Friedrich Casimir Medicus and Franz von Paula Schrank'. *Annals of Science* 69(3) (2012): pp. 413–433.

Porter, Roy. 'Introduction'. In *Science, Culture and Popular Belief in Renaissance Europe* (Manchester: Manchester University Press, 1991): pp. 1–15.

Pounds, Norman John Greville. *An Economic History of Medieval Europe* (London: Routledge, 1994).

Pumfrey, Stephen, Paolo L. Rossi, and Maurice Slawinski, eds. *Science, Culture and Popular Belief in Renaissance Europe* (Manchester: Manchester University Press, 1991).

Ranke, Leopold von. *Geschichten der romanischen und germanischen Völker: von 1494 bis 1535*, vol. 1 (Leipzig: Reimer, 1824).

Raphael, Renée. 'Producing Knowledge about Mercury Mining: Local Practices and Textual Tools' *Renaissance Studies* 34(1) (2020): pp. 95–118. https://doi .org/10.1111/rest.12584.

Rapp, Francis. *Les origines médievales de l'Allemagne moderne: de Charles IV à Charles Quint (1346–1519)* (Paris: Aubier, 1989).

Ravier, Benjamin. 'Voir et concevoir: les théâtres de machines (XVIe–XVIIIe siècles)' (PhD, Université Paris 1, 2013).

Reich, Ulrich. 'Philipp Melanchthon (1497–1560) und sein Einsatz für die Mathematik'. In *Rechenmeister und Mathematiker der Frühen Neuzeit, Schriften Des Adam-Ries-Bundes Annaberg-Buchholz* vol. 25 (Annaberg: TU Bergakademie Freiberg, 2017): pp. 1–14.

Reichert, Frank. 'Zur Geschichte der Feststellung und Kennzeichnung von Eigentums–und Herrschaftsgrenzen in Sachsen' (Diplomarbeit Technische, Universität Dresden, 1999).

Reichert, Frank. 'Die Kurfürstlich-sächsischen Markscheider Georg Öder die Jüngeren sen. und jun.' In *Fürstliche Koordinaten. Landesvermessung und Herrschaftsvisualisierung um 1600, Schriften zur Sächsischen Geschichte und Volkskunde* vol. 46 (Leipzig: Leipziger Universitätsverlag, 2014): pp. 147–188.

Reimers, Toni. 'Wurzeln des Markscheidewesens im Spiegel gelehrter Schriften: Eine mathematikhistorisch-bibliographische Analyse'. *Siegener Beiträge zur Geschichte und Philosophie der Mathematik* vol. 14 (2021): pp. 93–127.

Reiss, Timothy J. *Knowledge, Discovery and Imagination in Early Modern Europe: The Rise of Aesthetic Rationalism, Cambridge Studies in Renaissance Literature and Culture* (Cambridge: Cambridge University Press, 1997). https://doi.org/10.1017/ CBO9780511549465.

Reith, Reinhold. 'Know-How, Technologietransfer und die *Arcana Artis* im Mitteleuropa der Frühen Neuzeit'. *Early Science and Medicine* 10(3) (2005): pp. 349–377. https://doi.org/10.1163/1573382054615451.

Ridder-Symoens, Hilde de, and Walter Rüegg. *A History of the University in Europe: Universities in Early Modern Europe (1500–1800)*, vol. 2 (Cambridge: Cambridge University Press, 1996).

Riedel, Lothar. 'Adam und Johann Adam Schneider, Zwei Markscheidergenerationen in Marienberg'. *Sächsische Heimatblätter* 52 (3) (2006): pp. 201–209.

Roberts, Lissa, Simon Schaffer, and Peter Robert Dear, eds. *The Mindful Hand: Inquiry and Invention from the Late Renaissance to Early Industrialisation* (Amsterdam: Koninklijke Nederlandse Akademie van Wetenschappen, 2007).

Rochhaus, Peter. 'Adam Ries und die Annaberger Rechenmeister zwischen 1500 und 1604'. In *Rechenmeister und Cossisten der frühen Neuzeit*. Edited by Rainer Gebhardt (Annaberg: Schriften des Adam-Ries-Bundes 7, 1996: pp. 95–106.

Rochhaus, Peter. 'Zu den Rechenmeistern im sächsischen Erzgebirge während des 17. und 18. Jahrhunderts am Beispiel der Städte Annaberg, Johanngeorgenstadt und Schneeberg'. In *Verfasser und Herausgeber mathematisches Texte der frühen Neuzeit*. Edited by Rainer Gebhardt and Schriften des Adam-Ries-Bundes, vol. 14 (Annaberg: TU Bergakademie Freiberg, 2002): pp. 343–352.

Rodríguez, Roberto Matías. 'Ingenieria Minera Romana'. *II. Congreso de las Obras Públicas Romana* (Tarragona, 2004): pp. 157–189.

Rogge, Jörg. *Die Wettiner: Aufstieg einer Dynastie im Mittelalter* (Ostfildern: Jan Thorbecke Verlag, 2009).

Rosenberg, Gary D. 'The Measure of Man and Landscape in the Renaissance and Scientific Revolution'. In *The Revolution in Geology from the Renaissance to the Enlightenment* (Boulder: Geological Society of America, 2009): pp. 13–40.

Roux, Sophie. 'Forms of Mathematization (14th–17th Centuries)'. *Early Science and Medicine* 15(4–5) (2010): pp. 319–337. https://doi.org/10.1163/1573382 10X516242.

Rüdiger, Bernd. 'Zur Rolle der Lateinschulen bei der Vermittlung mathematikwissenschaftlichen Fortschritts im 16. Jahrhundert an Angehörige ausgewählter städtischer Berufe'. In *Visier- und Rechenbücher der frühen Neuzeit, Schriften des Adam-Ries-Bundes* vol. 19 (Annaberg: TU Bergakademie Freiberg, 2008): pp. 329–340.

Rüdiger, Bernd. 'Abraham Ries (1533–1604) und sein Werk: Angehöriger einer neuen Generation mit den Problemen und Lösungen Ihrer Zeit'. In *Kaufmanns-Rechenbücher und Mathematische Schriften der Frühen Neuzeit, Schriften Des Adam-Ries-Bundes* vol. 22 (Annaberg: TU Bergakademie Freiberg, 2011): pp. 23–36.

Rüdiger, Bernd. 'Adam Ries d. Ä. und dessen Söhne als Rechenmeister und Mathematiker sowie die Wandlungen im 16. Jahrhundert'. In *Rechenmeister und Mathematiker der Frühen Neuzeit*, Schriften Des Adam-Ries-Bundes vol. 25 (Annaberg: TU Bergakademie Freiberg, 2017): pp. 151–170.

Ruge, Sophus. *Geschichte der sächsischen Kartographie im 16. Jahrhundert* (Schauenburg: Lahr, 1881).

Sahmland, Irmtraut. 'Gesundheitsschädigungen der Bergleute: Die Bedeutung der Bergpredigten des 16. bis frühen 18. Jahrhunderts als Quelle arbeitsmedizingeschichtlicher Fragestellungen'. *Medizinhistorisches Journal* 23(3/4) (1988): pp. 240–276.

Schattkowsky, Martina. *Das Erzgebirge im 16. Jahrhundert: Gestaltwandel einer Kulturlandschaft im Reformationszeitalter* (Leipzig: Leipziger Universitätsverlag, 2013).

Schellhas, Walter. *Der Rechenmeister Adam Ries (1492 Bis 1559) und der Bergbau* (Freiberg: Wissenschaftliches Informationszentrum der Bergakademie, 1977).

Schillinger, Klaus. 'Zur Entwicklung der Vermessungsinstrumente im 16. Jahrhundert'. *Sächsische Heimatblätter* 34 (1988): pp. 44–49.

Schillinger, Klaus. 'The Development of Saxon Scientific Instrument-Making Skills from the Sixteenth Century to the Thirty Years War'. *Annals of Science* 47 (1990): pp. 277–289.

Schirmer, Uwe. 'Das spätmittelalterlich-frühneuzeitliche Erzgebirge als Wirtschafts- und Sozialregion (1470–1550)'. In *Das Erzgebirge im 16 Jahrhundert. Gestaltwandel einer Kulturlandschaft im Reformationszeitalter*. Edited by Martina Schattkowsky (Leipzig: Leipziger Universitätsverlag, 2013): pp. 45–76.

Schleiff, Hartmut, and Peter Konečný, eds. *Staat, Bergbau und Bergakademie: Montanexperten im 18. und frühen 19. Jahrhundert* (Stuttgart: Franz Steiner Verlag, 2013).

Schmidt, Ludwig. *Kurfürst August von Sachsen als Geograph: Ein Beitrag zur Geschichte der Erdkunde* (Dresden: Hoffmann, 1898).

Schmidt, Martin. *Die Wasserwirtschaft des Oberharzer Bergbaus* (Neuwied: Neuwieder Verlag, 1989).

Schmidt, Reinhard. 'Die Familie von Schönberg und das sächsische Oberbergamt'. In *Die Geschichte der Familie Von Schönberg* (Nossen: Von Schönberg'sche Stiftung, 2004): pp. 32–66.

Schmidt, Suzanne Karr. 'Pfinzing and Friends: Surveying Culture in Renaissance Nuremberg'. *Nuncius* 35(2) (2020): pp. 364–386. https://doi.org/10.1163/18253911-03502010.

Schneider, Ivo. *Johannes Faulhaber 1580–1635. Rechenmeister in einer Welt des Umbruchs,* Vita Mathematica (Basel: Birkhäuser, 1993). https://doi.org/10.1007/978-3-0348-7274-4.

Schneider, Ivo. 'Ausbildung und fachliche Kontrolle der deutschen Rechenmeister vor dem Hintergrund ihrer Herkunft und ihres sozialen Status'. In *Verfasser und Herausgeber mathematischer Texte der frühen Neuzeit, Schriften des Adam-Ries-Bundes,* vol. 14 (Annaberg: TU Bergakademie Freiberg, 2002): pp. 1–22.

Schotte, Margaret E. *Sailing School: Navigating Science and Skill, 1550–1800* (Baltimore: Johns Hopkins University Press, 2019).

Sennewald, Rainer. 'Die Stipendiatenausbildung von 1702 bis zur Gründung der Bergakademie Freiberg 1765/66'. In *Technische Akademie Freiberg, Festgabe zur 300. Jahrestag der Gründung der Stipendienkasse für die akademische Ausbildung im berg- und Hüttenfach zu Freiberg in Sachsen* (Freiberg: TU Bergakademie, 2002): pp. 407–429.

Shelby, Lonnie R. 'The Geometrical Knowledge of Mediaeval Master Masons'. *Speculum* 47(3) (1972): pp. 395–421. https://doi.org/10.2307/2856152.

Sibum, H. 'Les gestes de la mesure. Joule, les pratiques de la brasserie et la science'. *Annales* 53(4) (1998): pp. 745–774. https://doi.org/10.3406/ahess.1998.279696.

Sigrist, René. 'Scientific Standards in the 1780s: A Controversy over Hygrometers'. In *Jean-Andre Deluc: Historian of Earth and Man* (Geneva: Slatkine, 2011): pp. 148–183.

Smith, Pamela H. *The Body of the Artisan. Art and Experience in the Scientific Revolution* (Chicago: University of Chicago Press, 2004).

Smith, Pamela H. 'Science on the Move: Recent Trends in the History of Early Modern Science'. *Renaissance Quarterly* 62(2) (2009): pp. 345–375. https://doi.org/10.1086/599864.

Smith, Pamela H. 'Why Write a Book? From Lived Experience to the Written Word in Early Modern Europe'. *Bulletin of the German Historical Institute* vol. 47 (2010): pp. 25–50.

Sokoll, Thomas. 'Kameralismus'. *Enzyklopädie Der Neuzeit* (Stuttgart/Weimar: Metzler, 2007).

Stams, Werner. 'Die Anfänge der neuzeitlichen Kartographie in Mitteleuropa'. In *Kursächsische Kartographie bis zum Dreißigjährigen Krieg* (Dresden: VEB Deutscher Verlag der Wissenschaften, 1990): pp. 37–105.

Štefánik, Martin. 'Die Anfänge der Slowakischen Bergstädte. Das Beispiel Neusohl'. In *Stadt und Bergbau* (Cologne: Böhlau, 2004): pp. 295–312.

Stein, Karel. 'Zur Tätigkeit der Familie Öder in Böhmen'. *Sächsische Heimatblätter* 1 (1988): pp. 17–18.

Stiegler, Leonhard. 'Leibnizens Versuche mit der Horizontalwindkunst auf dem Harz'. *Technikgeschichte* 35 (1968): pp. 265–292.

Stoyan, Dietrich, and Thomas Morel. 'Julius Weisbach's Pioneering Contribution to Orthogonal Linear Regression (1840)'. *Historia Mathematica* 45(1) (2018): pp. 75–84. https://doi.org/10.1016/j.hm.2017.12.002.

Strano, Giorgio and Johnston, Stephen and Miniati, Mara and Morrison-Low, Alison, *European Collections of Scientific Instruments, 1550–1750* (Leiden: Brill, 2009).

Streefkerk, Chris, Werner, Jan and Frouke Wieringa. *Perfect gemeten: landmeters in Hollands noorderkwartier ca. 1550–1700* (Wormer, Netherlands: Stichting Uitgeverij, 1994).

Strieder, Jakob. 'Die deutsche Montan- und Metall-Industrie im Zeitalter der Fugger' *Abhandlungen und Berichte Deutsches Museum* 3(6) (1931): pp. 189–226.

Syndram, Dirk, and Martina Minning, eds. *Die Kurfürstlich-Sächsische Kunstkammer in Dresden. Das Inventar von 1587* (Dresden: Sandstein Verlag, 2010).

Taylor, Eva Germaine Rimington. 'The South-pointing Needle'. *Imago Mundi* 8(1) (January 1951): pp. 1–7. https://doi.org/10.1080/03085695108591973.

Tenfelde, Klaus, Christoph Bartels, and Rainer Slotta, eds. *Geschichte des deutschen Bergbaus. Der alteuropäische Bergbau: von den Anfängen bis zur Mitte des 18. Jahrhunderts* vol. 1 (Münster: Aschendorff, 2012).

Trunz, Erich. *Wissenschaft und Kunst im Kreise Kaiser Rudolfs II., 1576–1612* (Neumünster: Wachholtz, 1992).

Uhlig, Gottfried. *Geschichte des sächsischen Schulwesens bis 1600* (Dresden: Hellerau Verlag, 1999).

Valleriani, Matteo, ed. *The Structures of Practical Knowledge* (Cham: Springer, 2017).

Van Veen, Johan. *Dredge Drain Reclaim, The Art of a Nation* (The Hague: Martinus Nijhoff, 1955).

Vérin, Hélène. *La gloire des ingénieurs: l'intelligence technique du XVIᵉ au XVIIIᵉ siècle* (Paris: Albin Michel, 1993).

Victor, Stephen K. *Practical Geometry in the High Middle Ages: Artis Cuiuslibet Consummatio and the Pratike de Geometrie* (Philadelphia: American Philosophical Society, 1979).

Virol, Michèle. *Vauban: De la gloire du roi au service de l'Etat* (Seyssel: Champ Vallon, 2013).

Vogler, Bernard. *Le monde germanique et helvétique à l'époque des réformes: 1517–1618*, vol. 1 (Paris: Société d'enseignement supérieur, 1981).

Wakefield, Andre. *The Disordered Police State: German Cameralism as Science and Practice* (Chicago: Chicago University Press, 2009).

Walker, Mack. *German Home Towns: Community, State, and General Estate, 1648–1871. German Home Towns* (Ithaca: Cornell University Press, 1971).

Wandel, Lee Palmer. 'Maps for a Prince'. *Science in Context* 32(2) (2019): pp. 171–192. https://doi.org/10.1017/S0269889719000176.

Wappler, August F. 'Oberberghauptmann von Trebra und die ersten drei sächsischen Kunstmeister Mende, Baldauf und Brendel'. *Mitteilungen der Freiberger Altertumsverein* 41 (1905): pp. 69–178.

Wardhaugh, Benjamin. *Poor Robin's Prophecies: A Curious Almanac, and the Everyday Mathematics of Georgian Britain* (Oxford: Oxford University Press, 2012).

Watanabe-O'Kelly, Helen. *Court Culture in Dresden* (New York: Palgrave Macmillan, 2002).

Weber, Max. *General Economic History* (New York: Greenberg, 1927).

Weber, Wolfhard. *Innovationen im Frühindustriellen deutschen Bergbau und Hüttenwesen: Friedrich Anton von Heynitz* (Göttingen: Vandehoeck et Ruprecht, 1976).

Weber, Wolfhard. 'Bergbau und Bergakademie. Zur Etablierung des Bergstaates im 18. Jahrhundert'. *Nachrichtenblatt der Deutschen Gesellschaft für Geschichte der Medizin, Naturwissenschaft und Technik EV* 35(3) (1985): pp. 79–89.

Weißflog, Egon. 'Die Bergwaage des Abraham Ries'. *Jahrbuch des Adam-Ries-Bundes*, 2014, pp. 75–88.

Weißflog, Egon. 'Uhrmacher und Mechaniker im Umfeld von Adam und Abraham Ries'. In *Visier- und Rechenbücher der frühen Neuzeit, Schriften des Adam-Ries-Bundes*, vol. 19 (Annaberg: TU Bergakademie Freiberg, 2008): pp. 341–356.

Westermann, Ekkehard. *Bergbaureviere als Verbraucherzentren im Vorindustriellen Europa* (Stuttgart: Frank Steiner Verlag, 1997).

Wetzel, Michael. 'Der Erzgebirgische Kreis im Ausgestaltungsprozess des frühen Albertinischen Territorialstaates'. In *Das Erzgebirge im 16 Jahrhundert. Gestaltwandel einer Kulturlandschaft im Reformationszeitalter*. Edited by Martina Schattkowsky (Leipzig: Leipziger Universitätsverlag, 2013): pp. 33–44.

Whaley, Joachim. *Germany and the Holy Roman Empire. Maximilian I to the Peace of Westphalia, 1493–1648*, vol. 1 (Oxford: Oxford University Press, 2013).

Whaley, Joachim. *Germany and the Holy Roman Empire. The Peace of Westphalia to the Dissolution of the Reich 1648–1806*, vol. 2 (Oxford: Oxford University Press, 2013).

Wieden, Brage bei der, and Thomas Böckmann. *Atlas vom Kommunionharz in historischen Abrissen von 1680 und aktuellen Forstkarten* (Hannover: Verlag Hahnsche Buchhandlung, 2010).

Wilkening, Wilhelm. 'Erasmus Reinhold, der Verfasser der ersten deutschen Markscheidekunde'. *Mitteilungen aus dem Markscheidewesen* 67 (1960): pp. 13–15, 58–74.

Wilkening, Wilhelm. 'Aus dem Leben des Erasmus Reinhold'. *Mitteilungen aus dem Markscheidewesen* 70 (1963): pp. 11–12.

Williams, Gerhild Scholz. *Ways of Knowing in Early Modern Germany: Johannes Praetorius as a Witness to His Time* (New York: Routledge, 2017).

Willmoth, Frances. *Sir Jonas Moore: Practical Mathematics and Restoration Science* (Woodbridge: Boydell Press, 1993).

Wilsdorf, Helmut, Walther Herrmann, and Kurt Löffler. *Bergbau – Wald – Flösse: Untersuchungen zur Geschichte der Flösserei im Dienste des Montanwesens und zum montanen Transportproblem* (Berlin: Akademie-Verlag, 1960).

Wilski, Paul. *Lehrbuch der Markscheidekunde* (Berlin: Julius Springer, 1929).

Wilski, Paul. 'Über die Heutige Markscheidekunde'. *Österreichische Zeitschrift für Vermessungswesen* 31(4) (1933): pp. 61–66.

Witthöft, Harald. 'Die Metrologie bei Georgius Agricola. Von geistiger und materieller Kultur im 16. Jahrhundert'. *Der Anschnitt* 48(1) (1996): pp. 19–27.

Woitkowitz, Torsten. 'Der Landvermesser, Kartograph, Astronom und Mechaniker Johannes Humelius (1518–1562) und die Leipziger Universität um die Mitte des 16. Jahrhunderts'. *Sudhoffs Archiv* 92(1) (2008): pp. 65–97.

Wolf, Herbert. 'Die "Himmlische Fundgrube" und die Anfänge der deutschen Bergmannspredigt'. *Hessische Blätter für Volkskunde* 49/50(1) (1958): pp. 347–354.

Wunderlich, Herbert. *Kursächsische Feldmesskunst, artilleristische Richtverfahren und Ballistik im 16. und 17. Jahrhundert* (Berlin: Deutscher Verlag der Wissenschaften, 1977).

Yale, Elizabeth. 'Marginalia, Commonplaces, and Correspondence: Scribal Exchange in Early Modern Science'. *Studies in History and Philosophy of Science* 42(2) (2011): pp. 193–202. https://doi.org/10.1016/j.shpsc.2010.12.003.

Zaitsev, Evgeny A. 'The Meaning of Early Medieval Geometry: From Euclid and Surveyors' Manuals to Christian Philosophy'. *Isis* 90(3) (1999): pp. 522–553.

Ziegenbalg, Michael. 'Aspekte des Markscheidewesens mit besonderer Berücksichtigung der Zeit von 1200 bis 1500'. *Der Anschnitt* 30(2) (1984): pp. 40–49.

Ziegenbalg, Michael. 'An Interdisciplinary Cooperation: Painters of Landscape, Cartographers, Surveyors of Land and Mountain in the Renaissance'. *Histoire & Mesure* 8(3–4) (1993): pp. 313–324.

Ziegenbalg, Michael. 'Von der Markscheidekunst zur Kunst des Markscheiders'. *Berichte der Geologischen Bundesanstalt* 41 (1997): pp. 267–274.

Zilsel, Edgar. 'The Genesis of the Concept of Physical Law'. In *The Social Origins of Modern Science, Boston Studies in the Philosophy of Science* (2000): pp. 97–122.

Zilsel, Edgar. *The Social Origins of Modern Science* (Dordrecht: Kluwer, 2003).

Zimmermann, Christian von, ed. *Wissensschaftliches Reisen – reisende Wissenschaftler: Studien zur Professionalisierung der Reiseformen zwischen 1650 und 1800, Cardanus*, vol. 3, (Heidelberg: Palatina Verlag, 2003).

# Index

Page numbers in *italics* relate to Figures. Names of books are listed under the names of their authors.

281

Lightning Source UK Ltd.
Milton Keynes UK
UKHW021544050123
414796UK00008B/299

9 781009 267304